TAKING SIDES

Clashing Views on Controversial

Global Issues

SECOND EDITION

TAKING SIDES

Clashing Views on Controversial

Global Issues

SECOND EDITION

Selected, Edited, and with Introductions by

James E. Harf
University of Tampa

and

Mark Owen Lombardi
College of Santa Fe

WITHDRAWN

McGraw-Hill/Dushkin
A Division of The McGraw-Hill Companies

Photo Acknowledgment
Cover image: © 2004 by PhotoDisc, Inc.

Cover Art Acknowledgment
Charles Vitelli

Library of Congress Cataloging-in-Publication Data
Main entry under title:
Taking sides: clashing views on controversial global issues/selected, edited, and with introductions
by James E. Harf and Mark Owen Lombardi.—2nd ed.
Includes bibliographical references and index.
1. Globalization. 2. International relations. 3. Global environmental change. 4. Population. 5.
Emigration and immigration. I. Harf, James E., ed. II. Lombardi, Mark Owen, ed. III. Series
303
0-07-288077-5
ISSN: 1536-3317

Printed on Recycled Paper

To my daughter, Marie: May your world conquer those global issues left unresolved by my generation. (J. E. H.)

For Betty and Marty, who instilled a love of education and need to explore the world. (M. O. L.)

Preface

This volume reflects the changing nature of the international system. The personal reflections of its two editors bring this into sharp focus. We are both products of the separate environments in which we have been raised. Our 20-year age difference suggests that each of us brings to the study of international affairs a particular lens through which two distinct yet interconnected generations view contemporary world issues.

The elder editor began his formal schooling a few weeks after the United States dropped two atomic bombs on Japan. His precollegiate experience took place amid the fear that an evil enemy was lurking throughout the world, poised to conquer America and enslave its occupants. During his initial month in college, the Soviet Union launched the first space vehicle, *Sputnik*, forever changing the global landscape. Yet his 1961 international relations textbook assumed that, despite the bomb and the ability to exploit outer space, the driving forces behind national behavior had not changed very much. By the time he accepted his first university teaching position, the world had been locked in a protracted cold war for two decades. Yet evidence of a new global agenda of problems that were totally unrelated to the American-Soviet rivalry was beginning to appear. Soon this new set of issues would challenge the traditional agenda for the world's attention and also occupy the editor's research agenda during much of his career.

As the older editor now moves on to a new stage of his long career, the cold war exists no more. And the problems that have occupied his and others' attention for some time—population growth, resource scarcities, environmental degradation, and the like—are being transformed by the overarching presence of globalization. The Internet and other advances in communication and transportation signal an entirely new environment in which global actors design and carry out their policies. This was brought home in a rather graphic way in December 2000, as the older editor joined others in Kiev, Ukraine, to watch the newspapers debate the pending closure of the remaining operating nuclear power plant at Chernobyl. This editor's world has undergone three major changes, from a traditionalist, realist paradigm where military power dominated, to a global agenda where problems demanded the attention and cooperation of a kaleidoscope of actors, to yet another set of problems and a transformation of existing issues brought on by globalization. Superimposed on this agenda, and coming at a time when the older editor was assuming a new academic position, are the events of September 11, 2001, and the war on terrorism, which are changing long-held assumptions about war and violence.

The younger editor began his formal schooling just after the United States and the Soviet Union had settled "comfortably" into the era of mutual assured destruction (MAD). He entered his university studies as increased tensions and U.S.-Soviet rivalry threatened to tear apart the fragile fabric of détente. As he embarked on his teaching career, the communist empire was taking its first steps toward total disintegration. Amid this revolutionary change, deeper and more complex issues of security, technology, development, and ethnic conflict became more pronounced and increasingly present in the consciousness of actors and scholars alike. As he too embarked on a new academic career, September 11 woke Americans up to the harsh reality that the paradigms of violence and war were changing. Now at the midpoint of his academic life, this younger editor contemplates the speed and force of globalization and how the issues that this phenomenon has brought forth will affect the lens through which he observes world affairs, influencing his and other research agendas over the next few decades.

The decision of this book's publisher to produce a text on global issues reflects how the nature of international affairs has changed. Many textbooks on world politics and American foreign policy address longstanding, important concerns of the student of world events, employing for the most part the traditional paradigm of the past century. Simply put, this traditional scholarly view assumed a world where power, particularly military might, dominated the world picture. Nation-states, the only really important world actors, focused on pursuing power and then using it in conventional ways to become a larger presence on the international scene. The specific problems of the day might have changed, but the fundamental principles that guided the international system remained the same. Most books in these fields reflect this outlook.

But the field of international affairs is ever changing, and textbooks are beginning to reflect this evolution. This new volume takes into account the fact that the age of globalization has accelerated, transforming trends that began over three decades ago. No longer are nation-states the only actors on the global stage. Moreover, their position of dominance is increasingly challenged by an array of other actors—international governmental organizations, private international groups, multinational corporations, and important individuals—who might be better equipped to address newly emerging issues (or who might also be the source of other problems). This new agenda took root in the late 1960s, when astute observers began to identify disquieting trends: quickening population growth in the poorer sectors of the globe, growing disruptions in the world's ability to feed its population, increasing shortfalls in required resources, and expanding evidence of negative environmental impacts, such as a variety of pollution evils.

An even more recent phenomenon is the increase in ethnic pride, which manifests itself in both positive and negative ways. The history of post–cold war conflict is a history of intrastate, ethnically driven violence that has torn apart countries that have been unable to deal effectively with the changes brought on by the end of the cold war. The most pernicious manifestation of this emphasis on ethnicity is the emergence of terrorist groups who use religion and other aspects of ethnicity to justify bringing death and destruction on their perceived enemies. As national governments attempt to cope with this latest phenomenon, they too are changing the nature of war and violence. The global agenda's current transformation, brought about by globalization, demands that our attention turn toward the latter's consequences.

The format of *Taking Sides: Clashing Views on Controversial Global Issues* follows the successful formula of other books in the Taking Sides series. The book begins with an introduction to the emergence of global issues and the new age of globalization. It then presents 21 current global issues grouped into five parts. Population takes center stage in Part 1 because it not only represents a global issue by itself, but it also affects the parameters of most other global issues. Part 2 addresses a range of problems associated with global resources. Parts 3, 4, and 5 feature issues borne out of the emerging agenda of the twenty-first century, including the increased movement of peoples and ideas across national boundaries and new security issues in the post–cold war and post–September 11 eras.

Each issue has two readings, one pro and one con. The readings are preceded by an issue *introduction* that sets the stage for the debate and briefly describes the two readings. Each issue concludes with a *postscript* that summarizes the readings, suggests further avenues for thought, and provides additional *suggestions for further reading*. Also, Internet site addresses (URLs) have been provided on the *On the Internet* page that accompanies each part opener, which should prove useful as starting points for further research. At the back of the book is a listing of all the *contributors to this volume* with a brief biographical sketch of each of the prominent figures whose views are debated here.

Changes to this edition This second edition represents a significant revision. Five of the 21 issues are completely new: *Should the World Continue to Rely on Oil as a Major Source of Energy?* (Issue 6); *Is the Threat of a Global Water Shortage Real?* (Issue 10); *Is a Nuclear Terrorist Attack on America Likely?* (Issue 18); *Is There a Military Solution to Terrorism?* (Issue 19); and *Are Civil Liberties Likely to Be Compromised in the War Against Terrorism?* (Issue 20). In addition, for the issues on declining population growth rates in the developed world (Issue 4), worldwide food production (Issue 7), global warming (Issue 9), and the effects of globalization on the world community (Issue 14), one or both selections were replaced to bring the issues up to date. In all, there are 15 new selections.

A word to the instructor An *Instructor's Manual With Test Questions* (multiple-choice and essay) is available through the publisher for the instructor using *Taking Sides* in the classroom. A general guidebook, *Using Taking Sides in the Classroom,* which discusses methods and techniques for integrating the pro-con approach into any classroom setting, is also available. An online version of *Using Taking Sides in the Classroom* and a correspondence service for *Taking Sides* adopters can be found at http://www.dushkin.com/usingts/.

Taking Sides: Clashing Views on Controversial Global Issues is only one title in the Taking Sides series. If you are interested in seeing the table of contents for any of the other titles, please visit the Taking Sides Web site at http://www.dushkin.com/takingsides/.

James E. Harf
University of Tampa

Mark Owen Lombardi
College of Santa Fe

Contents In Brief

Contents

Author Michael Tobias argues that there can be no doubt that under virtually all plausible scenarios of future global population growth, such growth will continue at unprecedented levels and will cause accelerated damage to the planet. Max Singer, cofounder of the Hudson Institute, a public policy research organization, contends that the world's population will be declining in 50 years as a consequence of an increasing number of countries' achieving a higher level of modernity.

Robert S. McNamara, former president of the World Bank, argues that the developed countries of the world and international organizations should help the countries of the developing world to reduce their population growth rates. The late professor of economics and business administration Julian L. Simon maintains that international organizations seek to impose birth control methods on developing nations in the misguided name of "virtuous and humanitarian . . . motives."

Don Hinrichsen, an environmental reporter and a consultant for the United Nations Population Fund, argues that the planet's natural systems and ecological systems are having increasing difficulty withstanding the negative consequences of continuing population growth. Economics professor N. Gregory Mankiw maintains that, assuming each additional person on

Earth creates resources, if the amount created by each new individual is greater than the amount used by that individual, then the Earth's quality is enhanced by the presence of that individual.

Peter G. Peterson, chairman of the Blackstone Group, a private investment bank, argues that global aging will dominate the public-policy agenda of the developed world as it attempts to cope with a vast array of problems associated with such aging. Adjunct professor of demography Leon F. Bouvier and Jane T. Bertrand, chair of the Department of International Health and Development at Tulane University School of Public Health and Tropical Medicine, argue that a future developed society will be quite able to cope with declining growth rates.

The United Nations Environment Programme presents a comprehensive and gloomy assessment of the global environment at the turn of the millennium. Peter Huber, a senior fellow of the Manhattan Institute, contends that humankind is saving the Earth with the technologies that the "soft greens" most passionately oppose.

Hisham Khatib, deputy chairman of the Jordan Regulatory Commission, and his coauthors conclude that reserves of traditional commercial fuels, including oil, "will suffice for decades to come." Seth Dunn, a research associate and climate/energy team coleader at the Worldwatch Institute, argues that a new energy system is fast emerging—in part because of a series of revolutionary new technologies and approaches—that will cause a transition away from a reliance on oil as the primary energy source.

Sylvie Brunel, former president of Action Against Hunger, argues that "there is no doubt that world food production . . . is enough to meet the needs of" all the world's peoples. Lester R. Brown, former president of the Worldwatch Institute, maintains that little, if any, progress is being made to eradicate pervasive global hunger, despite increases in food productivity.

Adil Najam, an assistant professor in the Department of International Relations at Boston University, argues that consumption control among developed countries is a major key to avoiding continuing environmental crises. Lester R. Brown, Gary Gardner, and Brian Halweil, researchers at the Worldwatch Institute, examine 16 dimensions of population growth in the areas of food and agriculture, environment and resources, and economic impacts and quality of life, and they indicate that the effects of such growth are mainly adverse.

In the summary of its most recent assessment of climatic change, the Intergovernmental Panel on Climate Change concludes that an increasing set of observations reveals that the world is warming and that much of it is due to human intervention. Brian Tucker, a senior fellow of the Institute for Public Affairs, argues that there are still too many uncertainties to conclude that global warming has arrived.

PART 3   THE MOVEMENT OF PEOPLES   193

William J. Carrington, an economist at the Bureau of Labor Statistics, details what he considers an extensive global brain drain, or the migration of the more qualified citizens of the developing world to the richer countries. Jean M. Johnson and Mark C. Regets, senior analysts in the Division of Science Resources Studies at the National Science Foundation, contend that roughly half of all foreign doctoral recipients leave the United States immediately upon completion of their studies, while others leave some years later, creating a brain circulation rather than a total drain.

PART 4 THE FLOW OF INFORMATION AND IDEAS **249**

Journalist Thomas L. Friedman argues that globalization is built around three balances, each of which makes the system more individualized and democratic by empowering individuals on the world stage. Economists Christian E. Weller, Robert E. Scott, and Adam S. Hersh argue that over the past 20 years, globalization has brought about greater inequities, more poverty, and greater misery and has not increased development for the vast majority of the world's citizens.

Professor of sociology Paul Starr argues that the information revolution will help to create a period of individual freedom empowered by computer technology. Professor of law Jerry Kang argues that privacy, which is fundamental to the concept of freedom, is severely threatened by the current love of cyberspace. He suggests that the U.S. Congress and, by extension, individual states should take action to ensure individual privacy in cyberspace.

David Rothkopf, an adjunct professor of international and public affairs, discusses the creation of a global culture and argues that the Americanization of the world along a U.S. value system is good and should be encouraged in the interests of the United States. Jeffrey E. Garten, dean of Yale University's School of Management, argues that people around the world see globalization as a form of American imperialism. He contends that such fears have a strong basis in reality and cannot be ignored.

Journalist Al J. Venter examines newly emerging germs and asserts that they will expand the threat of germ warfare in the coming years. Professor of political science Ehud Sprinzak argues that a biological or chemical terrorist attack is highly unlikely and that governments, in justifying expanding defense spending and other measures, are exploiting people's fears of such an occurrence.

Professor of government Graham Allison contends that the United States must assume that terrorist groups like Al Qaeda have dirty bombs and nuclear weapons and that it must therefore act accordingly. He argues that to assume that such groups do not possess such weapons invites disaster. Jessica Stern, a lecturer in public policy, argues that Americans are in danger of overestimating terrorist capabilities and thus creating a graver threat than actually exists. She warns that the United States must not overreact in its policy response and that prudent security measures will greatly reduce such threats now and in the future.

General Wesley K. Clark argues that with the proper flexible strategy, the U.S. military can accomplish the objective of eliminating groups like Al Qaeda and help the United States win the war against terrorism. Journalist Andrew Stephen maintains that the problems behind terrorism are deep and that U.S. patriotism and bravado will not win the war against it.

Issue 20.   Are Civil Liberties Likely to Be Compromised in the War Against Terrorism?   350

Morton H. Halperin, a senior fellow of the Council on Foreign Relations, cites the Patriot Act, as well as the strategy employed by the George W. Bush administration to enact it, in sounding an alarm about the abuses of civil liberties that he feels will likely result from this new legislation. Kim R. Holmes, vice president and director of the Kathryn and Shelby Cullom Davis Institute for International Studies at the Heritage Foundation, and Edwin Meese III, Ronald Reagan Distinguished Fellow in Public Policy at the foundation, argue that the Bush administration's approach to fighting terrorism, including its legislative package, strikes the proper balance between liberty and security.

Issue 21.   Are Cultural and Ethnic Conflicts the Defining Dimensions of Twenty-First-Century War?   362

Political scientist Samuel P. Huntington argues that the emerging conflicts of the twenty-first century will be cultural and not ideological. He identifies the key fault lines of conflict and discusses how these conflicts will reshape global policy. John R. Bowen, a professor of sociocultural anthropology, rejects the idea that ethnic and cultural conflicts are decisive. He argues that political choices made by governments and nations, not cultural divides and intercultural rivalry, dictate much of global international affairs.

Introduction

Global Issues in the Twenty-First Century

James E. Harf
Mark Owen Lombardi

The Emergence of the Age of Globalization

The cold war era, marked by the domination of two superpowers in the decades following the end of World War II, has given way to a new era of globalization. This new epoch is characterized by a dramatic shrinking of the globe in terms of travel and communication, increased participation in global policy making by an expanding array of national and nonstate actors, and an exploding volume of problems with ever-growing consequences. While the tearing down of the Berlin Wall over a decade ago dramatically symbolized the end of the cold war era, the creation of the Internet graphically illustrates the emergence of the globalization era, and the fallen World Trade Center symbolizes a new paradigm for conflict and violence.

Early signs of this transition were manifested during the latter part of the twentieth century by a series of problems that transcended national boundaries. These problems caught the attention of policymakers who demanded comprehensive analyses and solutions. The effects of significant population growth in the developing world, for example, called for multilateral action by leaders in developed and developing nations alike. Acid rain, created by emissions from smokestacks, was the precursor to a host of environmental crises that appeared with increasing frequency, challenging the traditional political order and those who commanded it. Finally, much of the world began to sense that the planet is in reality "Spaceship Earth," with finite resources that are in danger of being exhausted and of which careful stewardship is increasingly necessary.

These global concerns remain today. Some are being addressed successfully, while others are languishing amid a lack of consensus about the nature of the problems and how to solve them. Also, with the shattering of the cold war system and the advent of the globalization age, new issues have emerged to broaden our conception of the global agenda. In the past 10 years, the freedom of people to move has expanded, and with this freedom are concerns about refugees and displaced persons within the developing world and about the alleged brain drain from the poorer nations to countries that are most able to provide a better life. More recently, the world has come to fear an expanded potential for terrorism, as new technologies combined with old but strong hatreds have conspired to make the world far less safe than it had been.

The increase in the movement of people pales in comparison to the accelerated movement of information and ideas, which has ushered in global concerns over cultural imperialism, individual privacy, and freedom. The ability both to retrieve and to disseminate information will have an impact in this century as great as, if not greater than, the impacts of the telephone, radio, and television in the twentieth century. The potential for global good or ill is mind-boggling. Finally, traditional notions of great-power security embodied in the cold war rivalry have given way to concerns about terrorism, genocide, nuclear proliferation, cultural conflict, and the expansion of international law.

Globalization heightens our awareness of a vast array of global issues that will challenge individuals as well as governmental and nongovernmental actors. Since the demise of the cold war world, analysts and laypersons have become free to define, examine, and explore solutions to such issues on a truly global scale. This text seeks to identify those issues that are central to the discourse on globalization. The issues in this volume provide a broad overview of the mosaic of global issues that will impact students' daily lives.

What Is a Global Issue?

We begin by addressing what a "global issue" is.[1] By definition, the word *issue* is characterized by disagreement along several related dimensions:

1. whether or not a problem exists;
2. the characteristics of the problem;
3. the preferred future alternatives or solutions; and/or
4. how the preferred futures are to be obtained.

These problems are real, vexing, and controversial because policymakers bring to their analyses different historical experiences, values, goals, and objectives. These differences impede and can even prevent successful problem solving. In short, the key ingredient of an issue is disagreement.

The word *global* in the phrase *global issue* is what makes the set of problems confronting the human race today far different from those that challenged our ancestors. Historically, problems were confined to a village, city, or region. The capacity of humans to fulfill their daily needs was limited to a much smaller space: the immediate environment. When human feet and the horse were the principal modes of transportation, people pursued their food, fuel, and other necessities near their homes. When local resources were exhausted, people moved elsewhere in search of another environment that was better able to supply them. With the invention of the locomotive and the automobile, however, humans were able to travel greater distances on a daily basis. National energy systems and agricultural systems emerged, freeing humans from a reliance on their local community. Problems then became national, as dislocations in the production or distribution system in one part of a country had repercussions in some other part.

With the advent of transoceanic transportation capabilities, national distribution systems gave way to global systems. The "breadbasket" countries of the world—those best able to grow food—expanded production to serve an ever-larger global market. Oil-producing countries with vast reservoirs eagerly increased production to satisfy the needs of a growing worldwide clientele. Countries that were best able to provide other resources to a larger part of the world entered the global trade system as well. These spatial changes were a consequence of significant increases in population and per capita consumption, both of which dramatically heightened the demand for resources.

Additionally, the larger machines of the Industrial Revolution were capable of moving greater quantities of dirt and of emitting greater amounts of pollutants into the atmosphere. With these new production and distribution capabilities came more prevalent manifestations of existing problems as well as new problems. Providers either engaged in greater exploitation of their own environment or searched further afield in order to satisfy additional resource requirements.

The character of these new problems is different from those of earlier eras. First, they transcend national boundaries and impact virtually every corner of the globe. In effect, these issues help make national borders increasingly meaningless. Environmental pollution, poisonous gases, and diseases such as acquired immunodeficiency syndrome (AIDS) do not recognize or respect national borders.

Second, these new issues cannot be resolved by the autonomous action of a single actor, be it a national government, international organization, or multinational corporation. A country cannot guarantee its own energy or food security without participating in a global energy or food system.

Third, these issues are characterized by a wide array of value systems. To a family in the developing world, giving birth to a fifth or sixth child may contribute to the family's immediate economic well-being. But to a research scholar at the United Nations Population Fund, the consequence of such an action multiplied across the entire developing world leads to expanding poverty and resource depletion.

Fourth, these issues will not go away. They require specific policy action by a consortium of local, national, and international leaders. Simply ignoring the issue cannot eliminate the threat of chemical or biological terrorism, for example. If global warming does exist, it will not disappear unless specific policies are developed and implemented.

Finally, these issues are characterized by their persistence over time. The human race has developed the capacity to manipulate its external environment and, in so doing, has created a host of opportunities and challenges. The accelerating pace of technological change suggests that global issues will proliferate and will continue to challenge human beings throughout the next millennium.

In the final analysis, however, a global issue is defined as such only through mutual agreement by a host of actors within the international community. Some may disagree about the nature, severity, or presence of a given issue, and these concerns become areas of focus only after a significant number of actors (states, international organizations, the United Nations, and others) begin to focus systematic and organized attention on the issue itself.

The Creation of the Global Issues Agenda

Throughout the first part of the cold war, the international community found itself more often being an observer of, rather than a player in, the superpower conflict for global domination being waged by the United States and the Soviet Union. Other forces and trends competed for the attention of world leaders during this period: Economic development exploded in many parts of the globe, while other areas floundered in poverty. Resource consumption expanded alongside economic growth throughout the developed world, creating both pollution and scarcity. Population growth rates increased dramatically in societies that were unable to cope with the consequences of a burgeoning population. New forms of violence replaced traditional ways of employing force. And nonstate actors became a larger part of the conflict problem.

As the major participants in the cold war settled into a system of routinized behavior designed to avoid catastrophic nuclear war, other actors in the international system began to worry about the disquieting effects associated with exploding population and consumption. Nation-states, international governmental organizations, and private groups sought to identify these growth-related problems and to seek solutions. The United Nations (UN) soon became the principal force to which those affected by such global behavior turned.

International community members soon began to join forces under the auspices of the UN to design comprehensive strategies. Such plans became known as international regimes—a set of agreements, structures, and plans of action involving all relevant global actors in addressing specific issues before them.

For example, in response to concerns of the Swedish government in the early 1970s about the discovery of a significant amount of dead fish in its lakes, the UN brought together relevant parties to discuss the problem, which they assumed was caused by acid rain. This 1972 conference, called the United Nations Conference on the Human Environment, represented the first major attempt of the international community to address the global ecological agenda. Representatives from 113 countries met in Stockholm, Sweden (the Soviet-bloc countries boycotted the conference for reasons unrelated to the environmental issue). Over 500 international nongovernmental organizations from such disparate interests as the environment, science, women, religion, and business also sent delegates. Members of the United Nations Secretariat and specialized international governmental organizations within the UN, such as the World Bank and the Food and Agricultural Organization (FAO), were present as well. Experts from all walks of life—scientists, environmentalists, and others—attended.

The conference in Stockholm ushered in a new age of thinking about issues in their global context and became the model for future conferences on global issues. Position papers were prepared for delegates, and preparatory committees hammered out draft declarations and plans of action. These served as the starting points for conference deliberations. The final plan of action included the establishment of a formal mechanism, the United Nations Environment Programme (UNEP), for addressing environmental issues in the future.

The most significant outcome of the conference was the creation of awareness among national government officials regarding the severity of ecological issues, the extent to which these issues cross national boundaries, and the need for international cooperation in order to solve these issues. The era of global issues identification, analysis, and problem solving was born.

Twenty years later, the global community reconvened in Rio de Janeiro, Brazil, at the United Nations Conference on Environment and Development, more popularly known as the Earth Summit. This meeting produced "Agenda 21," a report that outlined a plan of action for achieving development throughout the world while at the same time avoiding environmental degradation.

Another example of the international community's venture into this emerging agenda focused on the global population problem. In the early 1970s, the UN responded to the pleas of a variety of groups to consider the negative consequences of exploding birth rates in many developing countries. The reason for such concern grew out of the belief that such growth was deterring economic growth in those areas of the world that were the most poverty stricken.

The result was the United Nations World Population Conference held in Bucharest, Romania, in 1974, and attended by delegates from 137 countries as well as by representatives of numerous governmental and nongovernmental organizations. The principal result of the conference was the placement of population on the global agenda and the development of a formal plan of action. The international community has continued paying formal attention to the population issue and has convened every 10 years since Bucharest—in Mexico City, Mexico, in 1984 and in Cairo, Egypt, in 1994—to evaluate strategy and assume new tasks to lower growth rates as well as to address other population-related issues.

These kinds of conferences, where a myriad of experts join a wide range of governments, private groups, and individuals, is now characteristic of how the global community is addressing the emerging global agenda. But the international community was far less certain of the wisdom of this strategy in 1972. However, the success of the United Nations Conference on the Human Environment led to similar strategies in a host of other environmental areas and provided the blueprint for the analysis of unforeseen global issues. For most of the issues in this text, the international community is either formally addressing the problem or is considering its first steps.

The Nexus of Global Issues and Globalization

In 1989 the Berlin Wall fell, and with it a variety of assumptions, attitudes, and expectations of the international system. This event did not usher in a utopia, nor did it irrevocably change all existing ideas. It did, however, create a void. As a dominant cold war system that had been entrenched since the end of World War II was broken apart, alternative views, issues, actors, and perspectives began to emerge to fill the void left by the collapse of the old system. What helped to bring down the previous era and what has emerged as the new dominant international construct is what we now call globalization.

Throughout the 1990s, scholars and policymakers have struggled to define this new era. Some have analyzed it in terms of the victory of Western or American ideals, the dominance of global capitalism, and the spread of democracy. Others have defined it in terms of the multiplicity of actors now performing on the world stage, and how states and their sovereignty have declined in importance and impact vis-à-vis others, such as multinational corporations and nongovernmental groups like Greenpeace and Amnesty International. Still others have focused on the vital element of technology and its impact on communications, information storage and retrieval, and global exchange.

Whether globalization reflects one, two, or all of these characteristics is not as important as the fundamental realization that globalization is the dominant element of a new era in international politics. This new period is characterized by several basic traits that greatly impact the definition, analysis, and solution of global issues. They include the following:

- an emphasis on information technology;
- the increasing speed of information and idea flows;
- a need for greater sophistication and expertise to manage such flows;
- the control and dissemination of technology; and
- the cultural diffusion and interaction that comes with information expansion and dissemination.

Each of these areas has helped shape a new emerging global issues agenda. Current issues remain important and, indeed, these factors help us to understand them on a much deeper level. Yet globalization has also created new problems and has brought them to the fore such that significant numbers of actors now recognize their salience in the international system.

For example, the spread of information technology has made ideas, attitudes, and information more available to people throughout the world. Americans in Dayton, Ohio, had the ability to log onto the Internet and communicate with their counterparts in Kosovo to discover when NATO bombing had begun and to gauge the accuracy of later news reports on the bombing. Norwegian students can share values and customs directly with their counterparts in South Africa, thereby experiencing cultural attitudes firsthand without the filtering mechanisms of governments or even parents and teachers. Scientific information that is available through computer technology can now be used to build sophisticated biological and chemical weapons of immense destructive capability. Ethnic conflicts and genocide between groups of people are now global news, forcing millions to come to grips with issues of intervention, prevention, and punishment. And terrorists in different parts of the globe can communicate readily with one another, transferring plans and even money across national and continental boundaries with ease.

Globalization is an international system, and it is also rapidly changing. Because of the fluid nature of this system and the fact that it is both relatively new and largely fueled by the amazing speed of technology, continuing issues are constantly being transformed and new issues are emerging regularly. The nexus of globalization and global issues has now become, in many ways, the defining dynamic of understanding global issues. Whether it is new forms of terrorism, new conceptions of security, expanding international law, solving ethnic conflicts, dealing with mass migration, coping with individual freedom and access to information, or addressing cultural clash and cultural imperialism, the transition from a cold war world to a globalized world helps us understand in part what these issues are and why they are important.

Identifying the New Global Issues Agenda

The analysis of global issues by scholars and policymakers has changed. Our assumptions, ideas, and conceptions of what makes a problem a global issue has expanded. New technologies and actors within a changing global order have made the study of global issues a broader and more complex undertaking.

The organization of this text reflects these phenomena. Parts 1 and 2 focus on the continuing global agenda of the post–cold war era. The emphasis is on global population and environmental issues and the nexus between these two phenomena. Is world population out of control, or is it a regional phenomenon centered only in the poorer sectors of the globe? What, if anything, should the international community do about this problem? Does population affect the environment, and if so, how? Should humanity be concerned about resource availability, be it food, oil, water, air, or pristine land? Do the activities of humans account for the degradation of the environment? If so, is such change part of a natural evolution to be embraced, not feared?

Parts 3, 4, and 5 address the emerging agenda of the globalization age. Is human migration, forced or unforced, a positive development or a sign of growing human desperation for economic and social well-being? The new age of globalization has facilitated the flow of ideas worldwide. Is this dissemination of knowledge a force for good or a form of indoctrination by one culture over another? And will this greater flow of information promote human empowerment or restrict human freedom?

Has the end of the cold war brought a chance for global peace, or has it unleashed a set of forces that will pose greater and more insidious dangers than those faced during the last century? Has the threat of nuclear war been replaced by a new struggle involving terrorist groups using more dangerous weapons of mass destruction? Have the horrors of the Holocaust been revisited through more recent instances of ethnic cleansing and genocide? Can the international community prevent and police such conflicts through international forums? Finally, can the world overcome the debilitating effects of ethnic conflicts across the globe?

The revolutionary changes of the last few decades present us with serious challenges unlike any others in human history. However, as in all periods of historic change, we possess significant opportunities to overcome problems. The task ahead is to define these issues, explore their contexts, and develop solutions that are comprehensive in scope and effect. The role of all researchers in this field, or any other, is to analyze objectively such problems and search for workable solutions. As students of global issues, your task is to educate yourselves about these issues and become part of the solution.

Note

1. The five characteristics are extracted from James E. Harf and B. Thomas Trout, *The Politics of Global Resources* (Duke University Press, 1986), pp. 12–28.

On the Internet . . .

Population Reference Bureau

The Population Reference Bureau provides current information on international population trends and their implications from an objective viewpoint. The PopNet section of this Web site offers maps with regional and country-specific population information as well as information divided by selected topics.

http://www.prb.org

United Nations Population Fund (UNFPA)

The United Nations Population Fund (UNFPA) was established in 1969 and was originally called the United Nations Fund for Population Activities. This organization works with developing countries to educate people about reproductive and sexual health as well as about family planning. The UNFPA also supports population and development strategies that will benefit developing countries and advocates for the resources needed to accomplish these tasks. Explore this site to learn more about the UNFPA, including the organization's support of individuals' right to decide the number and spacing of their children without outside coercion.

http://www.unfpa.org

Population Connection

Population Connection (formerly Zero Population Growth) is a national, non-profit organization working to slow population growth and to achieve a sustainable balance between Earth's people and its resources. The organization seeks to protect the environment and to ensure a high quality of life for present and future generations. Population Connection's education and advocacy programs aim to influence public policies, attitudes, and behavior on national and global population issues and related concerns.

http://www.populationconnection.org

The CSIS Global Aging Initiative

The Center for Strategic and International Studies (CSIS) is a public policy research institution that approaches the issue of the aging population in developed countries in a bipartisan manner. The CSIS is involved in a two-year project to explore the global implications of aging in developed nations and to seek strategies on dealing with this issue. This site includes a list of publications that were presented at previous events.

http://www.csis.org/gai/

PART 1

Global Population Growth

*I*t is no coincidence that many of the global issues in this volume emerged at about the same time as the world population exploded. No matter what the issue, the presence of a large and fast-growing population exacerbates the issue and transforms its basic characteristics. And the recent graying of populations around the globe has the potential for significant impact as well. The ability of the global community to respond to any given issue is diminished by certain population conditions, be it an extremely young consuming population in a developing country in need of producers; a large working-age group in a nation without enough jobs; a growing senior population for whom additional services are needed; or a rapidly growing, resource-consuming population in an area without sufficient food and water.

Thus, we begin this text with a series of issues directly related to various aspects of world population, which serves as both a separate global agenda and as a context within which other issues are examined.

- Is World Population Growth Out of Control?

- Should the International Community Attempt to Curb Population Growth in the Developing World?

- Does Population Growth Significantly Harm the Earth's Quality of Life?

- Will Declining Population Growth Rates in the Developed World Lead to Major Problems?

ISSUE 1

Is World Population Growth Out of Control?

YES: Michael Tobias, from *World War III: Population and the Biosphere at the End of the Millennium* (Continuum, 1998)

NO: Max Singer, from "The Population Surprise," *The Atlantic Monthly* (August 1999)

ISSUE SUMMARY

YES: Author Michael Tobias argues that there can be no doubt that under virtually all plausible scenarios of future global population growth, such growth will continue at unprecedented levels and will cause accelerated damage to the planet.

NO: Max Singer, cofounder of the Hudson Institute, a public policy research organization, contends that the world's population will be declining in 50 years as a consequence of an increasing number of countries' achieving a higher level of modernity.

In the context of global issues, a broad debate rages among policymakers and scholars about the respective consequences of increasing population levels and consumption patterns on the Earth's biosphere. The debate is politically charged because recent population growth has been confined primarily to the world's poorer regions, while consumption excesses are typically identified with rich countries.

Although an objective analysis of any specific global problem will likely suggest that increases in both human numbers and per capita consumption patterns play important roles, the rhetoric of most observers typically points to one or the other of these two macroinfluences as the predominant determinant of the magnitude of the problem.

Clearly, population numbers are critical to any global issue, perhaps more critical than any other factor. They affect the issue in both degree and kind. Population levels may serve as a threshold, above which a nonissue suddenly is transformed into a problem. For example, demographers allude to the *demographic threshold,* suggesting that once the percentage of the population below age 15 reaches 50 percent of the total, a country is hard-pressed to provide all the necessary services—education, food, housing, etc.—to this young consuming group prior to their becoming producers of goods and services (a problem of *kind*). But the extent of the problem is also a function of numbers. It is far easier to feed a country of 10 million than one of 20 million (a problem of *degree*). Without high population levels, consumption patterns would have far less impact.

So numbers do matter. Population size is important. And population growth rates are even more significant. They contain a potential built-in momentum that, if left unchecked, creates a geometric increase in births. A simple example illustrates the point. Let us assume that a young mother in a developing nation gives birth to three daughters before the age of 20. In turn, these three daughters each have three daughters before they reach the age of 20. The result of such fertility behavior is a ninefold (900 percent) increase in the population (minus a mortality-rate influence that is much lower than that in the developed world) within 40 years, or two generations. Contrast this pattern with that of a young mother in the developed world, who is currently reproducing at or slightly above replacement level. In the developed world, daughters replace mothers, and granddaughters replace daughters. Add a third generation of fertility, and the developing-world family has increased 27-fold within 60 years, while the developed-world family's size will likely remain virtually unchanged.

Looking at the potential problem at the macro level, consider the impact of the average annual growth rate of 2 percent for low-income countries (excluding China and India) during most of the 1990s. Those countries must also experience an annual 2 percent growth rate in food production, schools, jobs, housing, hospital beds, and all other basics in order to maintain the same quality of life, let alone improve it.

Thus, problems relating to population growth may, in fact, be at the heart of most global issues, including issues of physical conflict and violence. If growth is a problem, control it, and half the ecological battle is won. Fail to control it, and global problem solvers will be swimming upstream against an ever-increasing current. The importance of out-of-control population growth is irrefutable. Its overarching impact is undeniable. The only area of dispute is whether or not the global community has turned the corner in its efforts to curb the high birth rates of the past 40 years throughout the developing world.

In the following selections, the authors reach competing conclusions regarding if and when growth has or will slow down in the developing world and then level off. For Michael Tobias, 40 years of global population control efforts have not yielded the desired results, and the resultant damage to the biosphere during this period has been unprecedented and will only accelerate in the future. Max Singer, on the other hand, argues that the twenty-first century will witness rapid downsizing in the numbers. Not only will population growth level off, he asserts, but the actual size of the population will begin to decline within 50 years.

Demographic Madness

Chomolungma

It is estimated that every second more than twenty-eight people are born and ten die; that every hour nearly eleven thousand newborns cry out. Each day more than one million human conceptions are believed to come about, accompanied by some 150 thousand abortions. Among those newborns, every day thirty-five thousand will die by starvation, twenty-six thousand of them children. Meanwhile, every twenty-four hours the pace of war against the planet increases, sometimes in major affronts, other times imperceptibly, at least by our limited perceptual standards. That war includes the loss of fifty-seven million tons of topsoil and eighty square miles of tropical forest and the creation of seventy square miles of virtually lifeless desert every day. At current birth and death rates, the world is adding a Los Angeles every three weeks. If average human growth rates were to continue at their present course (the so-called "constant fertility variant" or rate of natural increase) the world's population would reach at least ten billion by the year 2030, twenty billion by 2070, forty billion by 2110, and eighty billion by the year 2150. Most social scientists believe the figure will be, at worst, between eleven and twelve billion. But some have not ruled out a population between fifteen and twenty billion.

While I have felt the oppressive weight of people—whether I envision them raging across the countryside, teeming over the concrete, slaughtering other species, burning down forests, or raping one another—I am still not quite conversant with the paradox of my own complicity. That, by association, as one more member of that species, I am as guilty, as out of control and blind as the next man. While my wife and I have had no child, we are two westerners, equal in ecologically destructive impact, I imagine, to at least four hundred of our brethren in the developing world. My response to that estimated ratio should be colossal guilt. Yet, in truth, I cannot trace, or visualize the known burden of myself upon nature. The quantum leap from my personal conduct and lifestyle to an imperiled world is vastly diffused and muted. Yet, there is no doubting the insoluble dilemma, that incessant proximity to more and more members of our own kind, who are at the crux of this biological configuration, of absolutely tragic dimensions. Such bipedal, carnivorous masses are unique in the annals of earthly experience, and surely outweigh all feeble, academic efforts to grapple with demographic theory. Even the theological struggle to explain, and resist, evil falls short of the psychological mayhem, the omnipresent paradox of so many *Homo sapiens.* "There is a concept which corrupts and upsets all the others," writes Jorge Luis Borges. "I refer not to Evil whose limited realm is that of ethics, I refer to the infinite." So many people: six, twelve, fifteen billion people; ever dwindling privacy; the volatility of competing demands, desires, and information. The psyche is overrun by these many components of congestion, this malaise of contagion, these very doubts as to our own nature, and the meaning of so many separate human lives.

So far, in the span of my own life, we have more than doubled our number as a species. Over 90 percent of the growth of the human population has taken place in less than one tenth of 1 percent of our whole species' history. And because half the world's population is now under the age of twenty-four, a third younger than fifteen, it is clear that we have currently reached the base—not the top—of a demographic Mount Everest, or Chomolungma, as those Buddhists dwelling directly beneath the mountain call it.

Though this century has already witnessed an unprecedented human population explosion, we are presently poised to proliferate as never before. Three billion young people will enter their reproductive years in the next twenty years. To sustain even a *nonpetroleum*-based, well-fed economy at current levels of affluence, it has been pointed out that the U.S. population should exceed no more than one hundred million people. We are, as of late 1997, at a petroleum-based population of 270 million. I can think of no better image of this steadfast pyramid of human consumers, this total population dynamic, than that of a mountain like Chomolungma (29,032 feet), cresting above the surrounding Himalayan roof of the world. Where the analogy abruptly ends, however, is with the adjectives appropriate to such a sacred mountain as Chomolungma: wild, free, and magnificent, descriptions that no longer apply to most of the world. Ironically, already Chomolungma herself is showing signs of wear. The effects of overpopulation are starkly clear. Since 1951, thousands of climbers and trekkers to the mountain have left some forty thousand pounds of nondecomposing refuse strewn upon what is, in fact, one of the most fragile of ecosystems, an eight thousand meter peak.

Some social scientists continue to ask, "At what point does the real population crisis set in? Or does it ever set in?" In parts of Egypt, that crisis has meant that school children obtain no more than three hours per day of instruction, so as to make way for incoming shifts of other students; in the Pacific Caroline Islands, the Yapese people were said to have been so overpopulated, and consequently destitute, that "families lived miserably on rafts in the mangrove swamps . . . and that sometimes four hungry men had to make a meal from a single coconut." In 1722, 111 people were left on Easter Island. Polynesians had settled there as early as the third century A.D., and had attained a stable population size of seven thousand. But excessive deforestation meant that no more fishing boats could be constructed, and the eventually malnourished population split into two warring tribes that only succeeded, ultimately, in virtually wiping one another out. Water scarcity in Pakistan, the absolute lack of new arable land in Kenya and Madagascar, are all crisis-level circumstances.

There is absolutely no doubting the links between human population growth—the daunting number of consumers—and environmental turmoil. It is known that, at a minimum, 75 percent of all global demand for fuelwood is the specific function of increasing numbers of consumers. By the year 2030, at least eight billion people will reside in countries previously—but no longer—containing tropical rain forests from which their primary sustenance had once been extracted. Global reforestation is occurring at only a fraction of what would be necessary to ensure long-term fuelwood supplies as against the exponential growth of human numbers. Ninety-four percent of all population growth in the coming decades will likely occur in those developing countries with the most impoverished resource base. In Pakistan, even by the late 1970s there was fuelwood scarcity. Yet Pakistan's population is poised to more than triple in size in the coming two generations, while only 2 to 3 percent of the country's forest cover remains. Where population is increasing most rapidly—from Myanmar to Cameroon—ecological deterioration tends to be the worst; and infant and child mortality, maternal mortality, and the extent of malnutrition and hunger, tend to be the highest.

The trends have long declared themselves. At present, for example, while the world population is growing by 1.8 percent per year, net grain output is growing at less than 1 percent per year. By the end of this century, the amount of arable land per person on the planet will be less than half of what it was in 1951. According to one model, because of the cumulative environmental impacts of overpopulation, global grain harvests will plummet by as much as 10 percent every forty months, on average, in the coming decades, and between fifty and four hundred million additional people will die from starvation.

Nor has this net food deficit been mollified by the nearly eight billion tons of additional artificial fertilizer spewed into the atmosphere each year in the form of carbon dioxide emissions, at least two-thirds of which are the explicit result of overpopulation (i.e., more automobiles and electrification). While future fossil fuel generation in the developed countries may be modified by gains in efficiency and the implementation of best available technologies, in the developing world per capita CO_2 emissions are projected to double in the coming generation. For governments in sub-Saharan Africa, where agricultural production has declined by more than 2 percent annually during the past twenty-four years, condemning at least thirty million people to starvation and chronic malnutrition, and 320 million others, or 65 percent of the total sub-Saharan population to "absolute poverty," such indicators are bewildering.

The sheer habit of finitude plagues any mind that attempts to reconcile seemingly irreconcilable conflict. We use the zero almost as casually as common change. But by 10^5—or one hundred thousand—we are no longer capable of visualizing the size. And this is precisely the point at which Aristotle, writing in Book VII of his *Politics*, suggested the emergence of a crisis in numbers. Human population, he said, is only in balance when it is "both self-sufficient and surveyable." Beyond that surveyable, self-sufficient size, Aristotle advocated, by any standards, hideous alternatives—compulsory abortion, and the mercy killing of children born deformed, though he did not stipulate what kind of deformities. But what does "surveyable" mean in the twentieth century? A man surveys his property. But what about a place like Macao, the most densely human-populated corner of the planet, with 61,383 individuals per square mile over a region of ten square miles? In an essay he named "Despair Viewed under the Aspects of Finitude and Infinitude," (part of his larger work, *On Fear and Trembling*), Soren Kierkegaard psychoanalyzed an individual attempting to survey the unsurveyable. We can't see in the infrared, hear ultrahigh frequencies, nor single ourselves out from a cluster of 5.8 billion other people. The effort leads to loneliness, a kind of ecological existentialism, or insanity.

All of our physical capabilities, desires, and metaphors are conditioned by the need for stability, a precursor of so many of our formulations, general principles, hypotheses, and behavior. Hence, demographers have tended to establish rather arbitrarily a point of stabilization for the human population, much in the manner of Hegel, whose "dialectic" presupposed eventual synthesis and resolution in the human psyche. The World Bank and United Nations have each predicated their respective population projections according to this general impulse, appropriating a formulaic optimism as a means of asserting the inevitability of classic demographic transition. But the whole mental image of humanity's developmental triumphs, the very hope of family planning, is totally absorbed by blue sky. While there is some exceedingly good news (the increasing number of countries showing fertility replacement trends, for example), there is just as much reason for a sinking heart, if not outright panic.

Global Shock

The bomb is now ticking. Every organism dimly perceives the growing shadow. In Oceanic Vanuatu the 1990 population of 151 thousand is expected to reach 205 thousand by the end of the 1990s. And, with its current high TFR [total fertility rate] of 5.53, the island population is expected to double by the year 2015, a mere generation away. Ten years later, in 2025 Vanuatu may well contain 348 thousand people. To feed and house such an increase in so small a region can only have the most devastating environmental consequences, reminiscent of Easter Island. Vanuatu is a surveyable problem. A dot in the ocean. And yet, even in this one workable microcosm, no solution to the population-environment collision has come to light.

Presently, eight to ten Paris-size cities a year are being added to the world's urban population. There is no way, as yet, to conduct real time population audits of these new megalopolises; no method by which the ecological or psychological impacts may be adequately measured. Too much is happening, too fast. Thought of differently, every eight months there is another Germany, every decade, another South America. And while TFRs throughout the developing world (China excluded) have declined from 6.1 to 4.0 on average, and overall population growth from 2.1 to 1.7-1.8 percent per year, getting these numbers down any further is appearing next to impossible. In fact, because the base population is so much huger than ever before, runaway growth is outstripping the aforementioned declines. The numbers, then, are tragically deceptive. For the human race to check its growth even at 16.5 billion, it will have to reach a two-child (replacement) family by the year 2080. In the United States, at present, only 23 percent of the population actually conforms to a two-child norm.

A global demographic transition assumes that declining fertility will follow declining mortality, an assumption that is proving to be less and less tenable in most low-income countries like Indonesia and India. These countries are stricken with an average sixty-two-year life expectancy, and a mean infant mortality of ninety-four per one thousand, versus eight per one thousand in the industrialized countries. More than half of the poor populations live in ecological zones that are biodiversity hotspots, those particularly vulnerable to further economically induced disruption. Eighty-five percent of all future megacities—ecological tinderboxes—are in the developing world. Each year, for example, Egypt, Ethiopia, and Nigeria add more people to the world than all of Europe combined. What we are seeing is the breakup of any predictability, just at the time when more and more government agencies and administrations are depending upon that presumed underlying orderliness to fulfill donor country targets, meet international treaty mandates, pay back loans, and get reelected. But ecological and economic fragmentation are rendering obsolete all classical demographic transition theory. Postindustrial fertility declines required a century in northwest Europe, but only thirty-five years throughout much of the Pacific Rim and Cuba. In Sri Lanka, China, and Costa Rica, diminished fertility did not coincide with noteworthy industrial development, whereas in Brazil and Mexico zealous industrialization has had little impact on continually escalating fertility rates.

By the year 2000, projections show that 1.2 billion women of reproductive age will have three children on average, and the number of those children under fifteen will have risen to 1.6 billion. By 2025 reproductive-age women will number 1.7 billion and their offspring also 1.7 billion. As technology enables women to conceive later in life, TFR data—sampled, typically, at one time—will tend to be less and less reliable. Moreover, male TFRs are yet to be accounted for. What this all means is that projections for future stabilization—be they population-oriented or ecological—are fraught with uncertainty.

Yet, according to the U.N., the U.S. Bureau of the Census, and the World Bank, demographic transition will effect a global human population stabilization in the middle of the coming century. According to an executive summary of a U.N. document from its Population Division, the range of human stabilization is between six billion (clearly impossible) and nineteen billion for the year 2100. These revised figures are based upon a replacement fertility rate of 2.06. But if fertility stabilization were just minutely higher, at 2.17, the world's population would hit nearly twenty-one billion in the year 2150! What is clearly terrifying to consider is that so minuscule an increase, multiplied by the human population over time, given all the current trends and variables, seems not only plausible, but dreadfully conservative. Demographers play a roulette game of chance. Yet, few people in government, or in science, dare to speak about a human population of twenty-one billion. Only the Vatican continues to uphold a belief that the world could carry on, even with forty billion people. Most scientists consider such convictions absurd. What is certain is the fact that human fertility is more combustible than ever before; its extent, advantage, brain power, carnivorous habits, footprint, hedonism, and malevolence, far in excess of the whole biological past combined.

What if, as outlined earlier, even the 2.17 figure is nothing more than demographic dreaming? What if a global average TFR of 3.0 persists, as it has for the past decade? After all, Africa is projected to remain at a TFR of six, its population expected to triple to 1.6 billion by 2025. And as of 1995, fifty-three countries still showed child mortality rates at or above one hundred per thousand. (In Europe, the average is thirteen.) Twenty countries in 1999 will still be officially considered malnourished, providing less than 90 percent of those respective populations' caloric requirements, engendering "stunting" and "wasting," as the conditions are known. And remarkably, sixty-eight countries containing 18 percent of the world's population are currently either maintaining or actually encouraging a rise in their fertility rates. The major population centers—India, China, Indonesia, and the U.S., which until recently prided themselves on family planning—are all losing ground to wave after wave of new baby booms. Given the enormity of the world's present fertility base, the triggers seem to be transcending culture or technology. They are purely mathematical. In spite of all the talk of environmental mediation, international family planning, and the very best intentions, such mathematics indicate a war that has gotten totally out of our control.

Projecting Into the Future?

Population projections are based upon a host of assumptions. If the assumptions are off to any extent, so are the projections. The variables are exasperating. For example, population momentum ensures that all growth will continue for many decades before any assumed new norms are likely to set in. Human disaster tends to work in favor of more and more people. In many countries, economic success is also translating into higher fertility rates, as witnessed in Singapore and China. And, as for the much-touted principle of education, in Kerala it has had a major impact on fertility, but less so in California, Mexico, or Malaysia. And in poor Mali, the literacy rate has actually declined in the past two decades, not risen.

The biggest contradiction inherent to all stablization projections is the very assumption that the human population will ever stabilize; or that the human condition is inevitably headed toward more wealth, more leisure, more security, when in fact all indications suggest the opposite. The standard of living, the so-called quality of life package, is rapidly diminishing for more and more people—billions of people in fact. Against such trends, how can anyone trust that fertility and mortality will eventually fall into some heavenly harmony? Economic and educational advantages are slipping for more people. On what grounds, then, can demography assume anything about demographic resolution?

Population experts have presumed that the conditions making for the most progressive region within a country will spread to all other regions in due time.

In India and China these are doubtful deductions. For example, the replicability of the Tamil Nadu/Kerala TFRs, literacy rates, and IMRs [infant mortality rates] may not be possible in any other Indian states, in which case, if India continues to grow, it will double, triple, and quadruple its size within a time-frame that is projected, instead, to stabilize. Zimbabwe has achieved a rather progressive contraceptive prevalence rate of over 30 percent and yet its TFR is still a high 4.4. The region of Chogoria in central Kenya is a good percentile point below the rest of the country in terms of TFR, but its success does not appear to be spreading elsewhere throughout that nation.

The entire North American growth rate of an annual 0.9 is projected to decrease to 0.3 by 2025 as a result of diminished net migration from Latin America. Yet statistics from California, the country's most populous state, show a doubling of numbers (comparable to Uttar Pradesh) and an enormous increase in migration (not unlike the situation in India's Bihar state). The notion that the world's total fertility rate is projected to decline from 3.3 to 2.9 in seven years, and then to 2.4 by 2025 is based upon nothing more than vague institutional optimism. "For countries with high fertility, the trend is assumed to be downward, and substantial fertility decline is projected." Just like that!

Fifty countries currently show incomes under $610 per capita per year. Another fifty-five or so are lower-middle income, from $610 to $2399. Still another thirty-three countries are, by any Western standards, poor. These include the Czech Republic and Brazil. All of these nations demonstrate the very economic adversity that has traditionally lent itself to higher and higher birthrates, not declines. While declining fertility assumptions by the World Bank and others are said to have no predictive relationship to socioeconomic factors, those are precisely the factors that have held sway during the initial fertility transitions in the two most populous countries in the world, and now appear to be doing so, contrary to most expectations, in Kenya. And if one assumption is incorrect, many probably are. For example, a higher TFR value will skew the anticipated decline in the proportion of individuals under the age of fifteen, meaning a larger percentage of childbearing females, and thus a larger built-in population momentum than expected. Optimism on this stage of life is a house of cards. Thus, the editors of the World Bank Working Papers caution a "universal qualifer that population will follow the indicated path if the assumptions prove to be correct." And if the assumptions prove to be incorrect? Add a global recession and the mounting "trigger" of ecological collapse throughout the world, and all such population assumptions and projections go up in smoke.

To appreciate the sense of discrepancy that prevails within the demographic community, note that the World Bank Working Papers project that China will "remain the most populous country until 2120." Yet dozens of independent demographers all predict that India will overtake China within two generations. Like some macabre game of chicken: two conflicting sets of data race toward a brink, crossing over the domains of urban and rural health care, ecology, religion, anthropology, politics, the world market economy, the shifting geopolitical landscape, the engineering, biotechnology, and military sectors, even ecotourism and telecommunications. The falling market for coconuts, the rising value of tennis shoes, the Pope's travel itinerary, new emergent contraceptive technologies, the age of certain political leaders, the extent of groundwater pollution, the death of mackerel, daily calorie supply, the scope of infant immunization, secondary school enrollment, the rate of inflation, political freedom, and civil rights, all enter into the equation of demographic projection.

In a recent study of women in twelve countries, it was shown that breast-feeding beyond the resumption of menstruation and sexual relations, following a birth had a "considerable contraceptive effect." UNICEF says that breast-feeding could save 1.5 million lives a year, as well as sparing the 250 thousand children who are permanently blinded from vitamin A deficiency disease. Yet, few hospitals or doctors normally advocate breast-feeding as a form of contraception. And the same bias prevails with regard to the "rhythm method" which can be 97.5 percent effective, according to the World Health Organization. Even periodic abstinence has an 80 percent effectiveness rate, better than spermicides or sponges, about the same as a diaphragm, and just slightly less than the condom.

And yet, despite forty years of concentrated family planning worldwide, neither breast-feeding (which of course requires no genius, no literacy, no technology whatsoever) nor other natural methods have exerted the kind of population controls that many have hoped for. Experts argue it is because the educational component is lacking. Others attribute this missed opportunity to maternal morbidity, economic fallout, ecological stress, or lack of access to family planning services affecting at least three hundred million couples. During the decade of the 1990s, another one hundred million couples will need family planning services just to maintain the status quo. According to a WHO/UNFPA/UNICEF statement, every year thirteen million children die before their fifth birthday and five hundred thousand women die from pregnancy complications, but "with current technology, the majority of these deaths could be prevented."

There is no question that new technologies, and the greater application of existing ones, could alter the demographic outlook. UNICEF saves nearly two million children each year through the administering of measles vaccinations. Breast-feeding traditionally saves lives. But the demographic dangling modifier is more intractable than anyone ever imagined: Will such technologies and techniques increase or decrease the overall size of the human population, keeping in mind that it is precisely the advances in medical technology, the best-laid plans of humanitarianism, for example, that paved the way for India's and Africa's population booms?

Such data leaves the question wide open as to whether or not new technologies are likely to have any impact on curtailing fertility in the future.

Every factor influencing demographic projections confronts the same unknowable forces. Disaster breeds disaster. Medical technology devised and applied compassionately can inadvertently invoke even greater pain and suffering by keeping more people alive for longer periods and setting the groundwork for more and more children. Healthier children are understood to be the secret to few children. And yet, just the opposite can also be argued. There are no rules. Human nature, like every other force of nature, defies consistency.

And even where the highest TFRs might be expected eventually to come down (i.e., Rwanda from 6.2, Yemen from 7.2) other high TFRs are now going even higher—in Mozambique, Tanzania, and Niger, for example. And there are a host of other crucial quantitative observations that defy certitude or predetermination. For example, the World Bank estimates that by the year 2000 the developing world will suddenly start declining in growth—just like that. Why should it, coming right after the largest population growth rate in world history? And there are now the largest proportion of childbearing females ever! How can it be assumed that within one generation, twenty-two million fewer children will be born every year? The reason, of course, is that it simply "has to be" that way if fertility and mortality are going to even out, or harmonize, by the year 2150. This "harmony" is an arbitrary goal that says more about the human psyche and its perennial faith in the future, than it does about any hard science of numbers.

A good indication of the inherent vulnerability of such data is the total number of females in their childbearing years (fifteen to forty-four) as calculated in 1995 and again in 2150. In every significant country, there will be many more such women 155 years hence. Unless all other assumptions are fulfilled in the domains of health care, education, environmental remediation, and economic equality, these numbers indicate anything but stabilization.

In countries like Iran, Iraq, Pakistan, Egypt, India, Indonesia, Brazil, Mexico, Nigeria, the Philippines, Vietnam, and China, these inherent fertility faults in the smooth surface of mathematics—inherent volatilities of culture and economics, politics and religion, ecology and healthcare—provide no sure indication whatsoever of the planet's future. In China, 1998 will see 44 million girls added to the "teenage" sector, but in the year 2150, 53 million of them will be added. In India, overall there will be 260 million teenage girls in 1998, but in 2150, 370 million of them. In Malaysia, 1998 will see 3.5 million teenage girls, but the year 2150 will see over 8 million of them. In Pakistan, the gap is enormous: from a 1995 figure of 33 million teenage females to nearly 80 million females of childbearing age in 2150. Officials in many of these countries may insist all they like that there are no teen pregnancies, but the fact remains that millions of teenage girls are having babies and in nearly every country such teen pregnancies are on the rise, as they are dramatically in a place like Los Angeles.

Many of these young people, in over forty countries, in countless interviews and discussions with me, have expressed their common depression, sense of futility, and fatalism about the planet as they view it, adding fuel to their determination to enjoy life while they can. There are very few good arguments to dissuade them from having sex in a world where tigers and chimpanzees are going extinct. Even the prescriptions for "safe sex" lose cogency in a world so fraught with decay. Add to these intuitive syndromes one other astonishing fact: worldwide the average age of first menstruation is dropping! This trend corresponds with what must be seen as a global emphasis on earlier sex. Throughout Latin America, birthrates among teen mothers are extremely high—15 percent in El Salvador, for example. By the year 2000, there will be nearly 130 million teenagers in Latin America. Only 10 percent of them use any form of contraception, often erratically and hence, ineffectively. In Africa, 40 percent of all women have their first child before the age of eighteen. In at least one coastal district of Kenya, a country that has until recently placed a premium on female education, girls are getting married as early as the age of nine.

Only Brazil, among the fastest growing of large nations, will have a slightly lower fertile female count in the twenty-second century than it does now—by a few million. Yet, Brazil's population will have doubled from its 1990 figure of 150 million to over 300 million by then. Again, according to the "harmony" (stabilization) scenario, China's and India's populations are expected to become stable by the year 2150, at 1.8 and 1.85 billion respectively, while Indonesia is expected to reach 375 million at that time. *The Working Papers* see Indonesia's population doubling by the year 2110, yet some government officials within Indonesia see it happening by the year 2040. Even in a defined, relatively small area—Shenyang, China, for example—projections are at odds. Shenyang's Environmental Protection Bureau has declared the city's annual population growth rate to be 1.5 percent. The U.N., however, estimates a growth rate of exactly double that.

In sum, future contraceptive technologies and medical breakthroughs, increasing economic disparities, massive urbanization, emerging viruses susceptible to ecological disruption, a fast-growing population base among teenagers, and the unpredictability of fertility rebounds following disasters, all bode of population instability.

But, regardless of the imprecision and huge discrepancies of demographic projections, one unambiguous fact remains: the current size of the human population has wreaked unprecedented damage on the biosphere, and is going to accelerate that damage. Millions of plant and animal species will be driven to extinction. Hundreds of millions of innocent children, women, and men have been, and will be, slaughtered, in one form or another. A billion people are hungry. The ozone layer is thinning, with consequences that are lethal for every living organism. The air, water, and soil across the planet have been fouled. The forests in many countries are gone or nearly gone. And the mammary glands of every mother on Earth are now infiltrated with DDT and other harmful chemicals. These essential facts—truths that distinguish this century from any other in our history—are all the result of uncontrolled human fertility and thoughtless behavior. Even if we should somehow manage to merely double our population size by the next century, attaining, in other words, a number of twelve billion, the ecological damage will be catastrophic.

Oddly, despite this litany of woes, there is a new sense of ecological consciousness and stewardship throughout the world; a spirit of environmental community that is rapidly trying to mobilize in order to meet this disaster head-on. Whether a similar mobilization to meet the even more difficult crisis of as yet unborn children can be accomplished quickly, humanely, and firmly, will determine whether any kind of qualitative human future is even possible.

Max Singer

 **NO**

The Population Surprise

Fifty years from now the world's population will be declining, with no end in sight. Unless people's values change greatly, several centuries from now there could be fewer people living in the entire world than live in the United States today. The big surprise of the past twenty years is that in not one country did fertility stop falling when it reached the replacement rate—2.1 children per woman. In Italy, for example, the rate has fallen to 1.2. In Western Europe as a whole and in Japan it is down to 1.5. The evidence now indicates that within fifty years or so world population will peak at about eight billion before starting a fairly rapid decline.

Because in the past two centuries world population has increased from one billion to nearly six billion, many people still fear that it will keep "exploding" until there are too many people for the earth to support. But that is like fearing that your baby will grow to 1,000 pounds because its weight doubles three times in its first seven years. World population was growing by two percent a year in the 1960s; the rate is now down to one percent a year, and if the patterns of the past century don't change radically, it will head into negative numbers. This view is coming to be widely accepted among population experts, even as the public continues to focus on the threat of uncontrolled population growth.

As long ago as September of 1974 *Scientific American* published a special issue on population that described what demographers had begun calling the "demographic transition" from traditional high rates of birth and death to the low ones of modern society. The experts believed that birth and death rates would be more or less equal in the future, as they had been in the past, keeping total population stable after a level of 10-12 billion people was reached during the transition.

Developments over the past twenty years show that the experts were right in thinking that population won't keep going up forever. They were wrong in thinking that after it stops going up, it will stay level. The experts' assumption that population would stabilize because birth rates would stop falling once they matched the new low death rates has not been borne out by experience. Evidence from more than fifty countries demonstrates what should be unsurprising: in a modern society the death rate doesn't determine the birth rate. If in the long run birth rates worldwide do not conveniently match death rates, then population must either rise or fall, depending on whether birth or death rates are higher. Which can we expect?

The rapid increase in population during the past two centuries has been the result of lower death rates, which have produced an increase in worldwide life expectancy from about thirty to about sixty-two. (Since the maximum—if we do not change fundamental human physiology—is about eighty-five, the world has already gone three fifths as far as it can in increasing life expectancy.) For a while the result was a young population with more mothers in each generation, and fewer deaths than births. But even during this population explosion the average number of children born to each woman—the fertility rate—has been falling in modernizing societies. The prediction that world population will soon begin to decline is based on almost universal human behavior. In the United States fertility has been falling for 200 years (except for the blip of the Baby Boom), but partly because of immigration it has stayed only slightly below replacement level for twenty-five years.

Obviously, if for many generations the birth rate averages fewer than 2.1 children per woman, population must eventually stop growing. Recently the United Nations Population Division estimated that 44 percent of the world's people live in countries where the fertility rate has already fallen below the replacement rate, and fertility is falling fast almost everywhere else. In Sweden and Italy fertility has been below replacement level for so long that the population has become old enough to have more deaths than births. Declines in fertility will eventually increase the average age in the world, and will cause a decline in world population forty to fifty years from now.

Because in a modern society the death rate and the fertility rate are largely independent of each other, world population need not be stable. World population can be stable only if fertility rates around the world average out to 2.1 children per woman. But why should they average 2.1, rather than 2.4, or 1.8, or some other number? If there is nothing to keep each country exactly at 2.1, then there is nothing to ensure that the overall average will be exactly 2.1.

The point is that the number of children born depends on families' choices about how many children they want to raise. And when a family is deciding whether to have another child, it is usually thinking about things other than the national or the world population. Who would know or care if world population were to drop from, say, 5.85 billion to 5.81 billion? Population change is too slow and remote for people to feel in their lives—even if the total population were to double or halve in only a century (as a mere 0.7 percent increase or decrease each year would do). Whether world population is increasing or decreasing doesn't necessarily affect the decisions that determine whether it will increase or decrease in the future. As the systems people would say, there is no feedback loop.

⁓◉⁓

What does affect fertility is modernity. In almost every country where people have moved from traditional ways of life to modern ones, they are choosing to have too few children to replace themselves. This is true in Western and in Eastern countries, in Catholic and in secular societies. And it is true in the richest parts of the richest countries. The only exceptions seem to be some small religious communities. We can't be sure what will happen in Muslim countries, because few of them have become modern yet, but so far it looks as if their fertility rates will respond to modernity as others' have.

Nobody can say whether world population will ever dwindle to very low numbers; that depends on what values people hold in the future. After the approaching peak, as long as people continue to prefer saving effort and money by having fewer children, population will continue to decline. (This does not imply that the decision to have fewer children is selfish; it may, for example, be motivated by a desire to do more for each child.)

Some people may have values significantly different from those of the rest of the world, and therefore different fertility rates. If such people live in a particular country or population group, their values can produce marked changes in the size of that country or group, even as world population changes only slowly. For example, the U.S. population, because of immigration and a fertility rate that is only slightly below replacement level, is likely to grow from 4.5 percent of the world today to 10 percent of a smaller world over the next two or three centuries. Much bigger changes in share are possible for smaller groups if they can maintain their difference from the average for a long period of time. (To illustrate: Korea's population could grow from one percent of the world to 10 percent in a single lifetime if it were to increase by two percent a year while the rest of the world population declined by one percent a year.)

World population won't stop declining until human values change. But human values may well change—values, not biological imperatives, are the unfathomable variable in population predictions. It is quite possible that in a century or two or three, when just about the whole world is at least as modern as Western Europe is today, people will start to value children more highly than they do now in modern societies. If they do, and fertility rates start to climb, fertility is no more likely to stop climbing at an average rate of 2.1 children per woman than it was to stop falling at 2.1 on the way down.

In only the past twenty years or so world fertility has dropped by 1.5 births per woman. Such a degree of change, were it to occur again, would be enough to turn a long-term increase in world population of one percent a year into a long-term decrease of one percent a year. Presumably fertility could someday increase just as quickly as it has declined in recent decades, although such a rapid change will be less likely once the world has completed the transition to modernity. If fertility rises only to 2.8, just 33 percent over the replacement rate, world population will eventually grow by one percent a year again—doubling in seventy years and multiplying by twenty in only three centuries.

The decline in fertility that began in some countries, including the United States, in the past century is taking a long time to reduce world population because when it started, fertility was very much higher than replacement level. In addition, because a preference for fewer children is associated with modern societies, in which high living standards make time valuable and children financially unproductive and expensive to care for and educate, the trend toward lower fertility couldn't spread throughout the world until economic development had spread. But once the whole world has become modern, with fertility everywhere in the neighborhood of replacement level, new social values might spread worldwide in a few decades. Fashions in families might keep changing, so that world fertility bounced above and below replacement rate. If each bounce took only a few decades or generations, world population would stay within a reasonably narrow range—although probably with a long-term trend in one direction or the other.

The values that influence decisions about having children seem, however, to change slowly and to be very widespread. If the average fertility rate were to take a long time to move from well below to well above replacement rate and back again, trends in world population could go a long way before they reversed themselves. The result would be big swings in world population—perhaps down to one or two billion and then up to 20 or 40 billion.

Whether population swings are short and narrow or long and wide, the average level of world population after several cycles will probably have either an upward or a downward trend overall. Just as averaging across the globe need not result in exactly 2.1 children per woman, averaging across the centuries need not result in zero growth rather than a slowly increasing or slowly decreasing world population. But the long-term trend is less important than the effects of the peaks and troughs. The troughs could be so low that human beings become scarcer than they were in ancient times. The peaks might cause harm from some kinds of shortages.

One implication is that not even very large losses from disease or war can affect the world population in the long run nearly as much as changes in human values do. What we have learned from the dramatic changes of the past few centuries is that regardless of the size of the world population at any time, people's personal decisions about how many children they want can make the world population go anywhere—to zero or to 100 billion or more.

POSTSCRIPT

Is World Population Growth Out of Control?

The growth issue can be most simply structured as one of an insurmountable built-in momentum versus dramatic change in fertility attitudes and behavior. Tobias is correct when he asserts that the population explosion of the latter part of the twentieth century has the potential of future fertility disaster because of the high percentage of the population that is either in the middle of or about to enter the reproductive years.

The key word here is *potential.* Its relevance grows out of the built-in momentum that has caused the developing world's actual population to rise substantially in the last 35 years despite a decline in the growth rate. The most recent comprehensive analysis of population patterns by the United Nations suggests that population growth in the first half of the twenty-first century will increase by over 50 percent, or more than 3 billion people. This is higher than projections from just two years earlier, primarily because of higher projected fertility levels in countries that are slow to show signs of fertility decline. See the United Nations Population Division report "World Population Prospects: The 2000 Revision Highlights" (2001).

According to the UN study, all growth in the first half of the twenty-first century will take place in the developing world. The report acknowledges that population will also grow in the developed world during the first 25 years of the new century, but it predicts that population will decline to levels approximating those in 2000 by midcentury. On the other hand, despite lowering birth rates in the developing world, the built-in momentum results in a 65 percent projected growth (from 4.9 billion in 2000 to 8.1 billion in 2050) during the first half of the century.

Wolfgang Lutz, Warren Sanderson, and Sergei Scherbov take issue with the UN's forecasts in "The End of World Population Growth," *Nature* (August 2, 2001). One major objection is the UN's use of a uniform fertility rate of 2.1 percent. To Lutz et al., the evidence reveals that in many countries—including a significant number in the developing world—the growth rate has fallen below 2.1 percent. As a consequence, Lutz et al. predict a lower growth rate for the period as well as an earlier leveling-off point.

Tobias suggests that the growth rate could continue on its present course, yielding 10 billion people by the year 2030, 20 billion by 2070, and 80 billion by 2150. He acknowledges, however, that a leveling-off figure advocated by most scientists of between 11 and 12 billion is a more realistic figure, although some have focused on a number closer to 20 billion. He likely overstates the probability of massive growth rates unseen in human history, however, when he suggests that "we are presently poised to proliferate as never before." Tobias bases his assessment on the large percentage of the world's population that is under the age of 24 (50 percent) or younger than 15 (one-third).

Singer contends that today's youth will not produce at the same level as their parents and grandparents based on the evidence of the last quarter century. Two factors are at work here. The demographic transition is evident in those countries of the developing world that are experiencing economic growth. Birth rates have dropped significantly, leading to lowered growth rates. But while the rates are higher than those for the newly industrializing countries of the developing world, growth rates for the remaining developing countries have also dropped in a large number of cases. In the latter situation, policy intervention lies at the heart of the lowered rates. The latter effort has been spearheaded by the United Nations and includes the work of many nongovernmental organizations as well. In global conferences held every 10 years (1974, 1984, and 1994), the entire international community has systematically addressed the problem of fertility in the developing world.

For Singer, the key is a change in human values: "Values, not biological imperatives, are the unfathomable variable in population predictions." The United Nations and other organizations have bought into this idea and are pursuing many initiatives to change the views of individuals in developing countries about the desirability of having large numbers of children.

In a sense, both authors are correct in that each acknowledges declining birth rates in the developing world and an eventual leveling-off of growth there. Their disagreement does point out dramatic implications of slight variations in both the timing and the degree of fertility reduction among the poorer countries.

The United Nations serves as an authoritative source on various population data, whether historical, current, or future oriented (see the United Nations Population Information Network at http://www.un.org/popin/). One UN agency, the United Nations Population Fund (UNFPA), issues an annual "State of the World Population," as well as other reports. Two private organizations, the Population Reference Bureau (http://www.prb.org) and the Population Institute (http://www.populationinstitute.org), which are both based in Washington, D.C., publish a variety of booklets, newsletters, and yearly reports. Admittedly, these organizations tend to emphasize the continued urgency rather than the seeds of progress. Others focus on either success stories or the potential for success growing out of recent policy intervention. The Population Council (http://www.popcouncil.org), which is headquartered in New York City, falls into the latter category.

A succinct, centrist, and easily understandable analysis of the future of world population and its implications can be found in Leon F. Bouvier and Jane T. Bertrand, *World Population: Challenges for the Twenty-First Century* (Seven Locks Press, 1999). For an analysis of the different perspectives of the current population debate, see Frank Furedi, *Population and Development: A Critical Introduction* (St. Martin's Press, 1997). Finally, the annual *State of the World* volume from the Worldwatch Institute typically includes a timely analysis on some aspect of world population.

ISSUE 2

Should the International Community Attempt to Curb Population Growth in the Developing World?

YES: Robert S. McNamara, from "The Population Explosion," *The Futurist* (November–December 1992)

NO: Julian L. Simon, from *The Ultimate Resource 2* (Princeton University Press, 1996)

ISSUE SUMMARY

YES: Robert S. McNamara, former president of the World Bank, argues that the developed countries of the world and international organizations should help the countries of the developing world to reduce their population growth rates.

NO: The late professor of economics and business administration Julian L. Simon maintains that international organizations seek to impose birth control methods on developing nations in the misguided name of "virtuous and humanitarian . . . motives."

The history of the international community's efforts to lower birth rates throughout the developing world goes back to the late 1960s, when the annual population growth rate hovered around 2.35 percent. At that time, selected individuals in international governmental organizations, including the United Nations, were persuaded by a number of wealthy national governments as well as by international nongovernmental population agencies that a problem of potentially massive proportions had recently emerged. Quite simply, demographers had observed a pattern of population growth in the poorer regions of the world quite unlike that which had occurred in the richer countries during the previous 150–200 years.

Population growth in the developed countries of the globe has followed a rather persistent pattern during the last two centuries. Prior to the Industrial Revolution, these countries typically experienced both high birth rates and death rates. As industrialization took hold and advances in the quality of life for citizens of these countries occurred, death rates fell, resulting in a period of time when the size of the population rose. Later, birth rates also began to decline, in large part because the newly industrialized societies were better suited to families with fewer children. After awhile, both birth and death rates leveled off at a much lower level than during preindustrial times. This changing pattern of population growth is known as the *demographic transition.*

This earlier transition throughout the developed world differed, however, from the newer growth pattern in the poorer regions of the globe observed by demographers in the late 1960s. First, the transition in the developed world occurred over a long period of time, allowing the population to deal more readily with such growth. On the other hand, more recent growth in the developing world has taken off at a much faster pace, far outstripping the capacity of these societies to cope with the changes accompanying such growth.

Second, the earlier growth in the developed world began with a much smaller population base and a much larger resource base than the developing world did, again allowing the richer societies to cope more easily with the dislocations caused by such growth. The developing world of the 1960s, however, found percentages of increase based on a much higher base. Coping under the latter scenario proved much more difficult. Moreover, available resources to assist in this coping, such as arable land, were far more abundant in the earlier period than today.

Finally, industrialization accompanied population change in the developed world, again allowing for those societies to address resultant problems more easily. Today's developing world has no such luxury. New jobs are not available, expanded educational facilities are nonexistent, unsatisfactory health services remain unchanged, and modern infrastructures have not been created.

The international community formally placed the population issue—defined primarily as excessive birth rates in the developing world—on the global agenda in 1974 with the first major global conference on population, held in Bucharest, Romania. But most of the poor countries resisted the argument of the wealthier nations that they had a problem. Those that accepted the notion pointed to economic development, not to formal fertility reduction programs, as the solution. Both sides eventually came to accept the other's description of the cause and cure of the problem and pledged to work together. Each bought into the assumption that "the best contraceptive was economic development," but until development was achieved, national family planning programs would help lower growth rates. This compromise was not universally accepted by all poor nations; some saw the motives of the developed world in this matter to be less than pure, a kind of ethnic genocide in its extreme.

In the first of the following selections, Robert S. McNamara argues that high population growth is exacerbating an already dire set of conditions in the developing world and that the industrialized countries of the globe should embark on a massive assistance program to help the "have not" countries reduce fertility. In the second selection, Julian L. Simon argues that organizations such as the United Nations Population Fund seek to impose birth control methods on the developing world in the misguided name of "virtuous and humanitarian . . . motives" while attacking the motives of their opponents.

Robert S. McNamara  **YES**

The Population Explosion

For thousands of years, the world's human population grew at a snail's pace. It took over a million years to reach 1 billion people at the beginning of the last century. But then the pace quickened. The second billion was added in 130 years, the third in 30, and the fourth in 15. The current total is some 5.4 billion people.

Although population growth rates are declining, they are still extraordinarily high. During this decade, about 100 million people per year will be added to the planet. Over 90% of this growth is taking place in the developing world. Where will it end?

The World Bank's latest projection indicates that the plateau level will not be less than 12.4 billion. And Nafis Sadik, director of the United Nations Population Fund, has stated that "the world could be headed toward an eventual total of 14 billion."

What would such population levels mean in terms of alleviating poverty, improving the status of women and children, and attaining sustainable economic development? To what degree are we consuming today the very capital required to achieve decent standards of living for future generations?

More People, Consuming More

To determine whether the world—or a particular country—is on a path of sustainable development, one must relate future population levels and future consumption patterns to their impact on the environment.

Put very simply, environmental stress is a function of three factors: increases in population, increases in consumption per capita, and changes in technology that may tend to reduce environmental stress per unit of consumption.

Were population to rise to the figure referred to by Sadik—14 billion— there would be a 2.6-fold increase in world population. If consumption per capita were to increase at 2% per annum—about two-thirds the rate realized during the past 25 years—it would double in 35 years and quadruple in 70 years. By the end of the next century, consumption per capita would be eight times greater than it is today.

Some may say it is unreasonable to consider such a large increase in the per capita incomes of the peoples in the developing countries. But per capita income in the United States rose at least that much in this century, starting from a much higher base. And today, billions of human beings across the globe are now living in intolerable conditions that can only be relieved by increases in consumption.

A 2.6-fold increase in world population and an eightfold increase in consumption per capita by 2100 would cause the globe's production output to be 20 times greater than today. Likewise, the impact on non-renewable and renewable resources would be 20 times greater, assuming no change in environmental stress per unit of production.

On the assumptions I have made, the question becomes: Can a 20-fold increase in the consumption of physical resources be sustained? The answer is almost certainly "No." If not, can substantial reductions in environmental stress—environmental damage—per unit of production be achieved? Here, the answer is clearly "Yes."

Reducing Environmental Damage

Environmental damage per unit of production can— and will— be cut drastically. There is much evidence that the environment is being stressed today. But there are equally strong indications that we can drastically reduce the resources consumed and waste generated per unit of "human advance."

With each passing year, we are learning more about the environmental damage that is caused by present population levels and present consumption patterns. The superficial signs are clearly visible. Our water and air are being polluted, whether we live in Los Angeles, Mexico City, or Lagos. Disposal of both toxic and nontoxic wastes is a worldwide problem. And the ozone layer, which protects us all against skin cancer, is being destroyed by the concentration of chlorofluorocarbons in the upper atmosphere.

But for each of these problems, there are known remedies—at least for today's population levels and current consumption patterns. The remedies are costly, politically difficult to implement, and require years to become effective, but they can be put in place.

The impact, however, of huge increases in population and consumption on such basic resources and ecosystems as land and water, forests, photosynthesis, and climate is far more difficult to appraise. Changes in complex systems such as these are what the scientists describe as nonlinear and subject to discontinuities. Therefore, they are very difficult to predict.

A Hungrier Planet?

Let's examine the effect of population growth on natural resources in terms of agriculture. Can the world's land and water resources produce the food required to feed 14 billion people at acceptable nutritional levels? To do so would require a four-fold increase in food output.

Modern agricultural techniques have greatly increased crop yields per unit of land and have kept food production ahead of population growth for several decades. But the costs are proving to be high: widespread acceleration of erosion and nutrient depletion of soils, pollution of surface waters, overuse and contamination of groundwater resources, and desertification of overcultivated or overgrazed lands.

The early gains of the Green Revolution have nearly run their course. Since the mid-1980s, increases in worldwide food production have lagged behind population growth. In sub-Saharan Africa and Latin America, per capita food production has been declining for a decade or more.

What, then, of the future? Some authorities are pessimistic, arguing that maximum global food output will support no more than 7.5 billion people. Others are somewhat more optimistic. They conclude that if a variety of actions were taken, beginning with a substantial increase in agricultural research, the world's agricultural system could meet food requirements for at least the next 40-50 years.

However, it seems clear that the actions required to realize that capacity are not now being taken. As a result, there will be severe regional shortfalls (e.g., in sub-Saharan Africa), and as world population continues to increase, the likelihood of meeting global food requirements will become ever more doubtful.

Similar comments could be made in regard to other natural resources and ecosystems. More and more biologists are warning that there are indeed biological limits to the number of people that the globe can support at acceptable standards of living. They say, in effect, "We don't know where those limits are, but they clearly exist."

Sustainability Limits

How much might population grow and production increase without going beyond sustainable levels—levels that are compatible with the globe's capacity for waste disposal and that do not deplete essential resources?

Jim MacNeil, Peter Winsemaus, and Taizo Yakushiji have tried to answer that question in *Beyond Interdependence,* a study prepared recently for the Trilateral Commission. They begin by stating: "Even at present levels of economic activity, there is growing evidence that certain critical global thresholds are being approached, perhaps even passed."

They then estimate that, if "human numbers double, a five- to ten-fold increase in economic activity would be required to enable them to meet [even] their basic needs and minimal aspirations." They ask, "Is there, in fact, any way to multiply economic activity a further five to ten times, without it undermining itself and compromising the future completely?" They clearly believe that the answer is "No."

Similar questions and doubts exist in the minds of many other experts in the field. In July 1991, Nobel laureate and Cal Tech physicist Murray Gell-Mann and his associates initiated a multiyear project to try to understand how "humanity can make the shift to sustainability." They point out that "such a change, if it could be achieved, would require a series of transitions in fields ranging from technology to social and economic organization and ideology."

The implication of their statement is not that we should assume the outlook for sustainable development is hopeless, but rather that each nation individually, and all nations collectively, should begin now to identify and introduce the changes necessary to achieve it if we are to avoid costly—and possibly coercive—action in the future.

One change that would enhance the prospects for sustainable, development across the globe would be a reduction in population growth rates.

Population and Poverty

The developing world has made enormous economic progress over the past three decades. But at the same time, the number of human beings living in "absolute poverty" has risen sharply.

When I coined the term "absolute poverty" in the late 1960s, I did so to distinguish a particular segment of the poor in the developing world from the billions of others who would be classified as poor in Western terms. The "absolute poor" are those living, literally, on the margin of life. Their lives are so characterized by malnutrition, illiteracy, and disease as to be beneath any reasonable definition of human dignity.

Today, their number approaches 1 billion. And the World Bank estimates that it is likely to increase further—by nearly 100 million—in this decade.

A major concern raised by poverty of this magnitude lies in the possibility of so many children's physical and intellectual impairment. Surveys have shown that millions of children in low-income families receive insufficient protein and calories to permit optimal development of their brains, thereby limiting their capacity to learn and to lead fully productive lives. Additional millions die each year, before the age of five, from debilitating disease caused by nutritional deficiencies.

High population growth is not the only factor contributing to these problems; political organization, macroeconomic policies, institutional structures, and economic growth in the industrial nations all affect economic and social advance in developing countries. But intuitively we recognize that the immediate effects of high population growth are adverse.

Our intuition is supported by facts: In Latin America during the 1970s, when the school-age population expanded dramatically, public spending per primary-school student fell by 45% in real terms. In Mexico, life expectancy for the poorest 10% of the population is 20 years less than for the richest 10%.

Based on such analyses, the World Bank has stated: "The evidence points overwhelmingly to the conclusion that population growth at the rates common in most of the developing world slows development. . . . Policies to reduce population growth can make an important contribution to [social advance]."

A Lower Plateau for World Population?

Any one of the adverse consequences of the high population growth rates—environmentally unsustainable development, the worsening of poverty, and the negative impact on the status and welfare of women and children—would be reason enough for developing nations across the globe to move more quickly to reduce fertility rates. Taken together, they make an overwhelming case.

Should not every developing country, therefore, formulate long-term population objectives—objectives that will maximize the welfare of both present and future generations? They should be constrained only by the maximum feasible rate at which the use of contraception could be increased in the particular nation.

If this were done, I estimate that country family-planning goals might lead to national population-stabilization levels that would total 9.7 billion people for the globe. That is an 80% increase over today's population, but it's also 4.3 billion fewer people than the 14 billion toward which we may be heading. At the consumption levels I have assumed, those additional 4.3 billion people could require a production output several times greater than the world's total output today.

Reducing Fertility Rates

Assuming that nations wish to reduce fertility rates to replacement levels at the fastest possible pace, what should be done?

The Bucharest Population Conference in 1974 emphasized that high fertility is in part a function of slow economic and social development. Experience has indeed shown that as economic growth occurs, particularly when it is accompanied by broadly based social advance, birth rates do tend to decline. But it is also generally recognized today that not all economic growth leads to immediate fertility reductions, and in any event, fertility reduction can be accelerated by direct action to increase the use of contraceptives.

It follows, therefore, that any campaign to accelerate reductions in fertility should focus on two components: (1) increasing the pace of economic and social advance, with particular emphasis on enhancing the status of women and on reducing infant mortality, and (2) introducing or expanding comprehensive family-planning programs.

Much has been learned in recent years about how to raise rates of economic and social advance in developing countries. I won't try to summarize those lessons here. I do wish to emphasize, however, the magnitude of the increases required in family planning if individual countries are to hold population growth rates to levels that maximize economic and social advance.

The number of women of childbearing age in developing countries is projected to increase by about 22% from 1990 to 2000. If contraception use were to increase from 50% in 1990 to 65% in 2000, the number of women using contraception must rise by over 200 million.

That appears to be an unattainable objective, considering that the number of women using contraception rose by only 175 million in the past *two* decades, but it is not. The task for certain countries and regions—for example, India, Pakistan, and almost all of sub-Saharan Africa—will indeed be difficult, but other nations have done as much or more. Thailand, Indonesia, Bangladesh, and Mexico all increased use of contraceptives at least as rapidly. The actions they took are known, and their experience can be exported. It is available to all who ask.

Financing Population Programs

A global family-planning program of the size I am proposing for 2000 would cost approximately $8 billion, with $3.5 billion coming from the developed nations (up from $800 million spent in 1990). While the additional funding appears large, it is very, very small in relation to the gross national products and overseas development assistance projected for the industrialized countries.

Clearly, it is within the capabilities of the industrialized nations and the multilateral financial institutions to help developing countries finance expanded family-planning programs. The World Bank has already started on such a path, doubling its financing of population projects in the current year. Others should follow its lead. The funds required are so small, and the benefits to both families and nations so large, that money should not be allowed to stand in the way of reducing fertility rates as rapidly as is desired by the developing countries.

The developed nations should also initiate a discussion of how their citizens, who consume seven times as much per capita as do those of the developing countries, may both adjust their consumption patterns and reduce the environmental impact of each unit of consumption. They can thereby help ensure a sustainable path of economic advance for all the inhabitants of our planet.

Julian L. Simon

 **NO**

The Rhetoric of Population Control

Inflammatory Terminology and Persuasion by Epithet

Fear of population growth has been inflamed by extravagant language. Examples are the terms "population explosion," "people pollution," and "population bomb." These terms are not just the catchwords of popular wordsmiths, whose rhetoric one is accustomed to discount. Rather, they have been coined and circulated by distinguished scientists and professors. One example comes from demographer Kingsley Davis, who began a recent article in a professional journal, "In subsequent history the Twentieth Century may be called either the century of world wars or the century of the population plague." (Margaret Sanger also wrote of China that "the incessant fertility of her millions spread like a plague.") Davis also has said that "Over-reproduction—that is, the bearing of more than four children—is a worse crime than most and should be outlawed." Or biologist Paul Ehrlich: "We can no longer afford merely to treat the symptoms of the cancer of population growth; the cancer itself must be cut out." And it was in his Nobel Peace Prize speech, of all places, that agronomist Norman Borlaug spoke of "the population monster" and the "population octopus."

Writers vie with each other to find the ugliest possible characterization of human beings. From a Greenpeace leader, "[O]ur species [is] the AIDS of the Earth: we are rapidly eroding the immune system of the Earth." . . .

Such language is loaded, pejorative, and unscientific. It also reveals something about the feelings and attitudes of contemporary antinatalist writers. Psychiatrist Frederick Wertham pointed out that many of these terms have overtones of violence, for example, "bomb" and "explosion," and many show contempt for other human beings, such as "people pollution." Referring to expressions such as "these days of the population explosion and the hydrogen bomb" and "both nuclear weapons and population growth endanger mankind," he wrote, "The atomic bomb is the symbol, the incarnation, of modern mass violence. Are we justified in even speaking in the same vein of violent death and birthrate? And is it not a perverse idea to view population destruction and population growth as twin evils?"

There is no campaign of counter-epithets to allay the fear of population growth, perhaps because of a Gresham's law of language: Ugly words drive out sweet ones. Reasoning by epithet may well be part of the cause of the fear of population growth in the United States.

Not only epithets but also value-smuggling neologisms have been used against fertility. The term "childfree" is a neologism coined by NON—the National Organization for Non-Parents—as a replacement for "childless." Their intention is to substitute a positive word, "free," for a negative word, "less." This neologism is an interesting example of skillful propaganda. Whereas the term "less" is only slightly pejorative—you can have less of something good (love) *or* of something bad (acne)—the term "free" *always* seems better than "unfree," and one can only be free of something bad. If not having children makes you "free," then this clearly implies that children are bad. In a similar vein, you now hear people speak with pain of "wetlands lost," a phenomenon earlier referred to with pleasure as "swamps drained."

Phony Arguments, Crude and Subtle

Some of the antinatalist propaganda is subtle. While seeming to be only straightforward birth-control information, in reality it is an appeal to have fewer children. Planned Parenthood was responsible for such a campaign on television and radio. The campaign was produced by the Advertising Council as a "public service" and shown on television during time given free by the broadcasters as part of their quid pro quo to the public in return for their licenses; that is, it was indirectly paid for by taxpayers. The following is drawn from a letter written in complaint to the Advertising Council decision makers—the only letter they said that they had ever received.

> You may have seen an advertising campaign staged by Planned Parenthood that ran on radio, television, and in many national magazines. There were a number of specific ads in the campaign including one that was headlined "How Many Children Should You Have? Three? Two? One?"; another that adduced "Ten Reasons for Not Having Children" and, finally, the most offensive one was called the "Family Game": the game was staged on a great monopoly board and every time the dice of life were thrown and a child was born—rather like going to jail without passing "go"—the background audio announced the disasters that came in the wake of children—"there goes the vacation," or "there goes the family room. . . ."
>
> One of the ads enjoins young people to "enjoy your freedom" before, by having children, you let some of that freedom go. Such a theme . . . continues the view that the contribution children make to persons and to society is a purely negative one. In this view children are a loss: they take space, constrict freedom, use income that can be invested in vacations, family rooms, and automobiles. We find no consideration here of how children enhance freedom, and of how the advantages of freedom itself are realized when shared rather than prized as a purely personal possession. Finally, one of the ads encapsulates the spirit of the entire campaign: "How many children should a couple have? Three? Two? One? None?" Such an ad belies the

claim that the advertising avoids the designation of any specific number of children as "preferred." Why not 12? 11? 10? or six? five? four? In the same ad, in order to lead audience thinking, it is noted that the decision to have children "could depend on their concern for the effect population growth can have on society." The direction of the effect on society is implied, but nowhere is the effect analyzed, or even clearly stated.

In summary, the ads not only teach family planning but recommend population control. Moreover, they do this by defining the range of acceptable family size as between zero and three, by placing children as negative objects alongside the positive goods supplied by industry, by equating the bringing up of children with merely equipping them with these same goods, by viewing children as an essential constriction of human freedom, and by suppressing a view of life and children that might lead people to think that having more children is a positive and rewarding act. There are values, not just techniques embodied in those ads.

Not all antinatality rhetoric is that subtle. Some is crude name calling, especially the attacks on the Catholic church and on people with Catholic connections. An example is the bold black headline on the full-page ad that was run in national magazines: "Pope denounces birth control as millions starve." Another example is the dismissal of opposing views by referring to the happenstance that the opponent is Catholic. Consider, for example, the religion-baiting of Colin Clark—a world-respected economist who presented data showing the positive effects of population growth—by sociologists Lincoln and Alice Day: "Colin Clark, an internationally known Roman Catholic economist and leading advocate of unchecked population growth . . ." (Lincoln Day also likened me to a "defrocked priest.") And Jack Parsons writes, "Colin Clark, the distinguished Roman Catholic apologist . . . refrains from discussing optimization of population at all . . . an extraordinary omission." Gunnar Myrdal is not a Catholic and is a Nobel prize winner, and yet *he* called the concept of an optimum population level "one of the most sterile ideas that ever grew out of our science." But Parsons feels free to attribute religious motives to Clark's choice of technical concepts and vocabulary when Clark does not mention this "optimization" concept. And in the widely read text of Paul Ehrlich and others, *Population, Resources, and Environment*, we find a reference to Clark as an "elderly Catholic economist," an innovation in the name calling by referring to Clark's age as well as to his religion.

As a firsthand example in the same vein, my own views . . . were described by Paul Silverman, a biologist, before a packed auditorium on the first and greatest Earth Day, in 1970, as "inspired by Professor Simon's contact with the Bible. . . . Indeed, a new religious doctrine has been enunciated in which murder and abstinence from sex are not distinguishable."

Grabbing Virtue, Daubing With Sin

A rhetorical device of the antinatalists (as of all rhetoricians, I suppose) is to attribute to themselves the most virtuous and humanitarian of motives, while attributing to their opponents motives that are self-serving or worse. Biologist

Silverman again: " . . . people such as Paul Ehrlich and Alan Guttmacher and presumably myself . . . out of our great concern for the future of the world and the threat to the quality of life . . . have urged that voluntary means be adopted for bringing about restraints on the overburdening of our environment by overpopulation. . . . We must, we can, and we will achieve a fine and beautiful world for ourselves and our children to inherit. . . . We can realize a new quality of life, free from avarice which characterizes our current society." (A few minutes before, the same speaker had said, "If voluntary restraints on population growth are not forthcoming, we will be faced with a need to consider coercive measures"—similar to Ehrlich's "by compulsion if voluntary methods fail.") (See my 1991 book, selection 58, for more discussion of how the population-control organizations assume the moral high ground and deny it to others.)

Why Is Population Rhetoric So Appealing?

Let us consider some of the reasons that antinatality rhetoric has won the minds of so many people. (For more extended discussion, see my 1991 book, selection 52.)

Short-Run Costs Are Inevitable, Whereas Long-Run Benefits Are Hard to Foresee

In the very short run, the effects of increased births are negative, on the average. If your neighbor has another child your school taxes will go up, and there will be more noise in your neighborhood. And when the additional child first goes to work, per-worker income will be lower than otherwise, at least for a short while.

It is more difficult to comprehend the possible long-run benefits. Increased population can stimulate increases in knowledge, pressures for beneficial changes, a youthful spirit, and the "economies of scale." . . . That last element means that more people constitute bigger markets, which can often be served by more efficient production facilities. And increased population density can make economical the building of transportation, communication, educational systems, and other kinds of "infrastructure" that are uneconomical for a less-dense population. But the connection between population growth and these beneficial changes is indirect and inobvious, and hence these possible benefits do not strike people's minds with the same force as do the short-run disadvantages.

The increase in knowledge created by more people is especially difficult to grasp and easy to overlook. Writers about population growth mention a greater number of mouths coming into the world, and even more pairs of hands, but they never mention more brains arriving. This emphasis on physical consumption and production may be responsible for much unsound thinking and fear about population growth.

Even if there are long-run benefits, the benefits are less immediate than are the short-run costs of population growth. Additional public medical care is needed even before the birth of an additional child. But if the child grows up to discover a theory that will lead to a large body of scientific literature, the economic or social benefits may not be felt for a hundred years. All of us tend to put less weight on events in the future compared with those in the present, just as a dollar that you will receive twenty years from now is worth less to you than is a dollar in your hand now.

The above paragraphs do not imply that, on balance, the effect of increased population will surely be positive in any longer-run period. The fact is that we do not know for sure what the effects will be, on balance, in fifty or one hundred or two hundred years. Rather, I am arguing that the positive effects tend to be overlooked, causing people to think—without sound basis—that the long-run effects of population growth *surely* are negative, when in fact a good argument can be made that the net effect *may* be positive.

Intellectual Subtlety of Adjustment Mechanisms

In order to grasp that there can be long-run benefits from additional people, one must understand the nature of a spontaneously ordered system of voluntary cooperation, as described by Mandeville, Hume, Smith, and their successors. But this is a very subtle idea, and therefore few people understand its operation intuitively, and believe explanations based on it.

Not understanding the process of a spontaneously ordered economy goes hand-in-hand with not understanding the creation of resources and wealth. And when a person does not understand the creation of resources and wealth, the only intellectual alternative is to believe that increasing wealth must be at the cost of someone else. This belief that our good fortune must be an exploitation of others may be the taproot of false prophecy about doom that our evil ways must bring upon us. (See below on the prophetic impulse.)

Apparent Consensus of Expert Judgment

Antinatalists convey the impression that *all the experts* agree that population is growing too fast in the United States and that it is simply a *fact* that population is growing too fast. An example from Lester Brown: "There are few if any informed people who any longer deny the need to stabilize world population." Other examples come from Paul Ehrlich, "Everyone agrees that at least half of the people of the world are undernourished (have too little food) or malnourished (have serious imbalances in their diet)." And, "I have yet to meet anyone familiar with the situation who thinks India will be self-sufficient in food by 1971, if ever." And from a *Newsweek* columnist and former high State Department official: "Informed men in every nation now know that, next to population growth and avoidance of nuclear war, the despoiling of nature is the biggest world problem of the next 30 years."

These "everyone agrees" statements are just plain wrong. Many eminent experts did not agree with them when they were made (and now the consensus disagrees with them . . .). But such assertions that "everyone agrees" may well be effective in manipulating public opinion. Which nonspecialist is ready to pit his or her own opinion against that of all the "informed people"?

Population as a Cause of Pollution

Fear of population growth is surely heightened by the linking of population and pollution issues. It has come to seem that if one is for pollution control then one must be against population growth. And pollution control in itself appeals to everyone, for very substantial reasons.

To understand why the link-up of population control and pollution control has occurred with such force, we must understand the nature of the rhetoric on both sides of the argument. One can directly demonstrate that more people increase the flow of a pollutant—for example, more cars obviously make more emissions. The argument that more people may reduce pollution is less direct and not so obvious. For example, as more people create a worse auto-emission pollution problem, forces of reaction arise that may eventually make the situation better than ever before.

Furthermore, the ill effects of people and pollution can be understood deductively. More people surely create more litter. But whether the endpoint after a sequence of social steps will be an even cleaner environment can only be shown by an empirical survey of experiences in various places: Are city streets in the United States cleaner now than they were one hundred years ago? Such empirical arguments are usually less compelling to the imagination than are the simplistic deductive arguments.

Population, Natural Resources, and Common Sense

With respect to natural resources, the population-control argument apparently makes perfect "common sense." If there are more people, natural resources will inevitably get used up and become more scarce. And the idealistic, generous side of young people responds to the fear that future generations will be disadvantaged by a heavy use of resources in this generation.

Perhaps such a doomsday view of natural resources is partly accounted for by the ease of demonstrating that more people will cause some particular negative effects—for example, if there are more Americans there will be less wilderness. The logic of the rebuttal must be global and much more encompassing than the logic of the charge. To show that the loss of wilderness to be enjoyed in solitude is not an argument against more people, one must show that an increase in people may ultimately lead to a general expansion of the "unspoiled" space available to each person—through easier transportation to the wilderness, high-rise buildings, trips to the moon, plus many other partial responses that would not now be possible if population had been stationary 100 years ago. It is obviously harder to show how good is the sum effect of these population-caused improvements than it is to show how bad is the partial effect of a decrease in this or that wilderness area that one may enjoy in solitude. Hence the result is a belief in the ill effects of population growth.

Judgments About People's Rationality

At the bottom of people's concern about population growth often lies the belief that other people will not act rationally in the face of environmental and resource needs. Arguments about the need to stop population growth now often contain the implicit premise that individuals and societies cannot be trusted to make rational, timely decisions about fertility rates. This is the drunkard model of fertility behavior. . . .

One of the themes that runs through much of the population movement is that the experts and the population enthusiasts understand population economics better than other persons do. As John D. Rockefeller III put it, "The average citizen doesn't appreciate the social and economic implications of population growth."

It is not obvious why a politician or businessman—even though a very rich one—should have a clearer understanding of the costs of bearing children than "an average citizen." But Rockefeller has much power to convert his opinion into national action.

Forces Amplifying the Rhetoric

Media Exposure

Antinatality views get enormously more exposure than pro-natality or neutralist views. Paul Ehrlich has repeatedly been on the Johnny Carson show for an unprecedented hour; no one who holds contrary views gets such media exposure. This is also clear from a casual analysis of the titles of articles listed in the *Reader's Guide to Periodical Literature.*

Money

The leaders of population agencies that have vast sums of money at their disposal—UNFPA and USAID—take as their goal the reduction of population growth in the poorer countries. Scientists who work in population studies and who have a reasonable degree of career prudence are not likely to go out of their way to offend such powerful potential patrons. Individuals and organizations hitch all kinds of research projects to this money-star. Furthermore, various agencies such as UNFAO realize that their own budgets will be larger if the public and government officials believe that there are fearsome impending dangers from population growth, environmental disaster, and starvation. Therefore, their publicity organs play up these threats.

Standards of Proof and of Rhetoric

The standard of proof demanded of those who oppose the popular view is much much more exacting than is the standard of proof demanded of those who share the popular view. One example: Decades ago the scientific procedure of the *Limits to Growth* study was condemned by every economist who reviewed it, to my knowledge. Yet its findings are still acclaimed and retailed by the "population community." But if I say that the world food situation has been improving year by year, you will either say "Prove it," or "I won't believe it." . . .

Furthermore, anti-doomsday people are in a double bind rhetorically. The doomsdayers speak in excited, angry, high-pitched voices, using language such as *Famine 1975!* They say that such tactics are acceptable because "we are faced with a crisis . . . the seriousness of which cannot be exaggerated." The fears they inspire generate lots of support money from the UN, AID, and popular fund-raising campaigns in full-page advertisements.

Many anti-doomsday people, on the other hand, speak in quiet voices—as reassurance usually sounds. They tend to be careful people. And they are totally ignored. The great geologist Kirtley F. Mather wrote a book called *Enough and To Spare* in 1944 that reassured the public that resources would be plentiful; it was withdrawn from the University of Illinois library just twice—in 1945 and 1952—prior to my 1977 withdrawal. But there are literally armfuls of books such as Fair-field Osborn's 1953 *Limits of the Earth* that have been read vastly more frequently. Even a book published by a vanity press and written by a retired army colonel who has Malthus's first name as "Richard" and who believes that *Over-population* (the title of the book) is a plot of the "Kremlin gangsters" had been withdrawn ten times between 1971 and 1980 (when I checked), and untold more times between its 1958 publication and 1971, when the charge slip was changed.

Finally—The Piper

Many of those in favor of population control are frank to admit the use of emotional language, exaggerated arguments, and political manipulation. . . . They defend these practices by saying that the situation is very serious.

Would the environment have gotten cleaner without the exaggerated and untrue scary statements made by the doomsayers starting in the late 1960s? Perhaps they helped speed the cleanup of our air and water—perhaps. But without false alarms, Great Britain started on its cleanup earlier than did the United States, and went further, faster. And even granting some credit to the doom-sayers, were the benefits worth the costs? Billions of dollars were wasted preparing the airplane industry for $3-per-gallon gasoline, and tens of billions wasted worldwide on raw materials purchased in fear that metal prices would soar.

Much more expensive was the loss of public morale and the spirit of adventurous enterprise due to false environmental scares. And most costly, in my view, has been the inevitable loss of trust in science and in our basic institutions as people realized that they had been systematically fooled.

Exaggeration and untruth run up debts with the piper, who eventually gets paid. Philip Handler, president of the National Academy of Sciences, was a strong supporter of environmental and population control programs. But even he worried about the piper.

> It is imperative that we recognize that we know little and badly require scientific understanding of the nature and magnitude of our actual environmental difficulties. The current wave of public concern has been aroused in large measure by scientists who have occasionally exaggerated the all-too-genuine deterioration of the environment or have over-enthusiastically made demands which, unnecessarily, exceed realistically realizable—or even desirable—expectations. . . .
>
> The nations of the world may yet pay a dreadful price for the public behavior of scientists who depart from . . . fact to indulge . . . in hyperbole.

As far back as 1972, John Maddox, the long-time editor of *Nature*, warned that

> [T]here is a danger that much of this gloomy foreboding about the immediate future will accomplish the opposite of what its authors intend. Instead of alerting people to important problems, it may seriously undermine the capacity of the human race to look out for its survival. The doomsday syndrome [the name of his book] may in itself be as much a hazard as any of the conundrums which society has created for itself.

This may already have come to pass. In 1992 Theodore Roszak refers to himself as one of "those of us who presume to act as the planet's guardians" and who agrees that "the important thing is to spread the alarm" because "there is no question in my mind that these problems are as serious as environmentalists contend." He finds himself taken aback by physician Helen Caldicott's warning that "every time you turn on an electric light you are making another brainless baby" because nuclear power is the cause of anencephalic births along the Mexican border. "Despite my reservations, I do my best to go along with what Dr. Caldicott has to say—even though I suspect . . . that there is no connection between light bulbs and brainless babies." But he worries that "a fanatical antienvironmental backlash [is] now under way" because of the exaggerations and falsehoods.

Were it occurring, a fanatical backlash would not be a good thing; fanaticism of any kind will be destructive in these matters, where cool scientific consideration of the evidence is so desperately needed to assess dangers of all kinds. But this is what Handler and Maddox warned of—the piper that will finally be paid for untruths about population, resources, and environment.

This raises the question: To what extent is the current public belief that the U.S. economy and society are on the skids related to false doomsday fears that we are running out of minerals, food, and energy? And to the unfounded belief that the United States is an unfair plunderer of the world's resources, a supposed "exploitation" that people believe must bring grave consequences for the United States?

And when we count the costs of the doomsayers' excesses we must remember the tragedy of the human lives not lived because countries such as China and Indonesia prevent births in the name of the now-discredited doctrine that population growth slows economic development. It will fall to the future historian to balance the questionable benefits against those undeniable costs.

POSTSCRIPT

Should the International Community Attempt to Curb Population Growth in the Developing World?

There are at least two basic dimensions to this issue. First, ought the international community involve itself in reducing fertility throughout the developing world? That is, is it a violation of either national sovereignty (a country should be free from extreme outside influence) or human rights (an individual has the right to make fertility decisions unencumbered by outside pressure, particularly those from another culture)? Second, can the international community be successful in such policy intervention, or is it just a waste of money?

McNamara assumes that the answer to the first question is "yes" and addresses the second by linking population growth and increases in resource consumption. For McNamara, a 2.6-fold increase in population and an 8-fold increase in consumption per capita (by the end of the twenty-first century) would result in a 20-fold impact on nonrenewable and renewable resources. He cites the projected agricultural needs to demonstrate that the Earth cannot sustain such consumption levels. Additionally, McNamara suggests that other consequences include environmentally unsustainable development, worsening of poverty, and negative impacts on the welfare of women and children.

Those who oppose such intervention point to several different reasons. The first, originally articulated at the 1974 Bucharest conference, argues that economic development is the best contraceptive. The demographic transition worked in the developed world, and there is no reason to assume that it will not work in the developing world. Second, some who oppose intervention simply do not see the extreme negative environmental consequences of expanding populations. For them, the pronatalists (those favoring fertility reduction programs) engage in inflammatory discourse, exaggerated arguments, and scare tactics devoid of much scientific evidence. Simon suggests why such approaches are appealing at first but fall apart upon further analysis. For him, humans have always found a way to fulfill increased resource needs: technological fixes have been the norm since the start of the industrial age.

An excellent account of the 1994 Cairo population conference's answer to how the international community ought to respond can be found in Lori S. Ashford's "New Perspectives on Population: Lessons From Cairo," *Population Bulletin* (March 1995). Other sources that address the question of the need for international action include William G. Hollingsworth's *Ending the Explosion: Population Policies and Ethics for a Humane Future* (Seven Locks Press, 1996) and Elizabeth Liagin's *Excessive Force: Power, Politics, and Population* (Control Information Project for Africa, 1996).

ISSUE 3

Does Population Growth Significantly Harm the Earth's Quality of Life?

YES: Don Hinrichsen, from "6,000,000,000 Consumption Machines," *International Wildlife* (September/October 1999)

NO: N. Gregory Mankiw, from "Be Fruitful and Multiply," *Fortune* (September 7, 1998)

ISSUE SUMMARY

YES: Don Hinrichsen, an environmental reporter and a consultant for the United Nations Population Fund, argues that the planet's natural systems and ecological systems are having increasing difficulty withstanding the negative consequences of continuing population growth.

NO: Economics professor N. Gregory Mankiw maintains that, assuming each additional person on Earth creates resources, if the amount created by each new individual is greater than the amount used by that individual, then the Earth's quality is enhanced by the presence of that individual.

Intuitively, many people, perhaps even a significant majority, are likely to agree that population growth significantly harms the Earth's quality of life. After all, one need look no further than the highway during morning rush hour to find a poignant reminder of the effects of an ever-increasing population. "No vacancy" signs at popular destinations seem to appear with increasing frequency. The waiting lines at the checkout counter, the movie theater, and the amusement park appear to be evidence of a growing population. The list seems endless.

Scholars and policymakers in certain quarters mirror the initial intuition displayed by most people who find themselves in situations like those mentioned above. These activists point to many presumed consequences of population growth as evidence of a declining quality of life associated with such growth. Other researchers argue that a particular population characteristic, such as an absolute size or a certain rate of growth, is neither good nor bad. It is only when this characteristic is combined with something else—too few jobs, insufficient food, inadequate housing, overcrowded schools—that the population condition is vilified. In these circumstances, cries abound to "do something" so that the corresponding nonpopulation condition can be eliminated and one's quality of life can be improved.

Indeed, population growth rates in the developing world were such a cause for concern that the global community formally placed the issue on its agenda in 1974 by holding the first major global conference on population in Bucharest, Romania, and it has continued to expend substantial resources in addressing the problem. Simply put, it was the belief that population growth at both personal and national levels is strongly associated with poverty and a lack of development that spurred world leaders to take action.

It is instructive that the international community has recently revisited its original thesis about the link between growth rates and poverty. For example, an increasing "revisionist" literature that questions whether or not growth and poverty are as intertwined as previously thought is emerging. Also, if the 1974 population conference was informally called an economic development conference because of its emphasis on such development, the 1994 population conference, which was held in Cairo, might properly be termed a women's empowerment conference. One would have been hard-pressed at Cairo, for example, to find mention of the growth-poverty nexus. Instead, the more recent conference focused almost entirely on giving women control of their "minds, bodies and pocketbooks."

This viewpoint about the lack of an association between the rather specific condition of poverty and underdevelopment on the one hand and high population growth on the other repeats itself in more general terms in countless ways. However, the arguments take two basic forms. One, the impact of growth is difficult to measure precisely, and examples presented by its advocates are simply random impressions. Critics who utilize this reasoning argue that the world is much too complex to suggest such a simple paradigm. To them, poverty can just as easily be explained by the lack of capital, bad physical environment, unscrupulous leaders, or the "luck of the draw."

The line of reasoning in the second argument takes the dispute to a higher level of disagreement. To this group of critics, not only is increased population not harmful to one's quality of life, its consequences are essentially positive. For example, more people means more consumption, which means more jobs and more profits. More people also suggests more minds at work solving humankind's major problems, such as finding a cure for cancer. And more American workers means a more secure social security system.

The following selections are at opposite ends of the argument about population growth and quality of life. Don Hinrichsen uses the occasion of the birth of "Earth's six billionth human," the one putting the planet's present population over the 6 billion mark, to highlight the strains placed by continuing growth on the planet's vital systems. N. Gregory Mankiw, in contrast, advances the position that the one factor that is clearly associated with world population increases has been and continues to be higher standards of living for Earth's inhabitants.

Don Hinrichsen  **YES**

6,000,000,000 Consumption Machines

Sometime on October 12, 1999—most likely in China or India, according to demographic probabilities—the Earth's six billionth human [was] born.

As a consumer of water and food, forest products and clean air, animals and the ocean's bounty, this newborn will make but a tiny dent on natural resources during its sojourn on the planet. But put Baby Six Billion together with all the other human consumption machines already here, and alarm bells go off.

Can Earth's natural resources and ecological systems withstand the additive impact of this latest member of our species? Worse yet, what will happen in the year 2025, when Baby Eight Billion is projected to be born?

If this latest addition to the human family arrives in a developed country—say, the United States—he or she will automatically be in the top 20 percent of the human race, at least in terms of good housing, potable water, proper sanitation, a high school or college education, sound medical care, jobs, disposable income and leisure time. But Baby Six Billion will also be part of an elite that consumes in record numbers. In all, 270 million Americans use up nearly 10 billion metric tons of materials a year, 30 percent of the planet's total. And the world's one billion richest people—which also include Europeans and Japanese, among others—consume 80 percent of the Earth's resources.

If, on the other hand, Baby Six Billion is indeed born in the Third World, where three-quarters of humanity is already concentrated, he or she stands a good chance of being thrown into misery and deprivation. One-third of Earth's people—two billion of them—already subsist on just $2 a day or less. Half of all people on Earth have improper sanitation facilities. A quarter have no access to clean water. A third live in substandard housing, many in tin-roofed shacks with dirt floors. A sixth will never learn to read, and 30 percent who enter the global workforce will never get adequate job opportunities. The other five billion people on Earth make do with just 20 percent of the planet's resources.

Rising expectations and the inevitable quest for improved living conditions in the Third World are likely to exacerbate this assault on resources. The average American consumes 37 metric tons of fuels, metals, minerals, food and forest products each year. By contrast, the average Indian consumes less than one metric ton. According to the United Nations, if the entire population of the Earth were to have the same level of consumption as the average American or West European, it would take three Planet Earths to supply the necessary resources.

Regardless of where Baby Six Billion [was] born, he or she will contribute to the relentless collective consumption that continues to devour global resources at rates most experts say are nonsustainable. And in the process, the human newcomer—along with his 5,999,999,999 companions—will produce enormous quantities of waste.

Whether Earth has the ability to absorb more people and provide for their ever-growing needs is not a closed question. Some technocrats have argued that the Earth's greatest resource is the innate capacity of human beings to invent or engineer their way out of population and resource crises. If that is so, however, human ingenuity is not keeping pace with human consumption as measured in the degradation of virtually every natural system—from the chilly North Atlantic with its vital fisheries to the steamy rain forests of Amazonia with their incomparable array of plants and animals.

When all is said and done, human activities caused by population growth and consumption patterns are taking a heavy toll on our planet's life-support systems—and on Earth's other species, which are disappearing at record rates as human numbers rise. The following report looks at the collective effect of six billion consumption machines on six aspects of the natural world. It is a grim picture, with only flashes of hope.

Water

Squandering the Planet's Lifeblood

Water is the liquid of life. Without it, the blue planet would be a dead and barren wasteland. Fresh water is also the most finite of Earth's resources. There is no more water on Earth now than there was 2,000 years ago when the human population was less than 3 percent of its current size. But population growth and rising use have put the squeeze on available resources.

Today, 31 countries with a collective population of half a billion people are experiencing chronic water shortages for all or part of the year. But within just 25 years, that figure will explode to 50 countries and 3 billion people—35 percent of all the people projected to be living on Earth in 2025.

Experts cite two reasons for this drastic increase: population growth plus the increasing demands of agriculture, industry and urban areas. During this century, the world's population has tripled, while the amount of water withdrawn from the planet's finite total has increased by more than six times. Since 1940, annual use of water has grown twice as fast as global population.

While population growth and escalating consumption patterns mean there is less water available per person, water resources are increasingly fouled with all manner of wastes. These include raw sewage and garbage from urban areas, toxic industrial effluents and such agricultural runoffs as fertilizers, pesticides and animal wastes. The UN Food and Agriculture Organization (FAO) estimates that each year roughly 450 cubic kilometers of wastewater—an amount equal to the entire renewable freshwater resources available to Malaysia on a yearly basis—are discharged into rivers, streams and lakes. More than 13 times that amount of clean water is required just to dilute and transport this dirty water. If current trends continue, the FAO projects, the world's entire river flow will be needed just for pollution transport and dilution by the middle of the twenty-first century.

As a global average, agriculture accounts for the lion's share (70 percent) of water taken for human use. Farming also accounts for the largest amount (70 percent in the U.S. and Europe, 50 to 60 percent in developing countries) of pollution to surface and ground waters. Disease carried by dirty water kills more than 12 million people a year, mostly women and children. And nearly all these deaths take place in the Third World.

There is another sinister side to the water crisis. As of 1996, the world's human population was expropriating 54 percent of all the accessible fresh water contained in rivers, lakes and underground aquifers. By 2025, population growth alone will push this figure to 70 percent. As humankind withdraws more and more water to satisfy its unquenchable thirst, less is available to maintain vital wetlands, like the Everglades in Florida.

The wholesale loss and degradation of life-giving riverine, lake and wetland habitats translates to a dramatic decline in populations of other species. Globally, close to one-quarter of all freshwater fish species are either endangered, vulnerable or on their way to extinction. Southeast Asia's Mekong River alone reports a two-thirds drop in fish catch due to dams, deforestation and the conversion of nearly 4,000 square miles of mangrove swamps into rice paddies and fish ponds.

Caught between finite and increasingly polluted water supplies on one hand and rapidly rising demand from population growth and development on the other, many countries face uneasy choices. The World Bank warns that the lack of fresh water is likely to be one of the major factors limiting economic development in the decades to come. It is also likely to spawn wars.

Forests

Earth's Green Lungs Begin to Fade

The earth's green mantle of forests provides humankind with multiple benefits. Forests absorb carbon dioxide and produce oxygen, regulating climate. They anchor soils and prevent erosion. They regulate water flow and protect watersheds. And they provide habitat for countless species of plants and animals. Yet over the course of the past half century, this green mantle has been reduced to tattered remnants.

Currently, about 39.5 million acres of forest, an area roughly the size of Nepal, are cut, bulldozed or burned each year. According to the World Resources Institute (WRI), an environmental think tank based in Washington, D.C., half of the world's original forest cover has been lost, with most of the destruction taking place during the last four decades. WRI reports that only one-fifth of the world's remaining forests are classified as "frontier forests"—pristine areas that have not been disturbed or degraded by human activities.

In Europe, despite green belts and conservation areas, only a tiny patch of the continent's original forest remains, cloistered in Bialowieza National Park in southeast Poland, hard against the border with Belarus. Here 1,000-year-old linden, oak and hornbeam stand cathedral-like—silent reminders of what has been lost irrevocably. Old-growth forests in the United States have been decimated, too; in the contiguous 48 states, 99 percent of frontier forests are gone—an empty echo of what once was.

Most experts link the loss of such forests, directly or indirectly, to human population growth and the insatiable demands of people. Lester Brown of the Washington-based Worldwatch Institute, which monitors human use of resources, reckons that 75 percent of the historical growth of population and 75 percent of the loss in global forest cover has taken place in the twentieth century. "The correlation makes sense," reasons Brown, "given the additional need for farmland, pastureland and forest products as human numbers expand. But since 1950, the advent of mass consumption of forest products has quickened the pace of deforestation."

In the Third World, conversion of forest resources to meet everyday human needs is significant. Dirk Bryant, a senior researcher at WRI, estimates that fuelwood collection and overgrazing by domestic animals are now responsible for degrading about 14 percent of the world's remaining frontier forests, nearly all of which—disregarding northern Canada and Russia—are found in developing countries.

But the relentless and rapidly escalating consumption of forest products by rich countries is also responsible for whittling away much of the remaining pristine forests. The use of paper and paperboard per person has nearly tripled since 1960, with the developed countries of North America, Europe and Asia accounting for most of it. North America, Europe and Japan, with just 16 percent of the global population, consume two-thirds of the world's paper and paperboard and half of its industrial wood.

Researchers at Friends of the Earth in the United Kingdom have determined that humanity's demand for forest products is already 25 percent beyond the point of sustainable consumption. What this means is that given population and income growth in the developing world and continued demand for forest products in the industrialized world, the future of the world's frontier forests and all the ecosystem benefits they provide to humankind are in jeopardy.

Soils

From Bare Earth: Hunger Amid Plenty

The world's topsoils, the "bottom line" in food production, are increasingly eroded and degraded by the demands both of large-scale mechanized agriculture and the desperate needs of subsistence farmers. We could be entering what some experts call the "century of scarcity," as rising demand for food is paralleled by a corresponding drop in supply.

Food shortages may seem an incredulous idea to those who subscribe to the "horn of plenty" scenario of agricultural productivity. After all, since the end of World War II, food production has tripled while population has only doubled. And the daily calories available per person in the Third World have increased from an average of 1,925 in 1961 to 2,540 in 1992.

Yet the prospects are unsettling. Much of the expansion of food production since the post-war days is explained by the adoption of crop rotation, mass production, use of petroleum-based fertilizers, chemical pesticides and expanded irrigation. Since the early 1960s, the introduction of genetically superior, disease-resistant cultivated crops—a signature part of what is known as the Green Revolution—also contributed heavily to food-production gains. But many of these successes have been accompanied by a downside—widespread land abuse and inappropriate agricultural policies, including $228 billion worth of subsidies spent on price supports and outright payments.

The gains in food output are not universal either: There is still widespread hunger in the midst of this plenty. The world has 840 million chronically malnourished people, mostly women and children, while an additional one billion suffer from protein malnutrition. Also, despite slower rates of population growth over the past decade, grain supplies per capita have actually fallen worldwide.

Declines in food production are particularly critical in many poor countries. Between 1985 and 1995, food production lagged behind population growth in 64 out of 105 developing countries. Africa, where food production per person fell in 31 out of 46 nations, fared the worst of all. It now produces nearly 30 percent less food per person than it did in 1970.

The change in direction in food availability in these areas is due primarily to two trends. On the one hand, rapid population growth and changing diets have increased demand. On the other, higher population densities in traditional agricultural areas, fragmentation of small farmsteads, poor land management and inappropriate agricultural and economic policies have suppressed supply.

Together, population growth, rapid urbanization and land degradation have also combined to reduce the amount of food-producing land available for each person on Earth. In developing countries as a whole, the average amount of arable land per person fell from about 0.3 hectares (a hectare equals 2.47 acres) in 1961 to less than 0.2 hectares in 1992.

On top of these alarming developments, nearly 2 billion hectares of crop and grazing land—an area larger than the United States and Mexico combined—suffer from moderate to severe soil degradation. The main causes are soil erosion, loss of nutrients, damage from inappropriate farming practices (including poorly built irrigation systems) and the misuse of agricultural chemicals. In the Philippines, for instance, nearly one-quarter of all cropland has been severely degraded.

According to WRI projections, by 2025 about 3 billion people, 35 percent of the global population, will live in land-short countries, with less than 0.07 hectares of fertile land per person. That is roughly the size of two tennis courts.

Air

Dark Skies, Changing Climates

Clean air is the life-giving resource most people take for granted. Yet increasingly, as human population spirals and consumption rises, the air we breathe is becoming both an agent of illness and the vehicle for modifying Earth's climate.

Few experts dispute the simple fact that more people means more air pollution. Even with the availability of vastly improved technologies to limit pollution, population growth translates directly into more use of energy, more cars on the road, more factories and hence more dirty urban air.

In turn, that often results in severe health problems. Today, more than one billion people suffer from dangerously high air-pollution levels. Most of those live in sprawling Third World cities where industries and power plants have few, if any, pollution controls and where traffic jams are a perpetual feature of urban life. Up to 700,000 of those people die every year from the air they breathe.

Cities such as Bangkok, Manila and Beijing are often entombed in a sickening pall spewed out from a rapidly growing fleet of vehicles and uncontrolled industrial emissions. In these cities and 17 others, air pollution—most commonly in the form of sulfur oxides, oxides of nitrogen, carbon monoxide and ozone—is one of the leading causes of respiratory infections and premature death. Just breathing the air in Mexico City has the same health effect as smoking three packs of cigarettes a day.

On the consumption side, the distribution of energy is uneven. Currently, the richest fifth of humanity consumes close to 60 percent of the world's energy, while the poorest fifth uses just 4 percent. The benefits of the fossil-fuel revolution, which drives industrial nations, have still not reached a full third of humanity—the two billion people who must burn fuelwood and organic waste for heating, cooking and lighting.

The other side of the atmospheric pollution problem is climate change, often called global warming. When carbon from burning of wood, coal, oil and other fossil fuels is released into the atmosphere, it combines with oxygen to form carbon dioxide, the gas responsible for two-thirds of human-induced changes in the world's climate. Atmospheric concentrations of carbon dioxide in 1997 reached 363.6 parts per million, the highest in more than 160,000 years. Altogether, carbon emissions are rising faster than the rate of population growth. In 1997, according to the Worldwatch Institute, global emissions of carbon totaled 6.3 billion tons. Since 1950, world carbon emissions have increased fourfold. Though western industrialized countries currently account for close to half this output, developing countries have increased their share dramatically in the past decade and are collectively responsible for 40 percent of global carbon emissions. China is now the world's second largest emitter, after the United States, with a 14 percent share.

Over the course of the next century, atmospheric concentrations are expected to double, triggering potentially devastating climatic changes on a regional and global scale. By 2100, according to the U.S. National Academy of Sciences, sea levels may rise by up to one meter, inundating vast swaths of coastal land, while average surface temperatures may increase by up to 3.5 degrees Celsius. Destabilization of the Earth's climate engine is expected to result in more intense heat waves, more severe droughts and floods, more devastating storms (tornadoes and hurricanes) and more frequent forest fires. These events, in turn, can add to the problem. The six months of extensive forest fires in Asia in 1997 and 1998 released more carbon into the atmosphere than Western Europe emits in an entire year.

Oceans

Trouble in Earth's Liquid Heart

Oceans, where life first evolved 3.5 billion years ago, cover 70 percent of the globe's surface. They wrap around the planet like an insulating blanket, making life possible on Earth today.

Oceans are the engines that drive the climate, defining weather and storing huge quantities of solar energy. They also make up the liquid heart of the planetary hydrological cycle, enabling roughly 430,000 cubic kilometers of water to evaporate every year.

But even this vast watery world is coming under increasing pressure from human activities. Just over half of humanity—some 3.2 billion people, according to some estimates—live and work within 120 miles of a sea coast, on just 10 percent of the Earth's land area. Two-thirds live within 250 miles of a coast.

These mounting human numbers and the development that follows in their wake have taken a grim toll on ocean resources nearby. Half the world's coastal wetlands, including salt marshes, for instance, have disappeared. And close to 70 percent of the world's beaches are eroding at rapid rates because of human impacts.

Coastal ecosystems, valuable because they function as nurseries for fish and other sea life, have been especially hard hit. Over the past century alone, 25 million hectares of mangrove forests—multi-rooted trees on the edge of the sea—have been destroyed or grossly degraded. Seagrass beds—underwater meadows in coastal shallows—have fared little better and are in retreat near virtually all inhabited coastal areas.

Coral reefs, the rain forests of the sea with perhaps 1 million species, are being pillaged as well. They are poisoned by sewage outfalls, overfished, dynamited, pummeled by ship's anchors, broken by recreational divers and bleached by unseasonally warm temperatures. Of the world's 230,000 square miles of reef-building corals, 60 percent could be lost within 40 years, marine biologists fear. Over 80 percent of the reefs in Southeast Asia alone are in peril.

One of the biggest threats to the integrity of ocean ecosystems is directly attributable to people and their insatiable demand for protein: the relentless hunt for fish. Of the world's 15 major oceanic fisheries, 11 are in decline. The catch of Atlantic cod has dropped 70 percent since 1970, while bluefin tuna stocks have declined by 80 percent over the same period.

A fivefold growth in seafood consumption since 1950 has pushed these and other fisheries to the brink and beyond. Between 1991 and 1995 the world's commercial fleets hauled in, on average, 84 million tons of seafood a year. Since seafood provides close to 20 percent of the world's total animal protein intake—up to 90 percent in the South Pacific and parts of Southeast Asia—the decline in fish catches is eroding food security for a number of poor countries in the tropics.

The overcapacity of the world's fishing fleets has itself become a threat to the integrity of ocean ecosystems. Currently, 5.8 million square miles of ocean bottom are trawled each year, the marine equivalent of strip-mining. Since bottom trawls are indiscriminate harvesters of marine life, the by-catch from these operations constitutes a horrendous waste of potential food. Every year, 10 pounds of fish and shellfish are discarded for every person on Earth—up to 40 million tons.

In too many places, the sea has also become a dumping ground for oil and a giant cesspool to collect the runoff of poisons from inland sources. Each year, for instance, effluents flowing from the Mississippi River system leave a lifeless dead zone 30 miles out into the Gulf of Mexico.

Animals

Plundering the Planet's Species

Human life cannot exist in the absence of complicated interactions of millions of species in biological systems. Yet we live in a period of the greatest loss of plant and animal species since the mega-extinctions of the Jurassic Period 65 million years ago.

Every year over the course of the coming decades, 50,000 plant and animal species are likely to disappear, ecologists warn. The percentage of birds, mammals, fish, reptiles and amphibians threatened with extinction is now in double digits, and the loss of insects and microorganisms is incalculable. Overall, human-induced habitat loss, killing by bushmeat hunters in the Tropics and the introduction of nonnative species, among other problems, have conspired to change the lineup of species on Earth.

Loss of biodiversity is not limited to wildlife. Since 1900, about three-quarters of the genetic diversity of agricultural crops have also disappeared, according to FAO estimates, along with half the wild gene pool upon which domestic cattle are dependent for improving their resistance to diseases, pests and changing environmental conditions.

Increasing population density and pressure for faster but unmanaged economic development are largely to blame. In a study of 50 countries in Asia and Africa, the United Nations Population Fund found that the loss of natural habitat was greatest in high-density areas and least in low-density areas. In the 10 countries that had lost the most habitat, population density averaged close to 200 people per square kilometer. In the 10 countries that had lost the least amount of habitat, the population density averaged just 29 people per square kilometer.

The outlook is particularly bleak in some of the most biologically rich countries of the Third World, where population growth and unsustainable exploitation of natural resources is savaging habitat in "biodiversity hotspots"— ecosystems with a superabundance of plant and animal species. So far, 24 of these hotspots containing half the planet's land species have been identified. Overall, five of the six most biologically diverse countries could see more than two-thirds of their original habitat destroyed or grossly degraded by the middle of the next century.

Meantime, the world's last great expanses of pristine, mostly uninhabited tropical forests now face imminent destruction. These large tracts of land— in the Guayana Shield region of northern South America, Amazonia, Africa's Congo and the island of New Guinea—are prime targets for logging. Together, they are about the size of the state of Alaska.

In other areas, the introduction of nonnative, or exotic, species contributes to extinction woes. Hawaii's native fauna and flora have been decimated by species brought in, deliberately or by accident, by people. On the U.S. mainland, exotics have been implicated in close to 70 percent of all fish extinctions this century. In Europe, much of the Black Sea's fauna has been eliminated by a combination of overfishing, pollution and exotics. Its commercially valuable fish species have declined from 26 to 5 in a decade.

On top of all that, an ominous new term has been added recently to the biologists' lexicon of threats to animals: "defaunation," also referred to as "the empty forest." From Laos to Congo, Brazil to Madagascar, impoverished people desperate to put food in the pot are killing whatever moves. Now, vast areas of tropical forest have been scoured nearly clean by hunters of bushmeat. For the first time, there are large areas of available habitat with few birds or mammals to live in them.

NO

N. Gregory Mankiw

Be Fruitful and Multiply

I confess: My wife and I are about to commit what some people consider a socially irresponsible act. Toward the end of the summer, we will bring our third child into the world.

A third child means, of course, that my wife and I are contributing to the world's population explosion, and to some people this makes our decision more than personal. To see how guilty I should feel, I just read the recent book by nature writer Bill McKibben, *Maybe One: A Personal and Environmental Argument for Single-Child Families*. McKibben tries to accomplish through persuasion what China has done through government fiat—make one child per couple the norm. The book wasn't published in time to change my family size, but even if it had been, it wouldn't have persuaded me to abstain from reproduction.

As McKibben is well aware, fear of overpopulation has a long and embarrassing history. Two centuries ago, Thomas Malthus argued that an ever-increasing population would continually strain society's ability to produce goods and services. As a result, mankind was doomed to forever live in poverty—a prediction that led Thomas Carlyle to label economics "the dismal science."

Fortunately, Malthus was far off the mark. Although the world population has increased about sixfold over the past two centuries, living standards are much higher. The reason is that growth in mankind's ingenuity has far exceeded growth in population. New ideas about how to produce and even about the kinds of goods to produce have led to greater prosperity than Malthus—or anyone else of his era—could have ever imagined.

The failure of Malthus' prediction, however, has not stopped others from repeating it. The most famous modern Malthusian is biologist Paul Ehrlich, whose 1968 book, *The Population Bomb,* warned of impending worldwide shortages in food and natural resources. Thirty years later, however, most natural resources are in abundant supply and are available at low prices. Even the famines that sometimes ravage less-developed countries are rarely due to overpopulation—civil war is a more common cause. Nonetheless, the fear of worldwide shortages because of overpopulation remains widespread.

Among the wealthy, reducing population growth is a popular cause. When David Packard, co-founder of Hewlett-Packard and father of four, died leaving $9 billion to his foundation, he specified that the foundation's highest priority should be lowering global birth rates. Packard's efforts may someday be dwarfed by those of investor Warren Buffett, father of three. Buffett claims he will give away most of his vast wealth, and according to some reports, the problem that most moves Buffett is the population explosion.

Those who fear overpopulation share a simple insight: People use resources. They eat food, drive cars, and take up space. Because resources are scarce, the only way to improve living standards, Malthusians argue, is to limit the number of people with whom we have to share these resources.

The rebuttal to this argument is equally simple: People create resources. They bring into the world their time, effort, and ingenuity. Before deciding whether world population growth is a curse or a blessing, we have to ask ourselves whether an extra person added to the planet uses more or less resources than he or she creates.

Environmentalists such as McKibben view humans as rapacious consumers who devour as much as they can get their hands on. About this, the environmentalists are largely right. But there is no problem as long as people pay for what they consume. In a market economy, the price system ensures that no one can consume resources without first creating some of equal or greater value.

Problems do arise when important resources fail to have prices attached to them. For instance, consider the issue that most concerns McKibben—global warming caused by the use of fossil fuels. Because people burning gasoline in their cars or oil in their furnaces do not pay for their impact on climate, they burn too much. The solution, according to McKibben, is fewer people. A more direct solution is a tax on fossil fuels.

Perhaps the most important resource without a price is society's pool of ideas. Every time a baby is born, there is some chance that he or she will be the next Newton, Darwin, or Einstein. And when that happens, everyone benefits. Although the government can easily protect the environment with a well-designed tax system, spurring the production of great ideas is much harder. The best way to get more geniuses is to have more people.

As a serial procreator, therefore, I make no apologies. When I welcome Peter Mankiw onto our planet, I will do so without a shred of guilt. I don't guarantee that he will find a cure for cancer or a solution to global warming, but there is always a chance. And in that chance lies the hope for our species.

POSTSCRIPT

Does Population Growth Significantly Harm the Earth's Quality of Life?

Research scholars and policymakers have not reached a conclusion about the role that population growth precisely plays in affecting our resource base or our quality of life in general. Nonetheless, observers of change, whether local or global, typically acknowledge that population plays a significant role. The problem is determining how much of the diminished quality is due specifically to population factors.

Sources abound that address the link between population and environmental factors. One important addition to this huge body of literature is U.S. State Department official Richard E. Benedick's article "Human Population and Environmental Stresses in the Twenty-First Century," in the Woodrow Wilson Center's *Environmental Change and Security Report* (Summer 2000). In it, Benedick proclaims that "inhabitants of planet Earth find themselves threatened by environmental dangers that would have been unimaginable" earlier. He explores his thesis with reference to three different links to population: forests, freshwater, and climate change.

Hinrichsen uses the same strategy in exploring the consequences of increased population growth for quality of life. People use resources. More people probably use more resources. Water, forests, soils, air, oceans, and animals are adversely affected by each additional person. Resource depletion is intimately tied up with quality of life. Hinrichsen may be right, but he has not been able to isolate the magnitude of population's effect on the systems that ensure a good quality of life.

Those who challenge the conventional wisdom, like Mankiw, suggest that quality of life is more about living standards than one's natural surroundings. The human race has managed to have its ingenuity grow faster than its numbers, Mankiw says, so all may benefit. Moreover, he argues that most natural systems are in abundant supply. He suggests that while it is true that people use resources, they also produce resources.

Lester R. Brown, Gary Gardner, and Brian Halweil outline the effects of population growth along 19 dimensions in *Beyond Malthus: Nineteen Dimensions of the Population Challenge* (Worldwatch Institute, 1999). Bjørn Lomborg's *The Skeptical Environmentalist: Measuring the Real State of the World* (Cambridge University Press, 2001) is enjoying huge sales and receiving an enormous amount of publicity, much of it critical, because of its questioning of most of the more conventional arguments (or "The Litany," in the words of the author) found across a variety of global issues. In a sense, *The Skeptical Environmentalist* can be viewed as the successor to the late Julian L. Simon's *The Ultimate Resource 2* (Princeton University Press, 1996).

ISSUE 4

Will Declining Population Growth Rates in the Developed World Lead to Major Problems?

YES: Peter G. Peterson, from "Reforms That Aging Industrial Countries Must Undertake," Luncheon Address to a CSIS Policy Summit on Global Aging (January 25–26, 2000).

NO: Leon F. Bouvier and Jane T. Bertrand, from *World Population: Challenges for the Twenty-First Century* (Seven Locks Press, 1999)

ISSUE SUMMARY

YES: Peter G. Peterson, chairman of the Blackstone Group, a private investment bank, argues that global aging will dominate the public-policy agenda of the developed world as it attempts to cope with a vast array of problems associated with such aging.

NO: Adjunct professor of demography Leon F. Bouvier and Jane T. Bertrand, chair of the Department of International Health and Development at Tulane University School of Public Health and Tropical Medicine, argue that a future developed society will be quite able to cope with declining growth rates.

While the poorer nations of the world have endured the consequences of over 30 years of huge birthrates, the richer sectors of the globe have witnessed the opposite phenomenon for an even longer period. The demographic transition began far earlier in the then newly industrialized countries of Western Europe and the United States. And for almost half a century, these nations have witnessed the third and final stage of this transition—low birth and death rates. The drop in the death rates in these countries has been a function of several factors. One major factor has been the dramatic decline in both infant mortality (within the first year of birth) and child mortality (within the first five years) due to women being healthier during pregnancy and nursing periods and to the virtually universal inoculation of children against principal childhood diseases. A second factor is that once people reach adulthood, they are living longer, in large part because of medical advances against key adult illnesses, such as cancer and heart disease.

Declining mortality rates yield an aging population in need of a variety of services—health care, housing, and guards against inflation, for example—provided, in large part, by the tax dollars of the younger, producing sector of society. As the "gray" numbers of society grow, the labor force is increasingly called upon to provide more help for this class. Moreover, the decline of infant and child mortality suggests that a potentially larger group will live to adulthood and become part of the producing class, with the potential to produce more children who will continue the cycle.

Declining birth and death rates in developed countries mean that significantly more services will be needed to provide for the aging populations of these countries, while at the same time, fewer individuals will be joining the workforce to provide the resources to pay for these services. However, some experts say that the new workforce will be able to take advantage of the skills of the more aged, unlike in previous eras. In order for national economies to grow in the information age, an expanding workforce may not be as important a prerequisite as it once was. Expanding minds, not bodies, may be the key to expanding economies and increasing abilities to provide public services.

However, the elderly and the young are not randomly distributed throughout society, which is likely to create regional problems. In the United States, for example, the educated young are likely to leave the "gray belt" of the north for the Sun Belt of the south, southwest, and west. Who will be left in the older, established sectors of the country that were originally at the forefront of the industrial age to care for the disproportionately elderly population? What will happen 30 or 40 years from now, when the respective sizes of the young and the elderly populations throughout the developed world will yield a much larger population at the twilight of their existence? Although the trend is most evident in the richer part of the globe, people are also living longer in the developing world, primarily because of the diffusion of modern medical practices. But unless society can accommodate the skills of older people, they may become an even bigger burden in the future for their national governments.

In the following selection, Peter G. Peterson presents the conventional argument about the doomsday condition of "gray America." He introduces the phrase "the Floridization of the developed world" to capture the essence of the problems associated with the changing age composition in industrial societies. In the second selection, Leon F. Bouvier and Jane T. Bertrand acknowledge that the developed world is on the "brink of an entirely new age" because of the leveling off of its population, but they do not assume that the future society will be less able to cope with the changing demographics. To them, the information age will be knowledge-intensive, not labor-intensive. That is, labor's traditional role in production—the hands and backs of a younger workforce—will be replaced by those who are best able to use their minds, no matter what age.

Peter G. Peterson

Reforms That Aging Industrial Countries Must Undertake

There is a great deal of talk in public policy circles about a variety of global challenges. I would argue that a far less discussed challenge, the aging of the world's population—by which I mean the unprecedented growth in the number of elderly and the unprecedented decline in the number of youth in the developed world—will do far more to shape our collective future than any of the other global challenges.

And unlike other challenges, the timing and magnitude of this global aging challenge is largely determined. The elderly of the next century have already been born; they can be counted and their cost to the public retirement systems can be projected. The projected costs would simply overwhelm the resources of even the greatest powers and make the costs of these other global challenges look like petty cash.

For those of you who are not geezers or who have looked at the subject of aging more on a domestic basis, permit me to start with a few global demographic tidbits:

Until the Industrial Revolution, the odds of a random encounter with an elderly person were 1 in 40. A few decades from now, in some developed countries, they will be 1 in 3. By the 2030's some developed countries may exceed a median age of 55—20 years older than the oldest median age on earth as recently as 1970. Global life expectancies have grown more over the past 50 years than over the previous 5000 years. Over the last century, longevity in the developed world grew from 45 years to over 75 years.

The 'Floridization' of the developed world   When will the percentage of elderly in Florida be true for [the] nation as a whole?

- U.S. 2023. Italy 2003. Japan 2005. Germany 2006.

An unprecedented burden on working-age people

- In most developed countries, the working-age portion of the population will shrink in absolute terms

- In Japan, between 2010 and 2020 there will be a 25% decline in the number of workers under the age of 30

Another demographic tidbit of great relevance in our pay-as-you-go system: over the next 25 years in the developed world, the number of elders or benefit receiving retirees is projected to grow 14 times faster than the number of taxpaying workers. As a result, the ratio of working taxpayers to non-working pensioners is projected to fall from today's ratio in the developed world of 3.1 to 1.5 by 2030 and in countries like Germany and Italy to 1 to 1 or even lower.

Another daunting statistic—but in other ways wonderful statistic—is the extraordinary growth in the so called old-old (over 85 years). A six-fold increase worldwide is anticipated over [the] next 50 years.

- These old-old consume 2–3 times more health care than the young-old. Over a recent decade their health care costs also grew 2½ times faster than the 65–69 year olds.
- None of these official projections contemplate any widely predicted major medical, biological, genetic breakthroughs. For example, U.S. longevity estimates for 2050 are already achieved by Japan.

Rapidly falling birthrates   As recently as 30 years ago, [the] average number of lifetime births per woman was about 5. Thereafter, a stunning drop in birthrate and completely unpredicted behavioral revolution—driven by: 1) Rising affluence, 2) Urbanization, 3) Female participation in work force, 4) New birth control technologies, 5) Legalized abortion.

These factors combined to cut the birth rate in half. It is fast approaching the replacement rate of 2.1. That is [the] global number. In the developed world, the birth rate [is] already down to 1.6. Italy, ironically enough as a leading Catholic nation has, at 1.19, the lowest rate among the developed nations. Been meaning to talk to Pope about that. The U.S., at about 1.9, is at the highest levels of developed countries and with our high immigration rates, we are the exception that expects a slight growth in population. Italy, on the other hand, projects a one-third drop in population in only 50 years.

Developed nations will confront a profound new phenomenon: current populations are projected to shrink significantly. Developing countries of course are still expanding, but they are also aging very rapidly. In 1950, 7 out of [the] 12 most populous nations in the world were developed countries. In 2050 there will be only 1—the United States. Nigeria, Pakistan, Ethiopia, Congo, Mexico and the Philippines replace the other developed nations.

There are many ways to measure fiscal impacts of current programs

1. First, the level of unfunded liabilities.

- These are benefits already earned by today's workers for which nothing has been saved
- Among developed countries, the pension portion of public retirement cost amounts to $35 trillion
- Add in health care, and the total is at least twice that much

- Nations' unfunded liabilities are six times greater than their official public debt
- In the U.S. we coin euphemisms, to put it gently, to convey the false impression these programs are funded. The term "trust fund" is an oxymoron. The trust fund does not contain assets, but liabilities—it is a convenient, but irrelevant, accounting and political fiction.

Thus, Americans are endlessly told, the Trust Fund will keep Social Security "solvent until 2034." If you believe that you will believe that Dennis Rodman has signed up for a course in assertiveness training. In truth, in these hand-to-mouth, pay-as-you-go systems, what really matters are annual operating deficits. They represent future benefits that need to be financed.

Over $8 trillion in aggregate deficits will accumulate between the years 2014 and 2034. Add on Medicare, and the deficit sum tops $15 trillion. These total to $23 trillion, or over $9 trillion in today's dollars. Against this set of melancholy facts, how more disingenuous can our political candidates be than to suggest to the American people that saving a fraction of a projected surplus over the next 10 years of $2–3 trillion saves Social Security?

2. Another way to assess fiscal effects is spending increase as a percentage of GDP. The amount spent on old-age entitlements varies among developed countries, although it always tops the amount spent on defense.

The Kerrey/Danforth Commission, a group of 20 Democratic and Republican Congressmen and Senators who agree on virtually nothing, were unanimous in their recognition that the entitlement monster could soon devour the federal budget. Absent change, by 2030, there would be not a penny left for anything, not even interest payments.

3. Still a third way of assessing fiscal effects is to think of what kind of tax increase would be required to sustain current pension promises. An unthinkable 25–40% of every worker's payroll would have to be taxed. In fact, some countries have payroll taxes that already exceed 40%.
4. Still another way of thinking of fiscal effects is to imagine trying to finance your way out of them. My analysis shows they would consume all the savings of the developed world. If it is true that industrial democracies can only respond to immediate crises—the so-called Pearl Harbor Syndrome—then such a crisis could easily occur in capital markets as various countries try to finance their way out of their problem.

When one hears these melancholy fiscal effects, the present system seems truly unsustainable. Herb Stein, the Nixon humorist and collector of oxymorons of government, quipped "When your horse dies, may I suggest you dismount." However, not withstanding all the melancholy fiscal news, we should not succumb to a lugubrious view of global aging. We should see it as a miraculous triumph of modern medicine. We should surely remember the extra years of life will be treasured not just by the elderly but their families as well. Indeed, would we have it any other way?

So, what is to be done? May I emphasize the theme I have been relentlessly repeating? This is a global problem that requires a global solution. I offer six strategies—all of them politically difficult, and some are far more likely in certain cultures than in others. Still, it is from some combination of these that the reforms will emerge.

Later retirement, longer work lives   This alone could generate enormous savings. Such a change in later retirement will require a sea change in the attitudes of labor unions and business as well. I would safely predict that just as racism was a bias in the work place and elsewhere that had to be transcended, ageism could easily be a dominant bias that will need to be transcended in the 21st century.

Ageism is the bias that proclaims a number of myths. Perhaps the most common is that most elderly do not want to do some work. This is simply untrue. The second is that most elderly don't need to work. Given [the] relatively small savings by large groups of elderly, this is almost certainly not true. In America, for example, the median 60 year old—and remember median means half have less—only has about $10,000 in net financial assets and that $10,000 includes stocks, bonds, CDs, savings and checking accounts, IRA and Keogh funds.

Still another myth is that most elderly aren't able to work. A classic case study of a British home improvement chain found that stores staffed by older workers had about one-sixth the turnover, 40 percent lower absentee rates, 60 percent less theft and 18 percent higher profits. And for those elderly that can't work, clearly a humane and durable safety net must be maintained by this richest of all nations. A final myth is that there is nothing for the elderly to do. Just look at the downsizing and high unemployment levels we are told. Quite apart from the awkward fact that paying very high taxes to support earlier retirement is hardly the healthy recipe for economic and employment growth, we need to get ourselves into a 21st century frame of mind. In a world of rapidly shrinking workforces we will need more workers, not fewer.

Encouraging more work from the non-elderly   This could be done either by:

    a. getting working-age citizens to work more, or
    b. by increasing the inflow of working-age immigrants.

Immigration will not only provoke serious cultural and political controversy within these other countries—consider Japan for example—but indeed between developed and developing nations, since we will obviously want to import their best educated and trained workers, which, ironically, these poor countries would have paid for.

The pro-natalism strategy   This would involve raising more numerous and productive children. It is the so-called 'Scandinavian approach' to battling demographic change. First, the empirical fact is that pro-natalist programs have worked only marginally well; raising a new taxpayer from birth is a long and costly process and may require more patience and wisdom than most governments now possess.

Beyond this, one might run into some strong cultural crosswinds if such a program is perceived as an attempt to roll back gains won by women in the workplace.

Reduce fiscal cost of elderly dependence by stressing filial obligation   This is an effort to increase the willingness of tomorrow's grown children to support their own parents. We in U.S. and Europe have much to learn about filial obligation from Asian countries like Japan and China, but to what extent can our very independent cultures be expected to import the "Confucian Ethic"?

Reduce the fiscal cost of elder dependency by targeting benefits on the basis of need   A 'means test' sounds mean—call it an affluence test. Such a policy would progressively reduce benefits for those above the median income level. In U.S. $200 billion of yearly benefits go to the most affluent quarter of all households.

Reduce the costs of elder dependency by increasing private and public savings   Attempts to increase public savings should begin by reducing the deficits or negative savings from the huge outyear elderly public spending and through more private savings by increasing and indeed requiring people to save and invest more of their own income—and thereby create genuinely funded plans.

However, make no mistake about it: such funding would involve the very large transition costs required to pay for two retirements—their parents' generation as well as their own. Thus, like [Federal Reserve Board] Chairman [Alan] Greenspan (I use his name on the reasonable assumption that he has a touch more credibility than I) I believe this will require genuine reform of the existing program—which is a euphemism for benefit cuts or tax increase. I clearly favor some gradual benefit reduction, further raising the retirement age, a "Diet COLA," affluence testing, or the like in some practical combination.

Looking beyond the fiscal challenges, *Gray Dawn* [Peterson's book on global aging] asks how global aging may transform our economy, our politics, our society, our military/security arrangements, and our very culture. Permit me to play Alice Toklas to your Gertrude Stein and pose some of these questions:

- What will happen to the structure and dynamism of the developed world's economies as they adjust to the new plentitude of elders, the new dearth of young workers, shrinking populations, and stagnating (or even shrinking) real output? In the U.S., for example, increases in the number of workers [have] accounted for about two-thirds of our work.
- Which sectors of the economy will prosper and which will decline in countries with shrinking populations? Likely winners in an aging world are: health care, asset management, and all manners of retirement "life style" businesses. Likely losers in some countries later in the next century [are] shrinking construction and infrastructure.
- On the face of unprecedented budgetary crowding out by metastasizing social benefit costs, will governments be able to maintain their national security and global commitments?
- And what of the politics? When nearly half of the adult population of developed countries (and perhaps two-thirds of the voters) are at or

beyond today's official retirement age, who will speak for the future? Is global aging destined to enthrone the senior lobby as an invincible political titan? Or does generational war loom in the future of the developed world?

- How will global aging reshuffle the ethics of life and death—incidentally, I know how convincingly a discussion on ethics comes from an investment banker—as technology and demographics combine to send health-care spending to unsustainable levels? Some people are going to have to give up some medical treatments that [have] some benefits. When medical progress collides with limited resources, who will live? Who will die? Who will decide?

- Does an aging society mean an aging culture? How will global aging transform culture, media, and entertainment as youth becomes less numerous and influential?

- How will it affect the global economy? Will widening pension deficits trigger a meltdown in global capital markets? Will the new Euro currency run aground on demographics as member nations respond to the aging challenge with widely divergent benefit reforms and fiscal policies?

- Take Italy for example: given their shaky political systems, and left-leaning traditions, will they try to run deficits far above the 3% of GDP EMU deficit ceiling and try to finance their way out of it? Recently, I was the official questioner of Mr. Duisenberg, head of the European Central Bank, and asked precisely that question. After the dinner, seven people came up to me and said that in all their experience at the Economic Club they had never heard such a non-answer.

- As one witty observer put it: America did it right. First it had its Civil War, then its dollar became a key currency. Europe appears to be on the verge of reversing this process: First it declares the Euro into being, then it has its civil war.

- Will it open or shut the door to more immigration from a still youthful developing world? And if the richer developed world does import large numbers of the most educated workers from poorer developing countries, what new bargain will these countries want in return?

- How will global aging affect the geopolitics of the next century? Will today's global divide between rich and poor nations become redefined as a divide between old and young nations? Will some developed nations someday come to depend on the capital surpluses of developing countries like China to keep themselves financially afloat? And if so, how will these new suppliers of capital use their newly acquired leverage? Will they, for example, insist on wholesale changes in global institutions? Will China, for example, insist that we reform Medicare in the same way that we recommended they change human rights policies as a condition for lending?

I hope I have presented you with a few things to talk about. But in the meantime, please don't put this *Gray Dawn* book down. I want you to be able to pick it up.

Leon F. Bouvier and Jane T. Bertrand  **NO**

The Demographic Challenges of the Twenty-First Century

The Developed World

Low fertility in developed nations seems here to stay. Whereas demographic "predictions" are hazardous, the evolution in fertility rates over the past 30 years in developed nations provides strong evidence of a trend that is unlikely to reverse itself in the current social and economic climate.

This trend began in 1973, when a momentous new demographic phenomenon began to unfold throughout most of the developed world: fertility fell below the level of 2.1 births per woman needed to replace the population in the long run *and remained there*. Country after country—Canada, the then two Germanys, the United Kingdom, the United States—all reported fertility rates under two births per woman in 1983 and still lower rates in the early 1990s. Actual population size did not begin to fall immediately because all of the nations experienced a rise in their fertility rates after World War II—some for just a few years, others (such as Australia, Canada, and the United States) for 10 to 15 years. As a result, the number of young couples of childbearing ages was quite large in the 1970s, so that even with extraordinarily low fertility *per woman*, births continued to outnumber deaths each year, reflecting population *momentum.* . . .

But by the late 1970s, the effect of that momentum had subsided in many European countries, and in some, deaths began to outnumber births by the mid-1980s. Population decline was noted in Austria and East and West Germany. Continued immigration and the higher fertility of immigrants warded off decline for a time in other countries, but by the early 1980s, France and the United Kingdom were losing population, as was Hungary.

In the younger developed nations—the United States, Canada, Australia, New Zealand—the trend toward actual population decline will occur later than in Europe because of the reverberating impact of the tremendous post–World War II baby boom generation and because immigration levels have remained high. Nevertheless, the U.S. Census Bureau has projected that its population would stop growing at about 309 million by 2050 and then begin to decline, given continued below-replacement fertility and if net immigration was limited to 450,000 per year.

Espenshade, Bouvier, and Arthur calculated that even with net immigration of 400,000 per year, the U.S. population would be reduced to 109 million before achieving zero growth in some 200 years, if the national fertility rate continued below replacement.

This low fertility pattern, first noted some 20 years ago, continues to this date in most European countries as well as the United States, Canada, Australia, New Zealand, and Japan. According to the most recently published data, Italy, Spain, Romania, and the Czech Republic have fallen to 1.2 and below, the lowest in the world. Other nations in this group are not far behind. (See table 1.)

Table 1

Total Fertility Rates in Selected Developed Countries, 1970, 1975, 1982, 1998

Country	1970	1975	1982	1998
Europe				
Austria	2.3	1.8	1.7	1.4
Belgium	2.2	1.7	1.6	1.6
Denmark	2	1.9	1.4	1.8
France	2.5	1.9	1.9	1.7
Germany, East	2.2	1.5	1.9	—
Germany, West	2	1.5	1.4	1.3*
Hungary	2	2.4	1.8	1.4
Italy	2.4	2.2	1.6	1.2
Netherlands	2.6	1.7	1.5	1.5
Norway	2.5	2	1.7	1.8
Spain	2.9	2.8	2	1.2
Sweden	1.9	1.8	1.6	1.6
Switzerland	2.1	1.6	1.6	1.6
United Kingdom	2.4	1.8	1.8	1.7
Other Areas				
Australia	2.9	2.2	1.9	1.8
Canada	2.3	1.9	1.7	1.6
Japan	2.1	1.9	1.7	1.4
United States	2.5	1.8	1.9	2

*This rate is for the combined Germanys.

Source: *World Population Data Sheets,* selected years (Washington, D.C.: Population Reference Bureau).

By 1995, Ireland (1.8 in 1998) and Portugal (1.4 in 1998) had joined the European low fertility groups and were accompanied by most of the new republics carved out of the erstwhile USSR. Thus, it is clear that fertility is remaining extremely low in all of Europe, Japan, and the former English colonies such as the United States, Canada, and Australia. It is also interesting to note that a slight increase in fertility has occurred in some Scandinavian countries. This may be a precursor of future trends and suggests that fertility could rise ever so slightly in other European countries in years to come.

Short-lived periods of low fertility have occurred in the past. During the Great Depression of the 1930s, fertility fell below replacement in most western European countries and in the United States. But the rates began to climb again by the late 1930s in the United States and during World War II in western Europe. Never before in modern history has fertility been so low in so many countries for such a long period as has been the case since 1973. And as of the late 1990s, the rates in most developed countries show little evidence of climbing back to the point where population growth, or at least population replacement, can be assured over the long run.

End of Growth in the Developed World

The significance of this new demographic pattern throughout most of the developed world suggests adaptation to a new era. The developed nations of the planet may well be poised on the brink of an entirely new age when the quantitative demands for labor will be modest, although the qualitative demands will be great.

We have seen harbingers of this momentous change. For example, John Naisbitt noted in his 1982 book, *Megatrends*: "In 1956 for the first time in American history, white-collar workers and technical-managerial positions outnumbered blue-collar workers. Industrial America was giving way to a new society where for the first time in history most of us worked with information rather than producing goods." Naisbitt named this the "information society." Sociologist Daniel Bell called it the "postindustrial society." These social observers, however, did not identify an end to population growth as an important ingredient in this emerging new society.

We are at the onset of a social upheaval as great as that touched off by the Industrial Revolution that began some 200 years ago. Interestingly, a demographic revolution occurred then also. With the improved living conditions resulting from the Industrial Revolution, mortality began to fall gradually from the end of the 18th century. Couples took a while to adjust to the fact that more of their children were surviving, so fertility generally did not begin to fall until late in the 19th century. During this interim period, population growth rose to over 1 percent a year in many western European countries, rates unprecedented in world history at that time. This growth helped supply workers needed for the cities' burgeoning industries, as well as immigrants to the "New World." Then western European couples realized that they could take advantage of improving social and economic opportunities by having fewer, but better educated, children. Today's population actors—at least those in developed nations—are once again responding appropriately to meet the needs of a new era.

This "information society," benefitting from high technology—and by 1990, from robotic technology and more recently from the Internet—will be knowledge-intensive rather than labor-intensive. The pattern of shifting the production of shoes and textiles, for example, to lower-wage developing countries is now being followed with the sophisticated "tools" of the information age. The 1983 decision by California-based Atari to move assembly-line work for its microcomputers to Hong Kong and Taiwan was but an early example of what has become standard practice. To some extent, that practice was behind the General Motors strike in 1998. The union argued that too many jobs were being sent to cheap labor countries such as Mexico. Even many service occupations may be eliminated in developed countries. Economist and Nobel prize winner Wassily Leontief has suggested that "the ability of the service sector to absorb displaced workers will diminish." He added, "As soon as not only the physical but also the controlling 'mental' functions involved in the production of goods and services can be performed without the participation of human labor, labor's role as an indispensable 'factor of production' will progressively diminish."

The Logical Result: Increased International Immigration

The current and future demographic scenarios are increasingly shaped by the prospect of increased international migration. Perhaps more than at any time in the past, residents of the poorest nations are making the decision to move across international borders sometimes legally, sometimes illegally in search of a better life. Humans have always been peripatetic, but with the emergence of nation-states and political barriers in the late 19th and early 20th centuries, migration had become relatively controlled. However, with recent spectacular advances in communications and transportation facilities, added to the staggering poverty in developing nations (aggravated by rapid population growth), more and more people are ignoring these constraints and are moving despite the possibly dire consequences. This is happening all over the world, not only between developing and developed nations but also between the less affluent and more affluent developed countries, as from Portugal to France and Italy to Germany, and between the poorest developing countries and their somewhat better-off neighbors, as from Egypt to Saudi Arabia and Colombia to Venezuela.

[Elsewhere] we quoted the eminent American demographer, Kingsley Davis. He asked, "One wonders how long the inequalities of growth between the major regions can continue without an explosion?" Davis's question is now being answered by millions upon millions of people who are giving up their long-accepted ways of life in search of something better. They may be legal immigrants whose decision to move resulted from considerable discussion and thought. They may be refugees forced to abandon their homeland because of political strife. They may be illegal immigrants who surreptitiously enter a country and lead guarded lives for fear of deportation.

It is evident that life in the 21st century will be greatly influenced by these three demographic phenomena: (1) the historically unprecedented levels of fertility decline in certain developed nations; (2) sustained population growth in the developing world over the past 40 years, which has created a "surplus population" in many nations; and (3) massive international migration by individuals in search of what they hope might be a better life.

POSTSCRIPT

Will Declining Population Growth Rates in the Developed World Lead to Major Problems?

The issue of the changing age composition in the developed world was foreseen a few decades ago, but its heightened visibility is relatively recent. This visibility culminated in the UN-sponsored Second World Assembly on Ageing, which was held in Madrid, Spain, in April 2002. The resulting plan of action commits governments to addressing the problem of aging and provides them with a set of 117 specific recommendations covering three basic areas: older individuals and development, advancing health and well-being into old age, and ensuring enabling and supportive environments.

With the successful demographic transition in the industrial world, the percentage of citizens above age 60 is on the rise, while the labor force is decreasing. In 1998, 19 percent of the developed world fell into the post-60 category (10 percent worldwide). Children under age 15 also make up 19 percent of the developed world's population, while the labor force is at 62 percent. With birthrates hovering around 1 percent or less and life expectancy increasing, the percentages will likely continue to grow toward the upper end of the scale.

For Peterson, the costs of global aging will not only outweigh the benefits, but the capacity of the developed world to pay for these costs is questionable at best. The economic burden on the labor force, he contends, will be "unprecedented." Peterson offers six major strategies to find a "global solution" to this global problem. He previously elaborated on this theme in "Gray Dawn: The Global Aging Crisis," *Foreign Affairs* (January/February 1999).

The national news media highlighted Peterson's "crusade" for government action on the global aging crisis in Phillip J. Longman, "The World Turns Gray: How Global Aging Will Challenge the World's Economic Well-Being," *U.S. News & World Report* (March 1, 1999). A particularly outspoken opponent of Peterson's thesis is Phil Mullan. In his book *The Imaginary Time Bomb: Why an Ageing Population Is Not a Social Problem* (I. B. Tauris, 2000), Mullan contends that the idea of an aging developed world has become "a kind of mantra for opponents of the welfare state and for a collection of alarmists."

For Bouvier and Bertrand, there seems to be a potential silver lining on the horizon. Although future increases in immigration will counterbalance the decline of the indigenous population, they assert, the real advance will be the decoupling of productivity expansion and workforce increases. The information age is knowledge-intensive, and becoming more so, not labor-intensive. Bouvier and Bertrand maintain that knowledge will continue to explode and that it will not be simply the reserve of the workforce. The aged will presumably be just as capable of contributing to the knowledge base.

One author who accepts Bouvier and Bertrand's thesis is the noted scholar of management Peter Drucker. In "The Future That Has Already Happened," *The Futurist* (November 1998), Drucker predicts that retirement age in the developed world will soon rise to 75, primarily because their greatest skill, knowledge, will become even more of an asset. He maintains that knowledge resources will become the most important commodity.

The Center for Strategic and International Studies (CSIS) in Washington is at the forefront of research and policy advocacy on the issue of global aging in the developed world. An important book on the fiscal problems facing the developed world because of aging is from the CSIS's Significant Issues Series: Robert Stowe England, *The Fiscal Challenge of an Aging Industrial World* (Center for Strategic and International Studies, 2002). An earlier report from the CSIS and Watson Wyatt Worldwide is "Global Aging: The Challenge of the New Millennium" (1999). This document's presentation of the raw data is particularly useful.

On the Internet . . .

UNEP World Conservation Monitoring Centre

The United Nations Environment Programme's World Conservation Monitoring Centre Web site contains information on conservation and sustainable use of the globe's natural resources. The center provides information to policymakers concerning global trends in conservation, biodiversity, loss of species and habitats, and more. This site includes a list of publications and environmental links.

http://www.unep-wcmc.org

renewingindia.org

Renewingindia.org is a comprehensive Web site promoting new and renewable sources of energy in India. Here you will find case studies, articles, links, and a list of organizations and businesses involved in renewable energy technologies, as well as an online discussion group.

http://www.renewingindia.org

The Hunger Project

The Hunger Project is a nonprofit organization that seeks to end global hunger. This organization asserts that society-wide actions are needed to eliminate hunger and that global security depends on ensuring that everyone's basic needs are fulfilled. Included on this site is an outline of principles that guide the organization, information on why ending hunger is so important, and a list of programs sponsored by the Hunger Project in 11 developing countries across South Asia, Latin America, and Africa.

http://www.thp.org

Global Warming Central

The Global Warming Central Web site provides information on the global warming debate. Links to the best global warming debate sites as well as key documents and reports are included. Explore the recent news section to find the latest articles on the subject.

http://www.law.pace.edu/env/energy/globalwarming.html

International Association for Environmental Hydrology

The International Association for Environmental Hydrology (IAEH) is a worldwide association of environmental hydrologists dedicated to the protection and cleanup of freshwater resources. The IAEH's mission is to provide a place to share technical information and exchange ideas and to provide a source of inexpensive tools for the environmental hydrologist, especially hydrologists and water resource engineers in developing countries.

http://www.hydroweb.com

PART 2

Global Resources and the Environment

T *he availability of resources and the manner in which the planet's in-habitants use them characterize another major component of the global agenda. Many people argue that renewable resources are being consumed at a pace that is too fast to allow for replenishment. Many also contend that nonrenewable resources are being consumed faster than we are able to find suitable replacements.*

The production, distribution, and consumption of resources also leave their marks on the planet. This section examines whether these impacts are permanent, too degrading to the planet, too damaging to our quality of life, or simply beyond a threshold of acceptability.

- Is Environmental Degradation Worsening?

- Should the World Continue to Rely on Oil as a Major Source of Energy?

- Will the World Be Able to Feed Itself in the Foreseeable Future?

- Is Global Environmental Stress Caused Primarily by Increased Resource Consumption Rather Than Population Growth?

- Is the Earth Getting Warmer?

- Is the Threat of a Global Water Shortage Real?

ISSUE 5

Is Environmental Degradation Worsening?

YES: United Nations Environment Programme, from *Global Environment Outlook 2000* (Earthscan, 1999)

NO: Peter Huber, from "Wealth Is Not the Enemy of the Environment," *Vital Speeches of the Day* (April 1, 2000)

ISSUE SUMMARY

YES: The United Nations Environment Programme presents a comprehensive and gloomy assessment of the global environment at the turn of the millennium.

NO: Peter Huber, a senior fellow of the Manhattan Institute, contends that humankind is saving the Earth with the technologies that the "soft greens" most passionately oppose.

T his issue focuses on a global arena that has occupied much newsprint and air time in the last 30 years: the global environment. The production, distribution, and consumption of the resources that are vital to our daily lives take place in the context of a set of constraints placed on every one of us by the environment in which we live. Its influence crosses every kind of human endeavor, from our ability to grow enough food to our capacity to enjoy clean air and water.

It is not surprising, therefore, that early in the post–World War II period the international community came to appreciate that modern technology had given the human race the ability to "move mountains" literally and figuratively, and so humans did, for better and for worse. Problems that were previously confined to a small geographical area—soot from steam engines, pollution downstream and downwind from local factories, the clearing away of a small tract of land for a village's agricultural needs, the Dust Bowl of the 1930s—now manifested themselves in larger and more pervasive ways. Because factories started erecting larger smokestacks, the air pollution that had rained down on the towns adjacent to the local factories was now carried much further downwind, even across national boundaries. Moreover, this pollution was transformed during its journey into a much more insidious monster: acid rain.

It was the effects of this last pollutant that forced the international community to consider whether or not other human activities had a similar price associated with them. In the early 1970s the United Nations answered a plea from the Scandinavian countries to investigate the mysterious death of millions of fish in their lakes and streams. The formal gathering, termed the 1972 Conference on the Human Environment and held in Stockholm, Sweden, ushered in a new global strategy for addressing what was to become a new set of global issues.

Since Stockholm, the international community has been much more sensitive to the potentially adverse effects of modern human behavior on the Earth's natural and ecological systems. The Stockholm model has been replicated well over 20 times since 1972 in the global community's efforts to ensure that the human race survives the various onslaughts on its planet.

No serious observer now disputes that human activity leaves its mark on the Earth's various systems. The question is whether or not the impact is within an acceptable range and whether or not the planet can recover within a reasonable period of time. The term *Spaceship Earth* has become part of our lexicon, and its visual image, the brilliant photograph of Earth from an early NASA mission, has been forever embedded on many people's minds. The implication of this new addition to our vocabulary is that everyone on this planet must collectively endure the consequences of human activity across the globe. As a result, it is easy for some people to be seduced by cries of "the sky is falling, the sky is falling," uttered by doomsayers who see the demise of the planet in every expanded human endeavor. Others, far fewer in number, respond as small voices in the wilderness, accusing the doomsayers of having "Chicken Little syndrome." Between these two extremes, though, are the majority of the international community's leaders. These officials have embarked on a series of steps in each environmental area to ascertain whether or not a problem really exists and, if so, the nature and degree of the problem, what a preferred future alternative should look like, and how policymakers might successfully address the problem and achieve the desired end state.

The two selections for this issue address the general question of the state of the environment. The first source is a major report of the United Nations Environment Programme (UNEP), the official UN agency created at Stockholm in 1972 to monitor the environment. This report was prepared with the participation of over 850 individuals and more than 30 environmental institutes worldwide, as well as with every other relevant UN organization. As such, it represents the current official scorecard for the state of the planet's national and ecological systems. The picture it paints is one of grave concern. The second selection is much different in both its message and method of delivery. Peter Huber offers an alternative view of the effects of much modern human activity in a speech before the Harvard Club. To Huber, science and technology have combined to allow the human race to make a rather positive footprint on much of the planet.

 **YES**

Synthesis

Two over-riding trends characterize the beginning of the third millennium. First, the global human ecosystem is threatened by grave imbalances in productivity and in the distribution of goods and services. A significant proportion of humanity still lives in dire poverty, and projected trends are for an increasing divergence between those that benefit from economic and technological development, and those that do not. This unsustainable progression of extremes of wealth and poverty threatens the stability of the whole human system, and with it the global environment.

Secondly, the world is undergoing accelerating change, with internationally-coordinated environmental stewardship lagging behind economic and social development. Environmental gains from new technology and policies are being overtaken by the pace and scale of population growth and economic development. The processes of globalization that are so strongly influencing social evolution need to be directed towards resolving rather than aggravating the serious imbalances that divide the world today. All the partners involved—governments, intergovernmental organizations, the private sector, the scientific community, NGOs [nongovernmental organizations] and other major groups—need to work together to resolve this complex and interacting set of economic, social and environmental challenges in the interests of a more sustainable future for the planet and human society.

While each part of the Earth's surface is endowed with its own combination of environmental attributes, each area must also contend with a unique, but interlinked, set of current and emerging problems. *GEO-2000* [*Global Environmental Outlook 2000*] provides an overview of this range of issues. This synthesis provides a summary of the main conclusions of *GEO-2000*.

The State of the Environment: A Global Overview

Climate Change

In the late 1990s, annual emissions of carbon dioxide were almost four times the 1950 level and atmospheric concentrations of carbon dioxide had reached their highest level in 160 000 years. According to the Intergovernmental Panel on Climate Change, 'the balance of evidence suggests that there is a discernible human influence on global climate'. Expected results include a shifting of climatic zones, changes in species composition and the productivity of ecosystems, an increase in extreme weather events and impacts on human health.

Through the United Nations Framework Convention on Climate Change and the Kyoto Protocol, efforts are under way to start controlling and reducing greenhouse gas emissions. During the Third Conference of the Parties in Buenos Aires in 1998, a plan of action was developed on how to use the new international policy instruments such as emission trading and the Clean Development Mechanism. However, the Kyoto Protocol alone will be insufficient to stabilize carbon dioxide levels in the atmosphere.

Stratospheric Ozone Depletion

Major reductions in the production, consumption and release of ozone-depleting substances (ODS) have been, and continue to be, achieved by the Montreal Protocol and its related amendments. The abundance of ODS in the lower atmosphere peaked in about 1994 and is now slowly declining. This is expected to bring about a recovery of the ozone layer to pre-1980 levels by around 2050.

Illegal trading, still a problem, is being addressed by national governments but substantial quantities of ODS are still being smuggled across national borders. The Multilateral Fund and the Global Environment Facility are helping developing countries and countries in transition to phase out ODS. Since 1 July 1999, these countries have, for the first time, had to start meeting obligations under the Montreal Protocol.

Nitrogen Loading

We are fertilizing the Earth on a global scale through intensive agriculture, fossil fuel combustion and widespread cultivation of leguminous crops. Evidence is growing that the huge additional quantities of nitrogen being used are exacerbating acidification, causing changes in the species composition of ecosystems, raising nitrate levels in freshwater supplies above acceptable limits for human consumption and causing eutrophication in many freshwater habitats. In addition, river discharges laden with nitrogen-rich sewage and fertilizer runoff tend to stimulate algal blooms in coastal waters, which can lead to oxygen starvation and subsequent fish kills at lower depths, and reduce marine biodiversity through competition. Nitrogen emissions to the atmosphere contribute to global warming. Consensus among researchers is growing that the scale of disruption to the nitrogen cycle may have global implications comparable to those caused by disruption of the carbon cycle.

Chemical Risks

With the massive expansion in the availability and use of chemicals throughout the world, exposure to pesticides, heavy metals, small particulates and other substances poses an increasing threat to the health of humans and their environment. Pesticide use causes 3.5 to 5 million acute poisonings a year. Worldwide, 400 million tonnes of hazardous waste are generated each year. About 75 per cent of pesticide use and hazardous waste generation occurs in developed countries. Despite restrictions on toxic and persistent chemicals such as DDT, PCBs and dioxin in many developed countries, they are still manufactured for export and remain widely used in developing countries. Efforts are under way to promote cleaner production, to limit the emissions and phase out the use of some persistent organic pollutants, to control waste production and trade, and improve waste management.

Disasters

The frequency and effects of natural disasters such as earthquakes, volcanic eruptions, hurricanes, fires and floods are increasing. This not only affects the lives of millions of people directly, through death, injury and economic losses, but adds to environmental problems. As just one example, in 1996–98 uncontrolled wildfires swept through forests in Brazil, Canada, China's north-eastern Inner Mongolia Autonomous Region, France, Greece, Indonesia, Italy, Mexico, Turkey, the Russian Federation and the United States. The health impacts of forest fires can be serious. Experts consider a pollution index of 100 $\mu g/m^3$ unhealthy; in Malaysia, the index reached 800 $\mu g/m^3$. The estimated health cost of forest fires to the people of Southeast Asia was US$1 400 million. Fires are also a serious threat to biodiversity, especially when protected areas are burnt. Early warning and response systems are still weak, particularly in developing countries; there is an urgent need for improved information infrastructures and increased technical response capabilities.

El Niño

Unusual weather conditions over the past two years are also attributed to the *El Niño* Southern Oscillation (ENSO). The 1997/98 *El Niño* developed more quickly and resulted in higher temperatures in the Pacific Ocean than ever recorded before. The presence of this mass of warm water dominated world climate patterns up to mid-1998, causing substantial disruption and damage in many areas, including temperate zones. Extreme rainfall and flooding, droughts and forest fires were among the major impacts. Forecasting and early warning systems, together with human, agricultural and infrastructural protection, have been substantially improved as a result of the most recent *El Niño*.

Land, Forests and Biodiversity

Forests, woodlands and grasslands are still being degraded or destroyed, marginal lands turned into deserts, and natural ecosystems reduced or fragmented, further threatening biodiversity. New evidence confirms that climate change may further aggravate soil erosion in many regions in the coming decades, and threaten food production. Deforestation continues at high rates in developing countries, mainly driven by the demand for wood products and the need for land for agriculture and other purposes. Some 65 million hectares of forest were lost between 1990 and 1995, out of a total of 3500 million hectares. An increase of 9 million hectares in the developed world only slightly offset this loss. The quality of the remaining forest is threatened by a range of pressures including acidification, fuelwood and water abstraction, and fire. Reduced or degraded habitats threaten biodiversity at gene, species and ecosystems level, hampering the provision of key products and services. The widespread introduction of exotic species is a further major cause of biodiversity loss. Most of the threatened species are land-based, with more than half occurring in forests. Freshwater and marine habitats, especially coral reefs, are also very vulnerable.

Freshwater

Rapid population growth combined with industrialization, urbanization, agricultural intensification and water-intensive lifestyles is resulting in a global water crisis. About 20 per cent of the population currently lacks access to safe drinking water, while 50 per cent lacks access to a safe sanitation system. Falling water tables are widespread and cause serious problems, both because they lead to water shortages and, in coastal areas, to salt intrusion. Contamination of drinking water is mostly felt in megacities, while nitrate pollution and increasing loads of heavy metals affect water quality nearly everywhere. The world supply of freshwater cannot be increased; more and more people depend on this fixed supply; and more and more of it is polluted. Water security, like food security, will become a major national and regional priority in many areas of the world in the decades to come.

Marine and Coastal Areas

Urban and industrial development, tourism, aquaculture, waste dumping and discharges into marine areas are degrading coastal areas around the world and destroying ecosystems such as wetlands, mangroves and coral reefs. Climatic changes also affect the quality of ocean water as well as sea levels. Low-lying areas, including many small islands, risk inundation. The global marine fish catch almost doubled between 1975 and 1995, and the state of the world's fisheries has now reached crisis point. About 60 per cent of the world's fisheries are at or near the point at which yields decline.

Atmosphere

There is a major difference between air pollution trends in developed and developing countries. Strenuous efforts have begun to abate atmospheric pollution in many industrialized countries but urban air pollution is reaching crisis dimensions in most large cities of the developing world. Road traffic, the burning of coal and high-sulphur fuels, and forest fires are the major causes of air pollution. People in developing countries are also exposed to high levels of indoor pollutants from open fires. Some 50 per cent of chronic respiratory illness is now thought to be associated with air pollution. Large areas of forest and farmland are also being degraded by acid rain.

Urban Impacts

Many environmental problems reinforce one another in small, densely-populated areas. Air pollution, garbage, hazardous wastes, noise and water contamination turn these areas into environmental hot spots. Children are the most vulnerable to the inevitable health risks. Some 30–60 per cent of the urban population in low-income countries still lack adequate housing with sanitary facilities, drainage systems and piping for clean water. Continuing urbanization and industrialization, combined with a lack of resources and expertise, are increasing the severity of the problem. However, many local authorities are now joining forces to promote the concept of the sustainable city.

Policy Responses: A Global Overview

As awareness of environmental issues and their causes develops, the focus of policy questions shifts towards the policy response itself: what is being done, is it adequate and what are the alternatives? *GEO-2000* includes a unique assessment of environmental policies worldwide.

Environmental laws and institutions have been strongly developed over the past few years in almost all countries. Command and control policy via direct regulation is the most prominent policy instrument but its effectiveness depends on the manpower available, methods of implementation and control, and level of institutional coordination and policy integration. In most regions, such policies are still organized by sector but environmental planning and environmental impact assessment are becoming increasingly common everywhere.

While most regions are now trying to strengthen their institutions and regulations, some are shifting towards deregulation, increased use of economic instruments and subsidy reform, reliance on voluntary action by the private sector, and more public and NGO participation. This development is fed by the increasing complexity of environmental regulation and high control costs as well as demands from the private sector for more flexibility, self-regulation and cost-effectiveness.

GEO-2000 confirms the overall assessment of *GEO-1*: the global system of environmental management is moving in the right direction but much too slowly. Yet effective and well tried policy instruments do exist that could lead much more quickly to sustainability. If the new millennium is not to be marred by major environmental disasters, alternative policies will have to be swiftly implemented.

One of the major conclusions of the policy review concerns the implementation and effectiveness of existing policy instruments. The assessment of implementation, compliance and effectiveness of policy initiatives is complicated and plagued by gaps in data, conceptual difficulties and methodological problems.

Multilateral environmental agreements (MEAs) have proven to be powerful tools for attacking environmental problems. Each region has its own regional and sub-regional agreements, mostly relating to the common management or protection of natural resources such as water supply in river basins and transboundary air pollution. There are also many global-level agreements, including those on climate change and biodiversity that resulted from the United Nations Conference on Environment and Development, held in Rio de Janeiro, Brazil, in 1992.

In addition to the binding MEAs, there are non-binding agreements (such as *Agenda 21*) and environmental clauses or principles in wider agreements (such as regional trade treaties). A major trend in MEAs over the years has been a widening focus from issue-specific approaches (such as provisions for shared rivers) to trans-sectoral approaches (such as the Basel Convention), to globalization and to the general recognition of the linkage between environment and development. Another trend is still unfolding: the step-by-step establishment of common principles (such as the Forest Principles) in different sectors.

The *GEO-2000* review of MEAs highlights two issues:

- the effectiveness of MEAs depends strongly on the institutional arrangements, the financial and compliance mechanisms, and the enforcement systems that have been set up for them;
- it is still difficult to assess accurately the effectiveness of MEAs and non-binding instruments because of the lack of accepted indicators.

Regional Trends

Africa

Poverty is a major cause and consequence of the environmental degradation and resource depletion that threaten the region. Major environmental challenges include deforestation, soil degradation and desertification, declining biodiversity and marine resources, water scarcity, and deteriorating water and air quality. Urbanization is an emerging issue, bringing with it the range of human health and environmental problems well known in urban areas throughout the world. Growing 'environmental debts' in many countries are a major concern because the cost of remedial action will be far greater than preventive action.

Although many African countries are implementing new national and multilateral environmental policies, their effectiveness is often low due to lack of adequate staff, expertise, funds and equipment for implementation and enforcement. Current environmental policies are mainly based on regulatory instruments but some countries have begun to consider a broader range, including economic incentives implemented through different tax systems. Although cleaner production centres have been created in a few countries, most industries have made little effort to adopt cleaner production approaches. However, some multinational corporations, large-scale mining companies and even local enterprises have recently voluntarily adopted precautionary environmental standards.

There is growing recognition that national environmental policies are more likely to be effectively implemented if they are supported by an informed and involved public. Environmental awareness and education programmes are expanding almost everywhere, while indigenous knowledge receives greater recognition and is increasingly used. Environmental information systems are still weak.

There is fairly high interest in many of the global MEAs, and several regional MEAs have been developed to support the global ones. The compliance and implementation rate is, however, quite low, mainly due to lack of funds.

Asia and the Pacific

Asia and the Pacific is the largest region and it is facing serious environmental challenges. High population densities are putting enormous stress on the environment. Continued rapid economic growth and industrialization is likely to cause further environmental damage, with the region becoming more degraded, less forested, more polluted and less ecologically diverse in the future.

The region, which has only 30 per cent of the world's land area, supports 60 per cent of the world population. This is leading to land degradation, especially in marginal areas, and habitat fragmentation. Increasing habitat fragmentation has depleted the wide variety of forest products that used to be an important source of food, medicine and income for indigenous people. Forest fires caused extensive damage in 1997–98.

Water supply is a serious problem. Already at least one in three Asians has no access to safe drinking water and freshwater will be the major limiting factor to producing more food in the future, especially in populous and arid areas. Energy demand is rising faster than in any other part of the world. The proportion of people living in urban centres is rising rapidly, and is focused on a few urban centres. Asia's particular style of urbanization—towards megacities—is likely to increase environmental and social stresses.

Widespread concern over pollution and natural resources has led to legislation to curb emissions and conserve natural resources. Governments have been particularly active in promoting environmental compliance and enforcement although the latter is still a problem in parts of the region. Economic incentives and disincentives are beginning to be used for environmental protection and the promotion of resource efficiency. Pollution fines are common and deposit-refund schemes are being promoted to encourage reuse and recycling. Industry groups in both low- and high-income countries are becoming increasingly sensitive to environmental concerns over industrial production. There is keen interest in ISO14 000 standards for manufacturing and in eco-labelling.

In most countries, domestic investment in environmental issues is increasing. A major thrust, particularly among developing countries, is on water supply, waste reduction and waste recycling. Environment funds have also been established in many countries and have contributed to the prominent role that NGOs now play in environmental action. Many countries are in favour of public participation, and in some this is now required by law. However, education and awareness levels amongst the public are often low, and the environmental information base in the region is weak.

Whilst there is uneven commitment to global MEAs, regional MEAs are important. They include a number of important environmental policy initiatives developed by sub-regional cooperative mechanisms.

One of the greatest challenges is to promote liberal trade yet maintain and strengthen the protection of the environment and natural resources. Some governments are now taking action to reconcile trade and environmental interests through special policies, agreements on product standards, enforcement of the Polluter Pays Principle, and the enforcement of health and sanitary standards for food exports.

Europe and Central Asia

Environmental trends reflect the political and socio-economic legacy of the region. In Western Europe, overall consumption levels have remained high but measures to curb environmental degradation have led to considerable improvements in some, though not all, environmental parameters. Sulphur dioxide emissions, for example, were reduced by more than one-half between 1980 and 1995. In the other sub-regions, recent political change has resulted in sharp though probably temporary reductions in industrial activity, reducing many environmental pressures.

A number of environmental characteristics are common to much of the region. Large areas of forests are damaged by acidification, pollution, drought and forest fires. In many European countries, as much as half the known vertebrate species are under threat and most stocks of commercially-exploited fish in the North Sea have been seriously over-fished. More than half of the large cities in Europe are overexploiting their groundwater resources. Marine and coastal areas are susceptible to damage from a variety of sources. Road transport is now the main source of urban air pollution, and overall emissions are high—Western Europe produces nearly 15 per cent of global CO_2 emissions and eight of the ten countries with the highest per capita SO_2 emissions are in Central and Eastern Europe.

Regional action plans have been effective in forging policies consistent with the principles of sustainable development and in catalysing national and local action. However, some targets have yet to be met and plans in Eastern Europe and Central Asia are less advanced than elsewhere because of weak institutional capacities and the slower pace of economic restructuring and political reform.

Public participation in environmental issues is considered satisfactory in Western Europe, and there are some positive trends in Central and Eastern Europe. Many countries, however, still lack a proper legislative framework for public participation although the Convention on Access to Environmental Information and Public Participation in Environmental Decision Making signed by most of the ECE [Economic Commission for Europe] countries in 1998 should improve the situation. Access to environmental information has significantly increased with the formation of the European Environment Agency and other information resource centres in Europe. The level of support for global and regional MEAs, in terms of both ratification and compliance, is high.

There has been significant success, particularly in Western Europe, in implementing cleaner production programmes and eco-labelling. Within the European Union, green taxation and mitigating the adverse effects of subsidies are important priorities. Legislation is being adopted on entirely new subjects. Examples include the Nitrates Directive, the Habitat Directive and the Natura 2000 plan for a European Ecological Network. Implementation is, however, proving difficult.

The transition countries need to strengthen their institutional capacities, improve the enforcement of fees and fines, and build up the capacity of enterprises to introduce environmental management systems. The major challenge for the region as a whole is to integrate environmental, economic and social policies.

Latin America and the Caribbean

Two major environmental issues stand out in the region. The first is to find solutions to the problems of the urban environment—nearly three-quarters of the population are already urbanized, many in mega-cities. The air quality in most major cities threatens human health and water shortages are common. The second major issue is the depletion and destruction of forest resources, especially in the Amazon basin. Natural forest cover continues to decrease in all countries. A total of 5.8 million hectares a year was lost during 1990–95, resulting in a 3 per cent total loss for the period. This is a major threat to biodiversity. More than 1,000 vertebrate species are now threatened with extinction.

The region has the largest reserves of cultivable land in the world but soil degradation is threatening much cultivated land. In addition, the environmental costs of improved farm technologies have been high. During the 1980s, Central America increased production by 32 per cent but doubled its consumption of pesticides. On the plus side, many countries have substantial potential for curbing their contributions to the build-up of greenhouse gases, given the region's renewable energy sources and the potential of forest conservation and reforestation programmes to provide valuable carbon sinks.

During the past decade, concern for environmental issues has greatly increased, and many new institutions and policies have been put in place. However, these changes have apparently not yet greatly improved environmental management which continues to concentrate on sectoral issues, without integration with economic and social strategies. The lack of financing, technology, personnel and training and, in some cases, large and complex legal frameworks are the most common problems.

Most Latin American economies still rely on the growth of the export sector and on foreign capital inflows, regardless of the consequences to the environment. One feature of such policies is their failure to include environmental costs. Economic development efforts and programmes aimed at fighting poverty continue to be unrelated to environmental policy, due to poor interagency coordination and the lack of focus on a broader picture. On the industrial side, some producers have adopted ISO 14 000 standards as a means of demonstrating compliance with international rules.

An encouraging aspect is the trend towards regional collaboration, particularly on transboundary issues. For example, a Regional Response Mechanism for natural disasters has been established with telecommunication networks that link key agencies so that they can make quick assessments of damage, establish needs and mobilize resources to provide initial relief to affected communities. There is considerable interest in global and regional MEAs, and a high level of ratification. However, the level of implementing new policies to comply with these MEAs is generally low.

North America

North Americans use more energy and resources per capita than people in any other region. This causes acute problems for the environment and human health. The region has succeeded, however, in reducing many environmental impacts through stricter legislation and improved management. Whilst emissions of many air pollutants have been markedly reduced over the past 20 years, the region is the largest per capita contributor to greenhouse gases, mainly due to high energy consumption. Fuel use is high—in 1995 the average North American used more than 1600 litres of fuel a year (compared to about 330 litres in Europe). There is continuing concern about the effects of exposure to pesticides, organic pollutants and other toxic compounds. Changes to ecosystems caused by the introduction of non-indigenous species are threatening biodiversity and, in the longer term, global warming could move the ideal range for many North American forest species some 300 km to the north, undermining the utility of forest reserves established to protect particular plant and animal species. Locally, coastal and marine resources are close to depletion or are being seriously threatened.

The environmental policy scene is changing in North America. In Canada, most emphasis is on regulatory reform, federal/provincial policy harmonization and voluntary initiatives. In the United States, the impetus for introducing new types of environmental policies has increased and the country is developing market-based policies such as the use of tradeable emissions permits and agricultural subsidy reform. Voluntary policies and private sector initiatives, often in combination with civil society, are also gaining in importance. These include voluntary pollution reduction initiatives and programmes to ensure responsible management of chemical products. The region is generally active in supporting and complying with regional and global MEAs.

Public participation has been at the heart of many local resource management initiatives. Environmental policy instruments are increasingly developed in consultation with the public and the business community. Participation by NGOs and community residents is increasingly viewed as a valuable part of any environmental protection programme.

Increasing accountability and capacity to measure the performance of environmental policies is an overarching trend. Target setting, monitoring, scientific analysis and the public reporting of environmental policy performance are used to keep stakeholders involved and policies under control.

Access to information has been an important incentive for industries to improve their environmental performance.

Despite the many areas where policies have made a major difference, environmental problems have not been eliminated. Economic growth has negated many of the improvements made so far and new problems—such as climate change and biodiversity loss—have emerged.

West Asia

The region is facing a number of major environmental issues, of which degradation of water and land resources is the most pressing. Groundwater resources are in a critical condition because the volumes withdrawn far exceed natural recharge rates. Unless improved water management plans are put in place, major environmental problems are likely to occur in the future.

Land degradation is a serious problem, and the region's rangelands—important for food security—are deteriorating, mainly as a result of overstocking what are essentially fragile ecosystems. Drought, mismanagement of land resources, intensification of agriculture, poor irrigation practices and uncontrolled urbanization have also contributed. Marine and coastal environments have been degraded by overfishing, pollution and habitat destruction. Industrial pollution and management of hazardous wastes also threaten socio-economic development in the region with the oil-producing countries generating two to eight times more hazardous waste per capita than the United States. Over the next decade, urbanization, industrialization, population growth, abuse of agrochemicals, and uncontrolled fishing and hunting are expected to increase pressures on the region's fragile ecosystems and their endemic species.

The command and control approach, through legislation, is still the main environmental management tool in almost all states. However, several new initiatives, such as public awareness campaigns, have been taken to protect environmental resources and control pollution. In addition, many enterprises such as refineries, petrochemical complexes and metal smelters have begun procedures for obtaining certification under the ISO 14 000 series. Another important approach to resource conservation has been a growing interest in recycling scarce resources, particularly water. In many states on the Arabian Peninsula, municipal wastewater is subjected at least to secondary treatment, and is widely used to irrigate trees planted to green the landscape.

Success in implementing global and regional MEAs in the region is mixed and commitment to such policy tools quite weak. At a national level there has, however, been a significant increase in commitment to sustainable development, and environmental institutions have been given a higher priority and status.

Polar Regions

The Arctic and Antarctic play a significant role in the dynamics of the global environment and act as barometers of global change. Both areas are mainly affected by events occurring outside the polar regions. Stratospheric ozone depletion has resulted in high levels of ultraviolet radiation, and polar ice caps, shelves and glaciers are melting as a result of global warming. Both areas act as sinks for persistent organic pollutants, heavy metals and radioactivity, mostly originating from other parts of the world. The contaminants accumulate in food chains and pose a health hazard to polar inhabitants. Wild flora and fauna are also affected by human activities. For example, capelin stocks have collapsed twice in the Arctic since the peak catch of 3 million tonnes in 1977. In the Southern Ocean, the Patagonian toothfish is being over-fished and there is a large accidental mortality of seabirds caught up in fishing equipment. On land, wild communities have been modified by introductions of exotic species and, particularly in northern Europe, by overgrazing of domestic reindeer.

In the Arctic, the end of Cold War tensions has led to new environmental cooperation. The eight Arctic countries have adopted the Arctic Environmental Protection Strategy which includes monitoring and assessment, environmental emergencies, conservation of flora and fauna, and protection of the marine environment. Cooperation amongst groups of indigenous peoples has also been organized. The Antarctic environment benefits from the continuing commitment of Parties to the Antarctic Treaty aimed at reducing the chance of the region becoming a source of discord between states. The Treaty originally focused on mineral and living resources but this focus has now shifted towards broader environmental issues. A similar shift is expected in the Arctic, within the broader context of European environmental policies. In both polar areas, limited financial resources and political attention still constrain the development and implementation of effective policies.

Peter Huber  **NO**

Wealth Is Not the Enemy
of the Environment

I thought I'd focus my remarks . . . on trying to prove to you how very green New York really is. And I don't mean Central Park. I mean the skyscrapers.

But let me begin by stepping back from the city and saying one or two contrarian things about the continent and the planet.

Begin with a little publicized environmental fact. The prevailing winds blow west to east across North America. On the continent itself, we burn enough fossil fuel to release some 1.6 billion metric tons of carbon a year into the air. Yet—as best anyone can measure these things, carbon dioxide levels drop as you move across the continent. One estimate in *Science* magazine puts America's terrestrial uptake of carbon at about 1.7 billion metric tons. What's going on?

Besides burning fossil fuels, we're building suburbs and roads and cities. Suburban sprawl is much in the news these days. Estimates again vary widely, but without doubt, a lot of what used to be farmland around cities is being transformed into suburb. Our cities occupy over twice as much land today as they did in 1920.

Which, on its own, would certainly be making the carbon balance even worse. If we're leveling forest to make room for sprawling cities, it has to be getting worse, right? But we aren't leveling forests. We're reforesting this continent.

U.S. forest cover reached a low of about 600 million acres in 1920. It has been rising ever since. Precisely how fast is hard to pin down. But all analyses show that America's forest cover today is somewhere between 20 million and 140 million acres higher than it was in 1920. At least 10 million acres have been reforested in the last decade alone.

Now how can we square these facts?

We burn huge amounts of fossil fuel, but atmospheric carbon levels drop. Our cities sprawl, but we have more forest. Am I making all this up? No. The facts are what they are. The interesting question is: how come?

Understanding these apparent paradoxes sharply defines, the two types of green I want contrast, hard green and soft.

Now let's work our way back to these facts from where we now sit, which is right here, in the heart of [Manhattan].

Here's something the Sierra Club and I agree on almost perfectly: the skyscraper as America's great green gift to the planet. It packs more people on to less land, which leaves more wilderness, undisturbed, in other places, where the people aren't. The city gets Wall Street, Saks, The Met, and the Times Square crowds, which leaves more fly-over country for bison and cougars. It's Saul Steinberg's celebrated *New Yorker* cover, painted green.

Conserve land and you conserve environment. Save "earth"—land itself— and you save the wilderness. There are other things to consider too, and I'll get to them. But if you're honestly interested in wilderness and environmental quality, land is the most important. When I say land, of course, I mean surface— including stream, river, lake and coastal waters, as well.

This is just basic ecology. Life on earth lives at the surface. The less humans disturb the surface—the land—the better for all other species. It's as simple as that.

That's why the Sierra Club and I basically agree about the city: though profligate in its consumption of most everything else, is very frugal with land. The one thing your average New Yorker does not occupy is 40 acres and a mule.

The original conservationists and Theodore Roosevelt [T. R.] had the focus just right. Recognize: to conserve wilderness, you have to conserve wilderness. Not trash. Not aluminum. Not zinc. Not energy. But rather forest, river, stream, coastal waters.

Today's soft say the two go hand in hand—conserve trash—stop using plastics—or conserve energy—ride a bike, not a car—and you conserve wilderness further up the line. But do you?

So how do humans destroy wilderness?

Lots of ways. We build cities and highways, of course. But in fact, they're quite secondary. Our cities, towns, suburbs, roads, and all the interstate highways currently cover under 3 percent of the U.S. land mass—under 60 million acres.

But for every acre of land we occupy ourselves, we use 6 acres for crops. Another 8 acres are designated as range—larders for our livestock. Most of the wilderness loss worldwide is to third-world agriculture.

However bucolic they may appear, farms aren't green. Endless miles of wheat are not bio-diverse prairie. Rangeland isn't pristine wilderness. Cattle denude it of native perennial grasses, to be replaced by sagebrush, mesquite, and juniper, or simply reduce it to desert. Acres of cassava are not acres of rain forest.

Shrinking agriculture is best, and we've done it.

Shrink the footprint of agriculture, and you can really save or restore a lot of wilderness. A 7 percent shrinkage in our agricultural footprint would return as much land to wilderness as wiping all our cities, suburbs, highways and roads off the map.

And shrinking agriculture is exactly what we've been doing, since about 1920. The result has been truly remarkable. Roughly 80 million more acres of North American cropland were harvested sixty years ago than are harvested today.

That explains one of the paradoxes I started with. Yes, our cities are sprawling. People are leaving the country and moving to the city. The cities themselves have negative population growth. They sprawl because they draw people off the rest of the land. We've flipped from a population that was 80 percent rural two centuries ago, to one that is 80 percent urban today. Sprawling cities absorb a bit of farmland nearby, but return a lot more land to the wilderness farther away.

Most "soft" greens—like the Sierra Club—have the right instincts about cities. They recognize that cities pack a lot of people into a small amount of space. That leaves more wilderness undisturbed.

But what they get right about cities, soft greens get wrong about almost everything else. Most of the policies soft greens prescribe are environmental disasters, because most force us to use more land.

Let me put this as bluntly as I can. From Al Gore to the Sierra Club, the green establishment strives to make the human footprint on the wilderness . . . bigger. If we do what they urge us to do with agriculture, our footprint on the land will at least double. If we do what they urge us to do with energy, it will more than double again.

We are saving the earth on this continent with the technologies that the soft greens most passionately oppose.

Until quite recently, humanity's advance meant retreat for the wilderness. The surface—land, river, and shallow coastal waters—supplied all our food, building materials, and fuel. The more we grew, the more land we seized.

In 1850, with one-tenth today's population, the United States cultivated as much cropland as we do today. Almost all of it was managed just as the modern soft would prefer: organic, pesticide-free, and genetically pure. For all that, the plows and the farmer wiped out the perennial grasses, crowded out the last ranges of the bison, eradicated the cougars and wolves, and created the dust bowl.

For much of this century, however, that process has been reversed. With genetic seed improvements, fertilizers, pesticides, and the relentless efficiencies of corporate farming, we almost trebled the land-to-food productivity of our agriculture.

Pesticides, packaging, preservatives, and refrigeration dramatically reduce the agricultural footprint, too.

One of the things that most concerned T.R. when he was pressing for forest conservation in 1905 was the future supply of lumber for railroad ties. Railroads were essential to the economy, and ties consumed enormous amounts of wood.

But that same year—1905—the railroads began coating their ties with creosote. That stopped the termites. The chemical preservative effectively slashed demand by two-thirds, by tripling their lifespan. Since then, wood preservatives and termite eradication have done far more to save forests in America than the recycling of newspapers.

Pesticides, preservatives, food irradiation, and plastic packaging all have comparable effects: they sharply reduce losses along the food chain, from farmer's field to dining room table, with commensurate reductions in agricultural sprawl.

Far more is still possible. The sun delivers an average energy density of roughly 180 watts per square meter of the United States. Wild plants currently convert about 0.35 w/m2 of that into stored energy, a dreadful 1:500 energy conversion efficiency.

Biotechnology could now double and redouble the agricultural productivity of land, in much the same way as silicon technology has doubled and redoubled the informational productivity of sand.

We've done even better with construction. A British frigate in 1790 required some 3,000 trees to build; the Exxon Valdez required none. Concrete, steel, and plastics substitute for hardwoods in our dwellings and furniture, which leaves the wood itself to the forest.

Far more substitution of this kind could be achieved. The total current U.S. wood harvest—90 percent of which is for construction, not energy—exceeds by several-fold the tonnage of all of the steel, copper, lead, nickel, zinc and other metals extracted in the United States. The raw materials for concrete are not at all scarce, and limitless amounts can be extracted from comparatively minuscule areas of land.

Soft greens endorse a few of these opportunities, but oppose most of them. They urge us to eat less meat, because letting cows eat our food before we do creates agricultural sprawl, but they oppose pesticides and preservatives, without which fungi, insects, and rodents do the cow's job instead. They urge us to improve the efficiency of motors and engines, but they oppose bio-technology, which improves the solar conversion efficiency of corn and cow. Sorts prefer natural, organic, free-range, wild—all of which require more land for less food.

The one solar technology that perennially fascinates soft greens is Pholtovoltaics (PV). Indeed, I owe my hard/soft taxonomy of green to Amory Lovins, of the Rocky Mountain Institute, who gained fame in the 1970s promoting what he called "soft" sources of energy like wind and solar over "hard" ones like uranium and coal.

Compared to a green plant, PV is certainly impressive technology. Selenium-doped silicon wafers mounted in glass or plastic can convert sunlight to useful energy better than chlorophyll in a typical leaf. About 60 times better: current PV technology can capture about 20 w/m2 in the United States.

New York, unfortunately, consumes 55 w/m2 of energy. So to power New York with PV, you have to cover every square inch of the city's horizontal surface with wafer—and then extend the PV sprawl over at least twice that area again, somewhere or other upstate.

PV technology could improve, of course. But the laws of nature and physics never allow anything approaching 100 percent capture; a reliable and economical 20 percent would be extraordinary, and would still require enormous amounts of ancillary material and space for energy storage, to keep the lights up when the sun is down.

The numbers for liquid fuels are even worse. The soft alternative here is "biomass." Shifting the prefix from "agri" to "bio" does make things sound greener from the get-go. "Tilling, not drilling. Biology, not geology. Living carbon, not dead carbon. Vegetables, not minerals." I am quoting here from the institute for local self reliance, which exhorts us to return to what it calls the "carbohydrate economy."

Which is to say: agriculture. We are to burn wood, garbage, bacterial mats, sunflower oil, buffalo gourd, peanut shells, chicken droppings, and tallow from lambs.

Farmers love the idea, which makes it doubly attractive to politicians. Green acres in your tank: natural, organic, free range, and wild. "Our goal," declare carbohydrate-economy pundits, "should be to get back to the ratio of 1920 when plant matter constituted one-third of all materials."

We know what the carbohydrate energy economy would look like: we've already lived it. In 1790, agriculture employed over 80 percent of the people; the founding fathers obtained 80 percent of their energy from wood. They had a "renewable" source of oil, too: the sperm whale. Their children and grandchildren almost fished it to extinction, until colonel Edwin Drake finally thought to mount his harpoon on a derrick in Titusville, Pennsylvania. Well into this century, vast areas of woods and forest were still being cut down, just to warm homes and cook food. The forests retreated steadily.

Any return to that way of life would mean resuming the retreat. Your car engine consumes far more calories than your muscles do.

No conceivable mix of solar, biomass, or wind technology could meet even half our current energy demand without (at the very least) doubling the human footprint on the surface of the continent. Sunlight is a thin, diffuse form of energy, and it's found nowhere but on the surface. Humanity burns the energy equivalent of about 1/7th of all the energy captured by all the green plants every year through photosynthesis. And that energy is already being used—to power the wilderness.

How did we escape from the carbohydrate economy in the first place? We began to dig. All U.S. oil wells and refineries together cover about 160,000 acres of land, or about one-fifth the area of King Ranch in southeastern Texas. All coal mines, including all strip mines, cover some two million acres, or well under 1 percent of the area occupied by U.S. crop land.

How did our energy footprint ever get so small?

Coal is dead trees—fossilized biomass. A coal mine yields something like 5000 w/m2 of land, an oil field more like 10,000. This makes them 5 to 10 thousand times more land frugal than existing PV technology, and 100 times more frugal than the highest PV or biomass efficiency even theoretically attainable.

There's no great mystery to why the numbers shake out that way: oil and coal are three-dimensional, sub-surface sources of energy; sunlight is all two-dimensional surface.

The environmental implications are enormous. For every acre occupied by dwellings and offices, we need about 14 acres of farm and range to produce the quite modest amount of energy—edible food—required by the people on that original acre. But we need a mere 0.3 acres of land to deliver the much greater quantities of energy people use as fuel.

In sum, it takes far less land to dig up energy than to grow it. That's what's so dreadfully wrong with all "renewables." By definition, renewables tap energy flow: solar and its immediate byproducts, plant growth or wind. But plant growth is either wilderness we should aim to conserve, or it's a farmer's field already under the plow, that the wilderness might otherwise reclaim.

Real estate on rooftops does offer some ecologically dead surface for solar capture, but there's nowhere near enough of it, and it isn't economical.

Technology won't ever close the gap. Biotechnology is young, which suggests great potential for improving the biomass energy economy. But conventional hard fuels are so much more concentrated forms of energy, and so much more profitable that they attract far more investment, which drives continuous innovation. Soft does improve, but hard improves even faster.

It just isn't a fair race, and it never can be. Sunlight is too thin, and it starts out spread across the surface—exactly where the wilderness is. Going after the sunlight means going after the energy that fuels the wilderness. It means overbuilding the wilderness itself.

Coal, oil, and gas, by contrast, represent ancient biomass already concentrated by nature, and buried in vast, three-dimensional reservoirs. From which it can now be extracted quite easily, through comparatively tiny incisions in the skin of earth. Given a choice, the wilderness will prefer the keyhole surgery every time.

A first soft green response is to suggest that renewables can be harvested interstitially, from within the sprawl that has already occurred. Mount PV arrays on existing rooftops. Process agricultural "waste" into usable fuels. Erect 400 foot-tall wind turbines amongst the rolling vineyards.

These are good rejoinders, so far as they go, but they don't go very far. Despite decades of subsidy and government promotion, "renewables" (other than conventional hydro) now generate barely 0.7 percent of our electricity. Add in the forest industry's on-site burning of waste wood, and renewables contribute perhaps 3 percent of all the energy we consume.

The economics just don't wash. Again, this isn't very surprising—it is, in fact, exactly the same reason that makes hard fuels kinder to the environment. It's just a lot cheaper to gather highly concentrated forms of energy, than dispersed ones.

A second soft rejoinder has nothing to do with the environment, really—it's a purely economic argument. We're going to run out of the stuff under the ground.

They've been saying that about energy since the '70s, just as they said it about food since the time of Thomas Malthus. But the fact is, we don't run out, and all serious economists understand why. Human ingenuity stretches resources and finds substitutes.

What terrifies the honest softs isn't that we're going to run out of fossils, but that we aren't. If we were going to run out any time soon, we wouldn't have to worry so much about global warming, would we? Running out would stop our emissions better than Al Gore possibly could.

In fact, we still have enormous reserves of fossil fuels, coal most especially. And beyond coal lie vast reservoirs of heavy oils in tar sands, which bio-engineered bacteria may make economically accessible. And uranium, and more.

Overall, the price of hard energy keeps dropping, which tells us all we can or need to know about the depths. Meanwhile, the price of land on the environmentally critical surface, and the value we attach to wilderness, keep rising.

The softs' last and most emphatic rejoinder is that hard fuels just devour surface in other, more insidious ways. Uranium becomes Chernobyl. Oil destroys Prince William Sound or—by way of global warming—the whole planet. Hard agriculture's real footprint includes all the land and water contaminated by fertilizers, pesticides, or wayward genes that kill monarch butterflies.

These are valid arguments. But to win the debate they have to be pushed further than they reasonably can be. Hard agriculture and hard fuels start out with footprints at least ten times smaller than the soft alternatives, and more often hundreds or thousands of times smaller. Their secondary sprawl, from pollution and such, has to be very bad indeed to make it as bad as soft alternatives.

And soft alternatives entail secondary sprawl of their own. Technologies that use more material and more surface are generally going to end up polluting more, as well.

The bicyclist emits greenhouse gas, as he puffs along the road—a lot of it, per pound of useful payload moved. Soft agriculture has run-offs, and relies on all-natural pesticides, bred into the pest-resistant crops themselves, and designed to kill predatory insects in much the same way as the chemicals Dow or Dupont would have us spray on less hardy plants. PV's are manufactured from toxic metals. More eagles have been killed by wind turbines than were lost in the Exxon Valdez oil spill. The Audubon Society labeled Enron's proposed wind farm in the Tehachapi mountains north of L.A. the "condor Cuisinart." The carbohydrate economy is the deforestation economy: third world practices make that quite clear.

All the while, hard technology keeps getting cleaner and more efficient. And in one very important sense, it's the hard fuels that are "renewable," not the soft ones.

Much of the land used when we extract fuels from beneath the surface is restored to wilderness after the fuels have been mined. Not so with soft "renewables": the surface-intensive technologies they depend on have to stay in place permanently—that's essential to the "renewing." Only the subsurface fuels are "renewable" so far as renewing the surface of the land is concerned.

Finally, when we have to allow for offsetting environmental benefits, too, soft alternatives certainly have their share, many of them purely aesthetic.

But hard technologies have their secondary benefits as well. Returning land to wilderness, as hard technologies do, can take care of a lot of pollution. Whatever impact pesticides have, freeing up 100 million acres to be reclaimed by forest will likely protect more birds than trying to bankrupt Dow Chemical through Superfund. The most beautiful way to purify water is probably the most effective too: maintain unfarmed, unlogged watersheds.

Carbon vividly illustrates this. As I mentioned at the outset, America is apparently sinking more carbon out of the air than it is emitting into it. What's doing the sinking? In large part, regrowth of forests on land no longer farmed or logged, and faster growth of existing plants and forests, which are fertilized by nitrogen oxides and carbon dioxide "pollutants."

The total forest ecosystem in the United States holds about 52 billion metric tons of carbon. Net growth rate of 3 percent a year is enough to consume all carbon emissions of the U.S. economy, and either in forests themselves or on surrounding grasslands and farms, that is about the net growth rate we seem to have.

Our carbon-energy cycle, begins with carbon sequestered in fossil fuels, and ends with carbon sequestered in trees. The sun supplied the energy that first converted atmospheric carbon to become fossil fuel, and today again, the sun supplies the energy that takes the carbon back out of the air and into the trees. Give it geologic time, and today's new biomass will eventually end up back in the depths once again, as new coal and oil. Well sealed landfills will help advance that process.

The soft green establishment would have us believe that wealth is the enemy of the wilderness. The facts are the opposite.

Wealth limits our residential sprawl: it urbanizes us.

Wealth makes possible hard technologies, which drastically reduce our agricultural and energy footprints on the land. Finally, and perhaps most importantly, wealth is the only means yet discovered to limit human fecundity. The richer people grow, the fewer children they bear.

The city itself is a center of negative population growth.

Western industrial nations have stabilized their population growth overall. And, of course, it is the wealthy nations, not the poor, that pour their land into wilderness conservation.

Wealth is green. Poverty isn't.

I started off by saying you're sitting in the greenest spot on earth, but I misspoke. The greenest spot is just south of here. It's Wall Street.

POSTSCRIPT

Is Environmental Degradation Worsening?

Since the human race developed the capacity to move mountains, both literally and figuratively, it has done so, for both good and ill. Thus, environmental degradation, once confined to the local community, has moved into the global arena, beginning with the 1972 Stockholm conference. Since then, various environmental problems have been the subject of much debate. Is there a problem? If so, what is its origin? How should it be addressed? For each environmental issue, the global community is at different stages in its fight to prevent, minimize, or eliminate some condition. For example, global warming remains a controversy despite an international agreement, the Kyoto Protocol of 1997. Greater progress has been made with respect to ozone layer depletion, in that 126 countries agreed to action in 1992. Acid rain has been the focus of regional action because of its regional character. And biodiversity issues have begun to be addressed, as have desertification, deforestation, and many others.

The point is that the international community must first reach some satisfactory level of consensus on the existence and nature of a problem before it can begin to find agreeable solutions. Thus, at any given moment in time, global policymakers might be at the problem identification stage (does it exist and, if so, what are its parameters?), the solution identification stage (choosing from a range of policy options), or the policy implementation stage (action undertaken to solve the problem).

The latter circumstance makes it difficult for the thoughtful reader to make some summary judgments about where we are in addressing questions of environmental degradation. Nonetheless, the *Global Environment Outlook 2000* report was an impressive undertaking. Admittedly, the individuals and groups who took part in the process brought to the table a concern about the environment and, presumably, a belief that it is in trouble. But given the complexity of the terrain and the distinctiveness of the issue areas, one is hard pressed to pass off the report as the work of a small group of individuals who brought their own agendas to the enterprise while ignoring paradigms that were inconsistent with their own beliefs. Accordingly, rejecting the essence of their findings, not to mention specific conclusions, requires an effort of the same magnitude.

But one need not reject the basic premise of the Huber article either. Much smaller in scope and ambition, it suggests that hard agriculture (substantial technological inputs) and hard fuels (nonrenewable sources) make an imprint "at least ten times smaller than the soft alternatives." For Huber, a major way of creating a much more environmentally friendly planet is to continue to utilize science in the pursuit of fulfilling food and energy needs, and environmental degradation might just take care of itself.

Marvin S. Soroos's *The Endangered Atmosphere: Preserving a Global Commons* (University of South Carolina Press, 1997) addresses the entire range of threats to the atmosphere posed by various external factors. *Global Environmental Politics,* 2d ed., by Gareth Porter and Janet Welsh Brown (Westview Press, 1996) examines environmental policy making across a wide range of issues. Lynton K. Caldwell's *International Environmental Policy: From the Twentieth to the Twenty-First Century,* 3rd ed., in collaboration with Paul S. Weiland (Duke University Press, 1996) represents the contribution of the seniormost scholar in the field of environmental policy making.

The Worldwatch Institute publishes periodic studies of the environment, including an annual *State of the World* volume, which examines the contemporary environmental scene of that year. The annual *World Resources* volume, which is published jointly by the World Resources Institute, the UN Environment Programme, the UN Development Programme, and the World Bank, is another good source. Finally, Stephen Collett's "Environmental Protection and the Earth Summit: Paving the Path to Sustainable Development," in Michael T. Snarr and D. Neil Snarr, eds., *Introducing Global Issues,* 2d ed. (Lynne Rienner, 2002), presents a succinct and easily readable description of current global environmental policy-making efforts.

Two Web sites are worthy of mention: The Environment News Service (http://www.ens-news.com) provides accounts of late-breaking environmental news, and The Environment: A Global Challenge (http:// library.thinkquest.org/ 26026/) provides 400 online articles on the environment.

ISSUE 6

Should the World Continue to Rely on Oil as a Major Source of Energy?

YES: Hisham Khatib et al., from *World Energy Assessment: Energy and the Challenge of Sustainability* (United Nations Development Programme, 2000)

NO: Seth Dunn, from "Decarbonizing the Energy Economy," in Lester R. Brown et al., *State of the World 2001* (W. W. Norton, 2001)

ISSUE SUMMARY

YES: Hisham Khatib, an honorary vice chairman of the World Energy Council, and his coauthors conclude that reserves of traditional commercial fuels, including oil, "will suffice for decades to come."

NO: Seth Dunn, a research associate and climate/energy team coleader at the Worldwatch Institute, argues that a new energy system is fast emerging—in part because of a series of revolutionary new technologies and approaches—that will cause a transition away from a reliance on oil as the primary energy source.

The new millennium witnessed an oil crisis almost immediately, the third major energy crisis in 30 years (1972–1973 and 1979 were the dates of earlier problems). The crisis of 2000 manifested itself in the United States via much higher gasoline prices and in Europe via both rising prices and shortages at the pumps. Both were caused by the inability of national distribution systems to adjust to the Organization of Petroleum Exporting Countries' (OPEC's) changing production levels. The 2000 crisis eventually subsided, but the issue is far from resolved.

These three major fuel crises are discrete episodes in a much larger problem facing the human race, particularly the industrial world. That is, oil, the Earth's current principal source of energy, is a finite resource that ultimately will be exhausted. And unlike earlier energy transitions, where a more attractive source invited a change (such as from wood to coal and from coal to oil), the next energy transition will be forced upon the human race in the absence of an attractive alternative. In short, we will be pushed out of our almost total reliance on oil and be forced to adopt a new system with a host of unknowns. What will the new system be? Will it be a single source or some combination? Will it be a more attractive source? Will the source be readily available at a reasonable price, or will a new cartel control much of the supply? Will its production and consumption lead to major new environmental consequences? Will it require major changes to our lifestyles and standards of living? When will we be forced to jump?

Before considering new sources of fuel, other questions need to be asked. Are the calls for a viable alternative to oil premature? Are we simply running scared without cause? Did we learn the wrong lessons from the earlier energy crises? More specifically, were these crises artificially created, or were they a consequence of the actual physical unavailability of the energy source? Were these crises really about running out of oil globally, or were they due to other phenomena, such as poor distribution planning by oil companies or the use of oil as a political weapon by oil-exporting countries?

For well over half a century now, Western oil-consuming countries have been predicting the end of oil. Using a model known as Hubbert's Curve (named after the U.S. geologist who designed it in the 1930s), policymakers have predicted that the world would run out of oil at various times; the most recent prediction is that oil will run out a couple of decades from now. Simply put, the model visualizes all known available resources and the patterns of consumption on a time line until the wells run dry. Despite such prognostication, it was not until the crisis of the early 1970s that national governments began to consider ways of both prolonging the oil system and finding a suitable replacement. Prior to that time, governments as well as the private sector encouraged energy consumption. Increases in energy consumption were associated with economic growth. After Europe recovered from the devastation of World War II, for example, every 1 percent increase in energy consumption brought a similar growth in economic output. To the extent that governments engaged in energy policy making, it was designed solely to encourage increased production and consumption. Prices were kept low, and the energy was readily available. Policies making energy distribution systems more efficient and consumption patterns both more efficient and lower were almost nonexistent.

Today the search for an alternative to oil continues. Nuclear energy, once thought to be the answer, may play a future role but at a reduced level. Both water power and wind power remain possibilities, as do biomass, geothermal, and solar energy. The question before us, therefore, is whether or not the international community has the luxury of some time before all deposits of oil are exhausted.

The two selections for this issue suggest different answers to this last question. While extolling the future virtues of renewable energy, Hisham Khatib and his colleagues do not see the world running out of fossil fuel, most principally oil, for a long time to come. On the other hand, Seth Dunn argues that a new energy system will be fast upon us, the culmination of a global effort to find a better alternative to the principal system of the past century.

Hisham Khatib et al.

 **YES**

Energy Security

The world has generally seen considerable development and progress in the past 50 years. Living standards have improved, people have become healthier and longer-lived, and science and technology have considerably enhanced human welfare. No doubt the availability of abundant and cheap sources of energy, mainly in the form of crude oil from the Middle East, contributed to these achievements. Adequate global energy supplies, for the world as a whole as well as for individual countries, are essential for sustainable development, proper functioning of the economy, and human well-being. Thus the continuous availability of energy—in the quantities and forms required by the economy and society—must be ensured and secured.

Energy security—the continuous availability of energy in varied forms, in sufficient quantities, and at reasonable prices—has several aspects. It means limited vulnerability to transient or longer disruptions of imported supplies. It also means the availability of local and imported resources to meet growing demand over time and at reasonable prices.

Beginning in the early 1970s energy security was narrowly viewed as reduced dependence on oil consumption and imports, particularly in OECD [Organization for Economic Cooperation and Development] and other major oil-importing countries. Since that time considerable changes in oil and other energy markets have altered the picture. Suppliers have increased, as have proven reserves and stocks, and prices have become flexible and transparent, dictated by market forces rather than by cartel arrangements. Global tensions and regional conflicts are lessening, and trade is flourishing and becoming freer. Suppliers have not imposed any oil sanctions since the early 1980s, nor have there been any real shortages anywhere in the world. Instead, the United Nations and other actors have applied sanctions to some oil suppliers, but without affecting world oil trade or creating shortages.

All this points to the present abundance of oil supplies. Moreover, in today's market environment energy security is a shared issue for importing and exporting countries. As much as importing countries are anxious to ensure security by having sustainable sources, exporting countries are anxious to export to ensure sustainable income (Mitchell, 1997).

From Hisham Khatib, Alexander Barnes, Isam Chalabi, H. Steeg, K. Yokobori, and the Planning Department, Organisation of the Arab Oil Producing Countries, *World Energy Assessment: Energy and the Challenge of Sustainability* (United Nations Development Programme, 2000). Copyright © 2000 by UNDP. Reprinted by permission.

However, although all these developments are very encouraging, they are no cause for complacency. New threats to energy security have emerged in recent years. Regional shortages are becoming more acute, and the possibility of insecurity of supplies—due to disruption of trade and reduction in strategic reserves, as a result of conflicts or sabotage—persists, although it is decreasing. These situations point to a need to strengthen global as well as regional and national energy security.... There is also a need for a strong plea, under the auspices of the World Trade Organization (WTO), to refrain from restrictions on trade in energy products on grounds of competition or differences in environmental or labour standards....

Energy Adequacy

... Most of the world's future energy requirements, at least until the middle of the 21st century, will have to be met by fossil fuels (figure 1). Many attempts have been made to assess the global fossil fuel resource base. Table 1 shows the results of two.

Figure 1

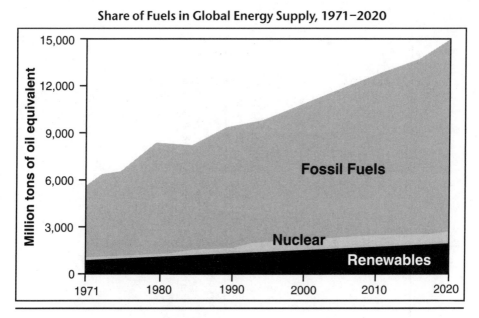

Share of Fuels in Global Energy Supply, 1971–2020

Source: IEA, 1998.

In 1998 world consumption of primary energy totalled almost 355 exajoules, or 8,460 million tonnes of oil equivalent (Mtoe)—7,630 Mtoe of fossil fuels, 620 Mtoe of nuclear energy, and 210 Mtoe of hydropower. To this should be added around 47 exajoules (1,120 Mtoe) of biomass and other renewables, for

Table 1

Global Energy Resource Base (Exajoules Except Where Otherwise Indicated)

Term	World Energy Council estimates		Institute for Applied Systems Analysis estimates			Consumption
	Proven reserves	Ultimately recoverable	Reserves	Resources	Resource base	1998
Conventional oil	6,300 (150)	8,400 (200)	6,300 (150)	6,090 (145)	12,390 (295)	142.8 (3.4)
Unconventional oil	—	23,100 (550)	8,190 (195)	13,944 (332)	22,050 (525)	n.a.
Conventional gas	5,586 (133)	9,240 (220)	5,922 (141)	11,718 (279)	17,640 (420)	85 (2.0)
Unconventional gas	—	—	8,064 (192)	10,836 (258)	18,900 (450)	n.a.
Coal and lignite	18,060 (430)	142,800 (3,400)	25,452 (606)	117,348 (2,794)	142,800 (3,400)	93 (2.2)
Uranium	3.4×10^9 tonnes	17×10^9 tonnes	(57)	(203)	(260)	64,000 tonnes

— Not available; n.a. Not applicable.

Note: Numbers in parentheses are in gigatonnes of oil equivalent.

a. Because of uncertainties about the method of conversion, quantities of uranium have been left in the units reported by the sources.

Source: WEC, 1998; IIASA, 1998.

VALUING THE COST OF ELECTRICITY SUPPLY SECURITY

The cost of electricity to a consumer—the consumer's valuation of the electricity supply (ignoring consumer surplus)—equals payments for electricity consumed plus the economic (social) cost of interruptions.

Supply insecurity causes disutility and inconvenience, in varying degrees and in different ways, to different classes of consumers—domestic, commercial, and industrial. The costs and losses (L) for the average consumer from supply interruptions are a function of the following:

- Dependence of the consumer on the supply *(C)*.
- Duration of the interruptions *(D)*.
- Frequency of their occurrence during the year *(F)*.
- Time of day in which they occur *(T)*.

That is, $L = C (D^d \times F^f, T^t)$, where *d*, *f*, and *t* are constants that vary from one consumer category to another.

The table shows estimates of the annual cost of electricity supply interruptions for the U.S. economy.

Economic cost of electricity supply interruptions for non-deferrable economic activities the United States, 1997

Consumer class and average duration of interruption	Cost to consumer per outage (U.S. dollars)	Cost to consumer per lengthy outage (U.S. dollars)	Estimated total annual losses (billions of U.S. dollars)
Residential (20 minutes)	0–20	50–250	0.9–2.7
Commercial (10 minutes)	25–500	5–20 (per minute)	2.9–11.7
Industrial (less than 30 seconds)	200–500 (small plant) 1,000–10,000 (large plant)	5,000–50,000 (per 8-hour day)	1.1–13.5

Note: Assumes nine outages a year for each class of consumer.

Source: Newton-Evans Research Company, 1998.

a total of 402 exajoules (9,580 Mtoe). The huge resource base of fossil and nuclear fuels will be adequate to meet such global requirements for decades to come.

Crude Oil

Proven oil reserves have increased steadily over the past 20 years, mainly because oil companies have expanded their estimates of the reserves in already discovered fields. This optimism stems from better knowledge of the fields, increased productivity, and advances in technology. New technologies have led to more accurate estimates of reserves through better seismic (three- and four-dimensional) exploration, have improved drilling techniques (such as horizontal and offshore drilling), and have increased recovery factors—the share of oil that can be recovered—from 30 percent to 40–50 percent (Campbell and Laherrere, 1998).

Huge amounts of untapp   ed unconventional oil also exist, augmenting conventional oil reserves.* Some 1.2 trillion barrels of heavy oil are found in the Orinoco oil belt in Venezuela. And the tar sands of Canada and oil shale deposits of the Russian Federation may contain 300 billion barrels of oil.

The U.S. Geological Survey assessed ultimate oil and gas reserves at the beginning of 1993 (IEA, 1998; WEC, 1998). The results, which tally with the World Energy Council (WEC) and International Energy Agency (IEA) figures (see table 1), point to ultimate conventional oil reserves of 2,300 billion barrels, with cumulative production until 1993 amounting to 700 billion barrels and unidentified reserves to 470 billion. No shortage of conventional liquid fuels is foreseen before 2020. Any deficiencies after that can be met by the ample reserves of unconventional oil.

Natural Gas

The U.S. Geological Survey also assessed ultimate natural gas reserves in 1993 (Masters, 1994). It estimated ultimate reserves at 11,448 trillion cubic feet (11,214 exajoules, or 267 gigatonnes of oil equivalent [Gtoe]), with cumulative production until 1993 amounting to 1,750 trillion cubic feet (1,722 exajoules, or 41 Gtoe). Cumulative world gas production through the end of 1995 was only 17.1 percent of the U.S. Geological Survey's estimate of conventional gas reserves.

Natural gas consumption is projected to grow 2.6 percent a year mostly as a result of growth in electricity generation in non-OECD countries. Despite this growth, cumulative production is expected to be no more than 41 percent of the . U.S. Geological Survey's estimate of conventional gas reserves by 2020. This points to a resource base large enough to serve global requirements for natural gas well into the second half of the 21st century.

Coal

Coal is the world's most abundant fossil fuel, with reserves estimated at almost 1,000 billion tonnes, equivalent to 27,300 exajoules, or 650,000 Mtoe (WEC, 1998). At the present rate of production, these reserves should last for more than 220 years. Thus the resource base of coal is much larger than that of oil and gas. In addition, coal reserves are more evenly distributed across the world. And coal is cheap. Efforts are being made to reduce production costs and to apply clean coal technologies to reduce the environmental impact.

Coal demand is forecast to grow at a rate slightly higher than global energy growth. Most of this growth will be for power generation in non-OECD countries, mostly in Asia. Although trade in coal is still low, it is likely to increase

---

*Conventional oil resources refer to those sources that are known and economically recoverable with present technologies. Unconventional oil resources refer to those sources that are less certain and/or are not economically recoverable with present technologies.—Eds.

slowly over time. Long-term trends in direct coal utilisation are difficult to predict because of the potential impact of climate change policies. Coal gasification and liquefaction will augment global oil and gas resources in the future.

Nuclear Energy

Although nuclear energy is sometimes grouped with fossil fuels, it relies on a different resource base. In 1998 nuclear energy production amounted to 2,350 terawatt-hours of electricity, replacing 620 Mtoe of other fuels. Uranium requirements amounted to 63,700 tonnes in 1997, against reasonably assured resources (reserves) of 3.4 million tonnes. Ultimately recoverable reserves amount to almost 17 million tonnes. Considering the relative stagnation in the growth of nuclear power, the enormous occurrences of low-grade uranium, and the prospects for recycling nuclear fuels, such reserves will suffice for many decades.

Renewables

Renewable energy sources—especially hydroelectric power, biomass, wind power, and geothermal energy—account for a growing share of world energy consumption. Today hydropower and biomass together contribute around 15 percent.

Hydroelectric power contributes around 2,500 terawatt-hours of electricity a year, slightly more than nuclear power does. It replaces almost 675 Mtoe of fuels a year, although its direct contribution to primary energy consumption is only a third of this. But it has still more potential. Technically exploitable hydro resources could potentially produce more than 14,000 terawatt-hours of electricity a year, equivalent to the world's total electricity requirements in 1998 (WEC, 1998). For environmental and economic reasons, however, most of these resources will not be exploited.

Still, hydropower will continue to develop. Hydropower is the most important among renewable energy sources. It is a clean, cheap source of energy, requiring only minimal running costs and with a conversion efficiency of almost 100 percent. Thus its annual growth could exceed the growth of global energy demand, slightly improving hydropower's modest contribution towards meeting world requirements.

Renewable energy sources other than hydro are substantial. These take the form mainly of biomass. Traditional biomass includes fuelwood—the main source of biomass energy—dung, and crop and forest residues. Lack of statistics makes it difficult to accurately estimate the contribution of renewables to the world's primary energy consumption. But it is estimated that the world consumed around 1.20 Gtoe in 1998. About two-thirds of this was from fuelwood, and the remainder from crop residues and dung. Much of this contribution is sustainable from a supply standpoint. But the resulting energy services could be substantially increased by improving conversion efficiencies, which are typically very low.

The contribution of biomass to world energy consumption is expected to increase slightly. It is mainly used as an energy source in developing countries.

While energy demand in these countries is steadily increasing, some of the demand is being met by switching from traditional to commercial energy sources.

Biomass energy technology is rapidly advancing. Besides direct combustion, techniques for gasification, fermentation, and anaerobic digestion are all increasing the potential of biomass as a sustainable energy source. The viability of wind energy is increasing as well. Some 2,100 megawatts of new capacity was commissioned in 1998, pushing global wind generating capacity to 9,600 megawatts. Wind power accounted for an estimated 21 terawatt-hours of electricity production in 1999. While that still amounts to only 0.15 percent of global electricity production, the competitiveness of wind power is improving and its growth potential is substantial. Use of geothermal energy for electricity generation is also increasing, with a present generating capacity of more than 8,300 megawatts.

The Resource Outlook

To summarise, no serious global shortage of energy resources is likely during at least the first half of the 21st century. Reserves of traditional commercial fuels—oil, gas, and coal—will suffice for decades to come. When conventional oil resources are depleted, the huge unconventional oil and gas reserves will be tapped as new extraction and clean generating technologies mature. Coal reserves are also huge: the resource base is more than twice that of conventional and unconventional oil and gas. Clean technologies for coal will allow greater exploitation of this huge resource base, mainly in electricity production, but also through conversion into oil and gas, minimising environmentally harmful emissions.

The uranium resource base is also immense, and it is unlikely, at least in the short term, to be tapped in increasing amounts. The ultimately recoverable uranium reserves will easily meet any nuclear power requirements during this century.

The renewable resource base is also promising. Only part of the global hydro potential has been tapped. Hydropower plants will continue to be built as demand for electricity grows and the economics of long-distance, extra-high-voltage transmission improve. Biomass has substantial potential and will continue to be used not only as a traditional fuel but also in increasingly sophisticated ways, through thermochemical and biochemical applications. New renewable sources, particularly wind power, will gradually increase the contribution of renewables to global energy supplies as the economies and technologies of these environmentally attractive sources continue to improve.

In short, the world's energy supplies offer good prospects for energy security in the 21st century. The fossil fuel reserves amount to 1,300 Gtoe and the fossil fuel resource base to around 5,000 Gtoe (see table 1), amounts sufficient to cover global requirements throughout this century, even with a high-growth scenario. That does not mean there will be no temporary or structural energy shortages, but as long as the energy resources are being explored and exploited, these shortages will not be due to resource inadequacy.

References

Campbell, C. J., and J. H. Laherrere. 1998. "The End of Cheap Oil." *Scientific American* 278: 60–65.

IEA (International Energy Agency). 1998. *World Energy Outlook.* Paris.

Masters, C. D. 1994. *World Petroleum Assessment and Analysis.* Proceedings of the 14th World Petroleum Congress. New York: John Wiley & Sons.

Mitchell, J. V. 1997. *Will Western Europe Face an Energy Shortage?* Strasbourg: Energy Council of France.

WEC (World Energy Council). 1998. *Survey of Energy Resources.* London.

World Bank. 1999. *World Development Report 1999/2000: Entering the 21st Century.* New York: Oxford University Press.

Seth Dunn

Decarbonizing the Energy Economy

The human harnessing of energy has long included the release of carbon atoms, dating at least as far back as a wood fire in the Escale cave near Marseilles, France, more than 750,000 years ago. Reliance on wood was a common feature of energy use in most settled parts of the world until the mid-nineteenth century. Growing population density and energy use in Great Britain meant that bulky and awkward-to-carry wood—in spite of its abundance—gradually lost out to coal, which was likewise abundant but more concentrated and easily transported. Though it was not noted at the time, the new fuel also happened to produce less carbon and more hydrogen per unit of energy—with one or two molecules of carbon per each one of hydrogen, versus a 10-to-1 ratio for wood. Thus began the first wave of "decarbonization."

Despite its negative health and environmental effects, coal remained King of the energy world for the remainder of the nineteenth century and well into the twentieth. But the automotive revolution eventually favored another new fuel. Oil boasted an even higher energy density, could be transported more easily and through pipelines, and emitted less soot. By the 1960s, it surpassed coal. With only one molecule of carbon for every two of hydrogen, oil marked the second wave of decarbonization.

At the end of the twentieth century, oil was still the world's leading energy source. Like its predecessors, however, it faced an up-and-coming challenger. Natural gas, which burned more efficiently, used a distributed network of pipes that made petroleum pipelines appear clumsy, and benefited from being known as the cleanest fossil fuel, began its ascent in the final decades of the second millennium. With one unit of carbon to four units of hydrogen, natural gas represents the third wave of decarbonization.

This molecular perspective on the energy economy is not typical: generally speaking, most experts prefer to discuss energy trends in terms of politics and prices, resources and reserves. Yet the trend toward decarbonization —progressively reducing the amount of carbon produced for a given amount of energy—is as illuminating and important as it is overlooked. Jesse Ausubel of

Rockefeller University goes so far as to argue that this pattern lies "at the heart of understanding the evolution of the energy system." Exploring this trend is therefore critical if we are to grasp where the energy economy might be headed in the future.

Yet even as the third wave begins its rise, a fourth has already appeared on the horizon. Initially building on the natural gas network for its distribution, and derived at first from natural gas to run high-efficiency fuel cells, hydrogen could within several decades displace natural gas. Eventually using its own full-fledged network, and created by the splitting of water into hydrogen and oxygen from solar, wind, and other forms of virtually limitless flows of renewable resources, the hydrogen economy will over time "free energy from carbon," as Nebojsa Nakićenović of the International Institute for Applied Systems Analysis (IIASA) has put it.

The first three waves of decarbonizing were driven by the search for more abundant and easily harnessed energy sources; local and regional environmental factors played a limited role in aiding the ascents of oil and natural gas. But the past century's discovery of carbon's role in changing Earth's climate, the role of humans in adding this carbon to the atmosphere, and the potential risks that accompany climate change have all made global environmental concerns a major new factor in the approaching fourth wave. Present and future generations thus face the challenge of deliberate decarbonization.

A key barrier to enhancing the decarbonization process is the lingering misperception among many governments and businesses that limiting carbon emissions necessitates reductions in energy use and economic growth. Yet the past quarter-century provides strong evidence that carbon and economic output can be "decoupled." . . . Between 1950 and 1999, the number of tons of carbon produced for each million dollars of gross world product was reduced by 39 percent, from 250 to 150 tons per million dollars of output, averaging an annual intensity improvement rate of 1 percent—but with a 2-percent annual average during the 1990s. Carbon intensity is becoming an important indicator of our movement toward a sustainable energy economy, particularly as evolving scientific understanding of climate change strengthens public concern and pressure for action. . . .

Beyond Fossil Fuels

"New" renewable energy sources are in the position of petroleum about a century ago: accounting for a fraction of world energy, but gaining footholds in certain regions and markets. It was concern about the future of oil that initially sparked the first modern wave of interest in renewable energy technologies in the 1970s and 1980s. The dynamic growth phase that began in the 1990s, however, features significant technological improvements and is driven in part by policies designed to address carbon emission reduction commitments.

The most spectacular recent growth in renewable energy use has occurred with wind power, which averaged a 24-percent annual increase in the 1990s and is now a $4-billion industry. Advances in wind turbine systems have dramatically lowered the generation cost of wind power over the last two decades, to the point where it is becoming cost-competitive with fossil-fuel-fired power generation in some regions. But its strong market entry also owes much to policy support, particularly electricity "in-feed laws" in Europe that provide generous fixed payments to project developers. Seven of the top 10 nations using wind power are European.

Solar photovoltaics (PVs), which convert sunlight into electricity, have also witnessed significant cost declines and market growth—a 17-percent annual average during the past decade. The global solar industry is estimated at $2.2 billion; BP Solar—the leading manufacturer, with a 20-percent share of the world market and estimated annual revenues of $200 million—has its cells in use in more than 150 countries and manufacturing facilities in the United States, Spain, India, and Australia. The PV market is experiencing a shift from primarily off-grid uses to grid-connected applications, which are the fastest-growing sector due to subsidy programs in Japan, Germany, and the United States. Also poised for growth is the off-grid rural market in developing nations, which is forecast to expand more than fivefold over the next 10 years.

Hydropower, geothermal power, and biomass energy have experienced slower but steady growth over the last decade, ranging from 1 to 4 percent annually. Increasingly, new hydropower systems in North America and East Asia are oriented toward small-scale applications. Geothermal power is on the rise in parts of Asia-Pacific and Latin America. Use of biomass energy—primarily agricultural and forestry residues—which accounts for as much as 14 percent of world energy and has both residential and commercial uses, is benefiting from modern applications in turbines and factories. Less-established technologies, such as harnessing wave and tidal energy, may yet prove viable.

Several recent publications have explored the possibility of renewable energy providing a significant share of world energy by mid-century. In the Shell Group Planning "Sustained Growth" scenario, renewables first capture niche markets and "by 2020 become fully competitive with conventional energy sources." Solar PV technologies experience cost reductions similar to those for oil in the 1890s, and between 2020 and 2030 developing countries turn aggressively to renewable energy.

During the next two decades, in the Shell study forecast, these technologies become widely commercial as fossil fuels plateau, with wind, biomass, and solar PVs achieving market penetration rates analogous to those of coal, oil, and gas in the past. By 2050, over 50 percent of primary energy supply comes from renewables, with 10 sources each holding a market share of 5–15 percent. The company responded to its own scenario by establishing in 1997 a Shell Renewables core business, which has earmarked $500 million over five years and has projects under way in solar, biomass, and wind energy in Europe, South America, the Middle East, Africa, and the Asia Pacific.

Renewable energy is prominent in other studies as well. The IIASA/WEC [World Energy Council] ecologically driven scenarios show renewables reaching a 40-percent share in 2050 and 80 percent in 2100. They note a "changing geography of renewables," as developing nations take a leadership role in harnessing the resources by the 2020s and account for two thirds of renewable energy use by 2050. In *Bending the Curve: Toward Global Sustainability,* the Stockholm Environment Institute (SEI) describes a 25-percent renewables scenario that "requires neither heroic technological assumptions nor economic disruption." Renewables promotion may, it argues, further the goals of economic development and job stimulation; the primary constraints to achieving the energy goal are institutional and political.

The SEI study lays out specific policies that could make its scenario a reality. Carbon taxes are coupled with reductions in other levies, as in the tax shifts of several European countries. Fossil fuel subsidies are phased out. New financing initiatives and economic incentives spur investment in renewable technologies. Expanded research, development, and demonstration create new technologies. Better information, capacity building, and institutional frameworks overcome barriers to investing in renewables. And global initiatives to transfer technologies and know-how make these sources the foundation of developing nations' energy economies.

A common question in discussions of the long-term role of renewable energy is whether these resources could conceivably meet worldwide energy requirements. Bent Sørensen at the Roskilde University in Denmark has delved into this question, using an array of economic, population, and energy data and projections to create scenarios that consider whether solar, wind, biomass, geothermal, and hydropower collectively could meet global energy demand by 2050. His scenarios achieve near-zero carbon emissions, and are more expensive than the current system only when environmental costs are neglected.

The Roskilde study concludes that a combination of dispersed and more centralized applications—placing solar PVs and fuel cells in buildings and vehicles; and wind turbines adjacent to buildings and on farmland, plus a number of larger solar arrays, offshore wind parks, hydro installations—would create a "robust" system capable of meeting the world's entire energy demand. But the study also stresses that significant additional technological and policy development would be required to realize the scenario.

The policy preconditions for a wholesale shift to renewable energy include a mix of free market competition and regulation, with environmental taxes correcting marketplace distortions; temporary subsidies to support the market entry of renewables; and the removal of hidden subsidies to conventional sources. Taxes would need to be synchronized internationally to avoid differential treatment of energy sources among countries, and to be adjusted if the market does not respond enough to the initial price change. And the energy transition would have to be kept on course by continuous "goal-setting" and monitoring.

In fact, the European Union (EU) has already established a target for renewables of 12 percent of total energy by 2010. National goals in Germany and Denmark have helped stimulate rapid wind power development that has in turn led to more ambitious goals. The Danish company BTM Consult, which projects that wind power could supply 10 percent of global electricity by 2020, argues that nations should set wind-specific goals, backed up by legally enforced mechanisms such as those now popular in Europe. The United States aims to increase wind power's share of electricity to 10 percent by 2010—compared with about 0.1 percent today—but it has yet to provide such policy support.

One of the variables shaping how fast an energy economy based on renewable resources emerges is the extent to which storage systems are developed that can harness the intermittent flows of these sources and store them for later use. Viable energy storage is essential for turning renewables into mainstream sources, and engineers have experimented with a long list of candidates, including batteries, flywheels, superconductors, ultracapacitors, pumped hydropower, and compressed gas. But the most versatile energy storage system, and the best "energy carrier," is hydrogen.

Entering the Hydrogen Age

The ultimate step in the decarbonization process is the production and use of pure hydrogen. As noted earlier, the gradual displacement of carbon by hydrogen in energy sources is well under way. Between 1860 and 1990, the ratio of hydrogen to carbon in the world energy mix increased more than sixfold.

Hydrogen—the universe's lightest and most abundant element—is known most commonly for its use as a rocket fuel. It is produced today primarily from the steam reformation of natural gas for a variety of industrial applications, such as the production of fertilizers, resins, plastics, and solvents. Hydrogen is transported by rail, truck, and pipeline and stored in liquid or gaseous form. Though it costs considerably more to produce than petroleum today, the prospect of hydrogen becoming a major carrier of energy has been revived due to advances in another space-age technology: the fuel cell.

An electrochemical device that combines hydrogen and oxygen to produce electricity and water, the fuel cell was first used widely in the U.S. space program and later in a number of defense applications such as submarines and jeeps. While these cells were traditionally bulky and expensive, technical advances and size and cost reductions have sparked interest in using them in place of internal combustion engines (ICEs), central power plants, and even portable electronics. Their initial costs are several times higher than these conventional systems, but are anticipated to drop sharply with mass production.

Fuel cells are nearing the market for both stationary and transportation uses. Ballard Power Systems and Fuel Cell Energy plan to deliver their first commercial 250-kilowatt units in 2001. DaimlerChrysler, which is devoting $1.5 billion to fuel cell efforts over the next several years, aims to sell 20–30 of its fuel cell buses to transit systems in Europe by 2002, and to mass-produce 100,000 fuel cell cars and begin selling them by 2004. Toyota and Honda have set 2003 commercialization dates for their fuel cell vehicles.

An important stimulus of the fuel cell market has been the state of California's requirement that 2 percent of new cars sold in 2003 be zero-emissions vehicles. The mandate has spurred new fuel cell investments and collaborations. The California Fuel Cell Partnership, composed of major car manufacturers, energy companies, and government agencies, intends to test 70 fuel cell vehicles by 2003, with energy companies delivering hydrogen and other fuels to refueling stations. In November 2000, the partnership unveiled its headquarters, which includes a refueling station and public education center, and the first fleet of vehicles in Sacramento.

Another region at the vanguard of the hydrogen transition is Iceland, where in February 1999 a $1-million joint venture to create the world's first hydrogen economy was launched by the government and other Icelandic institutions, DaimlerChrysler, Shell Hydrogen, and Norsk Hydro. The joint venture, Icelandic New Energy, emerged from a parliament-appointed study commission that recommended the initiative; it is now official government policy to promote the increased use of renewable resources—geothermal and hydroelectric resources provide 70 percent of the nation's energy—to produce hydrogen. The strategy is to begin with buses, followed by passenger cars and fishing vessels, with the goal of completing the transition between 2030 and 2040.

Hydrogen-powered buses are a logical first step because they can handle larger and heavier fuel cells and do not need to be refueled as often. Ballard has demonstrated fuel cell transit buses in Vancouver and Chicago, running on compressed hydrogen gas stored in tanks onboard the vehicles. Hydrogen refueling stations for buses and vans are also appearing in Germany, in the Munich airport and Hamburg, although these will initially supply vehicles with ICEs that use the fuel directly. The Hamburg station intends to eventually import hydrogen from Iceland.

The introduction of fuel cell cars faces three tough technical challenges: integrating small, inexpensive, and efficient fuel cells into the vehicles; designing tanks that can store hydrogen onboard; and developing a hydrogen refueling infrastructure. The design issue is being overcome by improvements in power density and reduced platinum requirements. The storage issue is being addressed through vehicle efficiency gains, tank redesign, and progress in storage technologies such as carbon nanotubes and metal hydrides. Although the direct hydrogen fuel cell vehicle is the simplest and most elegant approach, industry is devoting substantial research to having cars use onboard reformers that strip hydrogen from gasoline, natural gas, or methanol. From their perspective this approach may appear preferable to spending the money needed to develop a new refueling infrastructure, but the economics and environmental effects are less straightforward.

Studies indicate that by the time a critical mass of infrastructure and vehicles are in place, direct hydrogen will be more cost-effective than onboard reformers. Reformer-based fuel cell cars, furthermore, are unlikely to achieve the environmental performance of those using direct hydrogen. The Canada-based Pembina Institute, comparing the "well-to-wheel" greenhouse gas emissions of various hydrogen vehicle production systems over 1,000 kilometers of travel, found that reforming hydrogen from hydrocarbon fuels does provide an improvement over a gasoline-powered internal combustion engine, but the improvement varies widely, depending on the fuel. Gasoline and methanol, which are the preferred fuels for many transport and energy companies, offered the least improvement, with reductions of 22–35 percent. The hydrocarbon demonstrating the greatest climate benefits—natural gas, whose life-cycle emissions were 68–72 percent below that of the gasoline ICE—has been relatively ignored by industry.

One near-term solution to the "chicken-and-egg" infrastructure problem in many countries would be to use small-scale natural gas reformers at fueling stations, relying on existing natural gas pipelines to distribute the fuel. Marc Jensen and Marc Ross of the University of Michigan estimate that building 10,000 such stations—covering 10–15 percent of U.S. filling stations—would be enough to motivate vehicle manufacturers to pursue mass production of direct hydrogen fuel cell vehicles. This would require $3–15 billion in capital investment, which "can be weighed against the social and environmental benefits that will be gained as a fleet of hydrogen fueled vehicles grows." Ultimately, hundreds of billions of dollars will need to be invested over decades in a network of underground pipelines engineered specifically for hydrogen.

While natural gas is currently the most common source of hydrogen, and the reformation of coal and oil is also being explored, renewable energy sources are likely to eventually produce hydrogen most economically. Electrolysis of water can convert solar and wind energy into hydrogen; scientists have recently boosted the efficiency of solar-powered hydrogen extraction by 50 percent. Biomass can be gasified to produce the fuel. Other potential renewable hydrogen sources include photolysis—the splitting of water with direct sunlight—and common algae, which produces hydrogen when deprived of sunlight. Coupling renewable energy systems with hydrogen will address their inherent intermittency and easily meet energy demand, provided there are continued technical advances and cost reductions in fuel cells and electrolyzers.

Wise decisions made in today's early hydrogen economy could yield enormous economic and environmental benefits. Wrong turns toward an interim infrastructure, on the other hand, could strand millions of dollars in financial assets, lock in fleets of obsolete fuel cell cars, and add millions of extra tons of carbon emissions. There is an appropriate role for governments to play in collaborating with transport and energy companies to develop a direct hydrogen infrastructure through greater research into storage technologies and the identification of barriers and strategies to surmount them. A January 2000 report from the U.S. National Renewable Energy Laboratory concluded that "there are no technical showstoppers to implementing a near-term hydrogen fuel infrastructure for direct hydrogen fuel cell vehicles." The study did, though, point out engineering challenges and institutional issues, such as the need for codes and standards for hydrogen use.

If governments and industries can come up with forward-looking roadmaps, optimistic scenarios for hydrogen may materialize. In the Oxford Institute's Kyoto scenario, hydrogen becomes more competitive due to emissions policies that cause oil prices to rise, undercutting oil supply and reaching production of 3.2 million barrels per day of oil equivalent by 2010, and 9.5 million barrels by 2020. In their 1999 book *The Long Boom,* former Shell executive Peter Schwartz and colleagues describe a scenario in which fuel cells have displaced the internal combustion engine within two decades, and "by 2050 the world is running on hydrogen, or close enough to call it the Hydrogen Age"—with climate change concerns a major driver of the transition.

POSTSCRIPT

Should the World Continue to Rely on Oil as a Major Source of Energy?

The message of Khatib et al.'s assessment of foreseeable world energy supplies is "Don't panic just yet." The study reveals no serious energy shortage during the first half of the twenty-first century. In fact, the report suggests that oil supply conditions have actually improved since the crises of the 1970s and early 1980s. Khatib et al. go further in their assessment, concluding that fossil fuel reserves are "sufficient to cover global requirements throughout this century, even with a high-growth scenario."

Francis R. Stabler, in "The Pump Will Never Run Dry!" *The Futurist* (November 1998), argues that technology and free enterprise will combine to allow the human race to continue its reliance on oil far into the future. To be sure, Stabler's view of the future availability of gas is in the minority. One supporter is Julian L. Simon, who argues in *The Ultimate Resource 2* (Princeton University Press, 1996) that even God may not know exactly how much oil and gas are "out there." Another Stabler supporter is Bjørn Lomborg. In *The Skeptical Environmentalist: Measuring the Real State of the World* (Cambridge University Press, 2001), Lomborg argues that the world seems to find more fossil energy than it consumes, and he concludes that "we have oil for at least 40 years at present consumption, at least 60 years' worth of gas, and 230 years' worth of coal."

Simon and Lomborg's views are supported by Michael C. Lynch in an online article entitled "Crying Wolf: Warnings About Oil Supply," http://sepwww.stanford.edu/sep/jon/world-oil.dir/lynch.html.

Dunn follows the conventional wisdom in his article. He maintains that because oil is a finite resource, its supply will end some day, and that day will be sooner rather than later. In fact, Dunn suggests that new renewable energy sources are in the same position as oil a century ago—that is, they are "gaining footholds" in the energy market. Dunn has argued elsewhere that the global economy has been built on the rapid depletion of nonrenewable resources and that such consumption levels cannot possibly be maintained throughout the twenty-first century, as they were the previous century. See Christopher Flavin and Seth Dunn, "Reinventing the Energy System," in Lester R. Brown et al., *State of the World 1999* (W. W. Norton, 1999).

Finally, James J. MacKenzie provides a comprehensive yet succinct discussion on the peaking of oil in "Oil as a Finite Resource: When Is Global Production Likely to Peak?" a paper of the World Resources Institute (March 2000), which is available on the Web at http://www.wri.org/wri/climate/jm_oil_001.html.

ISSUE 7

Will the World Be Able to Feed Itself in the Foreseeable Future?

YES: Sylvie Brunel, from "Increasing Productive Capacity: A Global Imperative," in Action Against Hunger, *The Geopolitics of Hunger, 2000–2001: Hunger and Power* (Lynne Rienner, 2001)

NO: Lester R. Brown, from "Eradicating Hunger: A Growing Challenge," in Lester R. Brown et al., *State of the World 2001* (W. W. Norton, 2001)

ISSUE SUMMARY

YES: Sylvie Brunel, former president of Action Against Hunger, argues that "there is no doubt that world food production . . . is enough to meet the needs of" all the world's peoples.

NO: Lester R. Brown, former president of the Worldwatch Institute, maintains that little, if any, progress is being made to eradicate pervasive global hunger, despite increases in food productivity.

Visualize two pictures. The first snapshot, typical of photographs that have graced the covers of many of the world's magazines, reveals a group of people in Africa, including a significant number of small children, who show dramatic signs of advanced malnutrition and even starvation. The second picture (really several in sequence) shows an apparently wealthy couple finishing a meal at a rather expensive restaurant. The waiter removes their plates still half full of food, an untouched loaf of French bread, and assorted other morsels from the table and deposits them in the kitchen garbage can. These scenarios once highlighted a popular film about world hunger. The implication was quite clear: If only the wealthy would share their food with the poor, no one would go hungry. Today the simplicity of this image is obvious.

This issue addresses the question of whether or not the world will be able to feed itself by the middle of the twenty-first century. A prior question, of course, is whether or not enough food is grown throughout the world today to handle the current needs of all the planet's citizens. News accounts of chronic food shortages somewhere in the world have been appearing regularly for about 20 years. This time has witnessed graphic accounts in news specials about the consequences of insufficient food, usually in sub-Saharan Africa. Also, several national and international studies have been commissioned to address world hunger. An American study organized by President Jimmy Carter, for example, concluded that the root cause of hunger was poverty.

One might deduce from all this activity that population growth has outpaced food production and that the planet's agricultural capabilities are no longer sufficient. Yet the ability of most countries to grow enough food has not yet been challenged. During the 1970–2000 period, only one region of the globe, sub-Saharan Africa, was unable to have its own food production keep pace with population growth. All other regions of the globe experienced food increases greater than human growth.

This is instructive because, beginning in the early 1970s, a number of factors conspired to lessen the likelihood that all humans would go to bed each night adequately nourished: Weather in major food-producing countries turned bad; a number of countries, most notably Japan and the Soviet Union, entered the world-grain importing business with a vengeance; the cost of energy used to enhance agricultural output rose dramatically; and less capital was available to poorer countries as loans or grants for purchasing agricultural inputs or the finished product (food) itself. Yet the world has had little difficulty growing sufficient food, enough to provide every person with two loaves of bread per day as well as other commodities. Major food-producing countries have even cut back the amount of acreage devoted to agriculture.

Why, then, have famine and other food-related maladies appeared with increasing frequency? The simple answer is that food is treated as a commodity, not a nutrient. Those who can afford to buy food or grow their own do not go hungry. However, the world's poor have become increasingly unable to afford to create their own successful agricultural ventures or to buy enough food.

The problem for the next half century, then, has several facets to it. First, can the planet physically sustain increases in food production equal to or greater than the ability of the human race to reproduce itself? This question can only be answered by examining both factors in the comparison—likely future food production levels and future fertility scenarios. A second question relates to the economic dimension associated with an efficient global food distribution system: Will the poorer countries of the globe that are unable to grow their own food have sufficient assets to purchase it, or will the international community create a global distribution network that ignores a country's ability to pay? And third, will countries that want to grow their own food be given the opportunity to do so?

The selections in this issue address the question of the planet's continuing ability to grow sufficient food to feed its growing population. Sylvie Brunel contends that if world food production were distributed among all the world's peoples, then there would be plenty of food. Lester R. Brown is pessimistic about the likelihood of hunger being eradicated. He maintains that simply growing more food is not enough, a concept that world leaders have not yet grasped.

Sylvie Brunel  **YES**

Increasing Productive Capacity:
A Global Imperative

Can the earth feed its inhabitants? Despite the most alarmist predictions—those of Lester Brown in his State of the World published each year by the World Watch Institute of Washington; those of Paul Ehrlich, author of *The Population Bomb;* or those of the Club of Rome—there is no doubt that world food production, if equally distributed among all the world's peoples, is enough to meet the needs of them all. It is true that the increase in world agricultural production has slowed in recent years, a situation that has led to an immediate flood of alarmist predictions. The world, however, is not heading toward famine. And this is for a number of reasons.

The Increase in World Agricultural Production Continues to Outpace Population Growth

Only persons who are ill informed or of bad faith can argue that the trend is toward a decline in food production. They may even succeed in proving that claim. It is enough for them to select as the base year one in which harvests were particularly good and a second year in which they declined steeply in order to show a "disturbing" trend. Hervé Kempf demonstrated, for example, how a comparison of 1984 and 1991 would show an increase in cereal production of only 0.7 percent per year, which would be "disturbing," since it is far below the 1.7% annual rate of population increase. By selecting the preceding years, one can show, on the contrary, that world agriculture has never been more productive: A comparison between 1983 and 1990 shows an increase in cereal production of 2.7 percent per year. These two statistics are clearly equally deceptive, and Joseph Klatzmann, in a refreshing little book, repeatedly warned against the danger of blindly trusting statistical data taken out of context.

If we examine world agricultural production over a long period, it becomes clear that the production curve exceeds the population growth curve. While world population did indeed double in one generation, grain production increased more than threefold, from 600 million to approximately 1,900 million tons per year. Each human being has available in theory 20 percent more food than in the early 1970s, or 2,700 calories per person per day, which is far more than a person's estimated need of between 2,000 and 2,200 calories, depending on the sources. However, half of the current grain production does not directly benefit people: Approximately 20 percent is used to feed cattle, 5 percent is kept for seeds, and the remaining 25 percent is quite simply lost as a result of poor storage or destruction by rodents, insects, and so on, especially in developing countries. It is therefore not the impossibility of increasing agricultural production that threatens mankind, but rather the way in which this increase is achieved and for the benefit of whom.

Indeed, it is in fact not in the countries of the so-called Third World but rather in the developed countries that agricultural production has slowed, in other words precisely where the problems of hunger have been overcome. (At least they have been overcome in quantitative terms; in qualitative terms, obesity, on the one hand, and malnutrition caused by the economic and social marginalization of certain categories of persons, on the other, have become real societal problems.) The developed countries have chosen to voluntarily limit their agricultural production in order to adapt it to the level of demand at which production would be profitable, in other words, to the consumer market. The fact that there are some 800 million people suffering from malnutrition in the world in no way changes this calculation, since those persons are too poor to buy food.

The reduction or slowdown in the rate of increase in world food production is thus attributable mainly to the developed countries, for reasons that have nothing to do with ecological limitations. Pierre Le Roy estimates at 20 million tons the reduction in supply that results from Europe's policy of limitation of production (land left fallow), an amount that represents twice the total of all food imports by sub-Saharan Africa.

Food imports by the Third World are indeed increasing, rising from 20 million tons in 1960, or 2 percent of consumption, to 120 million tons in the mid 1990s, or 20 percent of consumption. Economic forecasts suggest that this dependency is likely to increase even further in the decades ahead and to rise to 160 million tons within two decades. The reasons for this growing dependence, which will create problems without precedent for the economies of poor countries that will face increasingly onerous food import bills, are both negative and positive.

The negative factor of continuing population growth and spreading urbanization in the countries of the South, where nearly half the population now lives in cities, explains why more and more people are consuming food that their farmers are incapable of providing. The positive factor of the increase in average living standards in the developing countries and the emergence of a middle class that consumes more meat and dairy products places increasing pressure on the demand for cereals, in particular secondary cereals for stock feed.

Two-Speed Agricultural Policies

Why cannot the Third World feed itself, even though self-sufficiency in food was the grand slogan of the 1970s and 1980s?

The "technical" impossibility of increasing agricultural production in the South is not the problem: The earth is far from reaching its maximum agricultural potential, and the Food and Agriculture Organization (FAO) has pointed out that the useful agricultural surface in developing countries (700 million hectares) could be doubled without encroaching on protected areas such as forests or areas in which people live. Latin America and Africa hold the greatest potential in this regard. In addition, the potential for increased production through more intensive farming methods remains considerable. Only 11 kilograms of fertilizer are used per hectare in Africa, compared with 66 kilograms in Latin America and 139 kilograms in Asia, and only 5 percent of land is irrigated in Africa (most of this in countries that are unable to take advantage of it, such as Sudan and Madagascar), compared to 37 percent in Asia and 14 percent in Latin America. This situation offers tremendous potential for growth.

But the political and economic choices made by the countries of the Third World have thus far been detrimental to agriculture, and in particular to small peasant farming. Investments in agriculture have been concentrated in regions in which purchasing power is greatest and are characterized by a concern to protect the income of farmers, which has been steadily declining. As a result, these investments are moving in the direction of a two-tiered world that is becoming increasingly unequal in terms of access to food.

On one hand, the developed countries enjoy rapid growth, and despite the fact that farmers represent on average no more than 3 percent of the active population, their food supply is abundant and diversified, prices are low, and import levels are low as a result of the massive support given to the agricultural sector (in the mid-1990s, Organization for Economic Cooperation and Development (OECD) countries each year spent more than two hundred billion dollars to support their agricultural sectors). In that part of the world, the concern is no longer the fear of shortage, but rather the quality of the food consumed. The agrofood industry, now powerful after being forced to steadily increase its output over the past decades in order to keep up with the steadily rising demand, is today facing another challenge, namely, shifting to production methods that focus less on quantity and more on the quality of the inputs used and on the quality of the final product. Producers are also concerned about the methods used to satisfy the demand of consumers, who now want food that is not only abundant but also varied and, above all, healthy. The successive scandals of mad-cow disease, salmonella poisoning in chickens, hormone-treated beef cattle; the rejection by consumer groups of genetically modified plants; animal feed that includes mud from cleaning stations; and the questions raised about the production of eggs by battery hens are indicative of a new era in which insistence on quality is now a greater challenge than the demand for quantity.

On the other side are the poor and vulnerable countries, where malnutrition is endemic and where a majority of the population still depends on agriculture. The food supply remains insufficient, however, because of the poor yields that result from the low level of technology used, the absence of incentives to produce because of economic policies that discourage agriculture, and unfavorable exchange rates that make the importation of agricultural imports expensive. It is therefore precisely in those countries that agricultural production needs to be increased. First, increased production would reduce the cost of food, particularly in the large urban centers, and thereby make it accessible to this large sector of the population that is too poor to eat properly. Second, increased production would reduce the food import bills of countries that are increasingly dependent on imports, mainly from the rich countries.

The Food Supply Is a Regional, Not a Global Problem

At the global level, the food supply is increasing for reasons that are both positive and negative. On the positive side, the agricultural sector in Eastern Europe, which needed to be restructured following the collapse of the Iron Curtain, is now on the road to recovery. On the negative side, the economic difficulties of East Asia have led to a decline in food imports by that region.

At the regional level, the structural overproduction that the world has been experiencing for the past twenty years hardly prevents sub-Saharan Africa and South Asia from experiencing hunger. Of the approximately 800 million malnourished people in the world, more than 200 million live in Africa (nearly 40 percent of the population) and 530 million in South Asia (or one person out of five).

What is therefore responsible for this disastrous and paradoxical situation at a time when, in order to reduce the supply of food, rich countries are destroying mountains of surplus food each year and forcing their farmers to leave a portion of their land uncultivated, through subsidies for land that is left fallow?

One answer is wars and conflicts, particularly in Africa, that disrupt agricultural production. A second, even more important answer is mass poverty, which prevents an entire sector of the world's population (one out of every five inhabitants of the Third World) from obtaining adequate food. That sector is incapable of producing enough food to meet its needs and lacks the means to purchase it, even when the food is available and can be bought. Worldwide, some 1.5 billion people live below the poverty level. Mass poverty is all the more serious, as it is always combined with ignorance: It is always the poorest classes that commit the most harmful errors of nutrition, since they lack the advantage of basic education. The errors of nutrition committed by pregnant women and children, who make up the primary groups at risk of hunger, should be the focus of particular attention, since these errors have disastrous consequences for the future of the entire society.

According to the United Nations Children's Fund (UNICEF), half of the world's malnourished children live in Asia, which has 100 million of the 200 million total, including 70 million in India. That country alone has two and a half times more malnourished children than all of sub-Saharan Africa.

Writing about South Asia . . . , Gilbert Etienne remarked on the extent to which the problem of hunger remains unresolved because of mass poverty and the slowdown in investments in agriculture. This is despite the notable progress achieved on the Asian continent.

The case of Africa gives cause for even greater concern:

1. Unlike the situation in other continents, the rate of malnutrition is not declining. Quite the opposite, in fact, because chronic malnutrition still affects nearly 40 percent of the region's population.

2. The high proportion of young people in the population, a sign of a vigorous population still characterized by high birthrates, has led to a high proportion of unemployed in relation to the number of those who are in a position to contribute to production. The burden on the economies of African countries is therefore particularly heavy, especially since most of these countries suffer from an acute lack of financial resources.

3. The continent's dependency on food from foreign countries is due to the low productivity of its own agriculture and the growing number of Africans now living in urban areas (more than one in three compared with one in ten a generation ago). This dependency is effectively addressed neither by imports (9 to 10 million tons per year), because of the lack of adequate financial resources, nor by food aid (approximately 2 million tons), which has been falling drastically for some years now. Consequently, the food needs of Africans are not being satisfied, since the widespread poverty of a sector of the urban population does not permit that sector to obtain food at market prices. At the same time many rural dwellers are unable to provide for themselves during the period between harvests, on account of the low productivity levels and inadequate access to food.

4. A large number of people in Africa are affected by war or internal conflict. Even in countries in which the populations could, in theory, be properly fed, there is an adverse impact on the food situation of the population because of the insecurity of the economic actors, the weakness of the State, and the destruction or confiscation of crops. In this regard, Africa is by far the continent most affected by conflicts, which also result in massive populations of refugees and displaced persons who depend on international aid for their survival.

5. Poverty and the pressure on land and resources of the high population growth rate are not matched by corresponding investments in agriculture that would bring about increases in yields. Africa is thus the continent in which the problems of deforestation, desertification, and soil erosion are most acute. It is also the continent in which access to drinking water and irrigation is still very limited.

The situation is not desperate, however: Despite the lack of investment in small peasant farming, agricultural production in Africa has risen by 2 percent per year since the early 1960s. Grain production has more than doubled, from 30 million to 66 million tons. This rate of increase is insufficient to meet the needs of a growing population (3 percent per year) because of the way in which it has been achieved (mainly by increasing the area of land under cultivation).

Nevertheless it shows that more intensive farming methods are needed in Africa and that this approach has the potential to significantly increase agricultural output. When the FAO states that the "load capacity of the land" in many countries has now been exceeded, it is basing its conclusions on the use of traditional production methods, such as use of the hoe more often than not, lack of fertilizers, lack of irrigation, and use of a diverse range of traditional varieties of grain to compensate for climatic and pedological constraints, with but modest results. The average yield for Africa remains 1,000 kilograms per hectare of millet and corn, which shows how much room for improvement there would be if African governments were to decide to treat their farmers a little better and to invest in their agricultural potential.

What are some of the ways in which agricultural production can be increased? It is interesting to note that food problems do not occur in countries that are at peace, that enjoy democracy, and in which farmers operate under conditions of relative legal and administrative security, even when these countries are densely populated and located in unfavorable climatic zones. It is better to live in Burkina Faso than in the Democratic Republic of Congo, even though the Congo is infinitely better endowed than Burkina Faso in terms of rainfall and available land. Similarly, the "white revolution" in Mali is revitalizing those regions that produce rice, millet, and cotton and enriching their farmers, even as hunger still plagues Madagascar, the former breadbasket of southern Africa, which has been making error after economic error over the last quarter of a century.

In order to bring about peace and security in Africa, a resumption of cooperation is necessary. However, the level of official development assistance has never been lower. Will the renegotiation of the Lomé Convention relaunch the partnership between Europe and Africa for the concerted development of agriculture?

Lester R. Brown

 **NO**

Eradicating Hunger: A Growing Challenge

In 1974, U.S. Secretary of State Henry Kissinger made a pledge at the World Food Conference in Rome: "By 1984, no man, woman, or child would go to bed hungry." Those attending the conference, including many political leaders and ministers of agriculture, came away inspired by this commitment to end hunger.

More than 26 years later, hunger is still very much part of the social landscape. Today, 1.1 billion of the world's 6 billion people are undernourished and underweight. Hunger and the fear of starvation quite literally shape their lives. A report from the U.N. Food and Agriculture Organization (FAO) describes hunger: "It is not a transitory condition. It is chronic. It is debilitating. Sometimes, it is deadly. It blights the lives of all who are affected and undermines national economies and development processes . . . across much of the developing world."

Kissinger's boldly stated goal gave the impression that there was a plan to eradicate hunger. In fact, there was none. Kissinger himself had little understanding of the difficult steps needed to realize his goal. Unfortunately, this is still true of most political leaders today.

In 1996, governments again met in Rome at the World Food Summit to review the food prospect. This time delegates from 186 countries adopted a new goal of reducing the number who were hungry by half by 2015. But as in 1974, there was no plan for how to do this, nor little evidence that delegates understood the scale of the effort needed. FAO projections released in late 1999, just three years after the new, modest goal was set, acknowledge that the objective for 2015 is not likely to be reached because "the momentum is too slow and the progress too uneven."

Assertions such as Kissinger's and those of other political leaders may make people feel good, but if they are not grounded in a carefully thought out plan of action and supported by the relevant governments, they ultimately undermine confidence in the public process. This in turn can itself undermine progress.

In its most basic form, hunger is a productivity problem. Typically people are hungry because they do not produce enough food to meet their needs or because they do not earn enough money to buy it. The only lasting solution is to raise the productivity of the hungry—a task complicated by the ongoing shrinkage in cropland per person in developing countries.

The war against hunger cannot be won with business as usual. Given the forces at work, it will no longer be possible to stand still. If societies do not take decisive steps, they face the possibility of being forced into involuntary retreat by continuing population growth, spreading land hunger, deepening hydrological poverty, increasing climate instability, and a shrinking backlog of unused agricultural technology. Eradicating hunger—never easy—will now take a superhuman effort.

A Hunger Report: Status and Prospects

As noted, 1.1 billion people are undernourished and underweight as the new century begins. The meshing of this number with a World Bank estimate of 1.3 billion living in poverty, defined as those living on $1 a day or less, comes as no surprise. Poverty and hunger go hand in hand.

The alarming extent of hunger in the world today comes after a half-century during which world food output nearly tripled. The good news is that the share of population that is hungry is diminishing in all regions except Africa. Since 1980, both East Asia and Latin America have substantially reduced the number and the share of their populations that are hungry. In the Indian subcontinent, results have been mixed, with the number of hungry continuing to increase but the share declining slightly. In Africa, however, both the number and the share of hungry people have increased since 1980.

The decline in East Asia was led by China, which brought the share of its people who are hungry down from 30 percent in 1980 to 11 percent in 1997. China's economic reforms initiated in 1978 led to a remarkable surge in agricultural output, one that boosted the grain harvest from roughly 200 kilograms per person a year to nearly 300 kilograms. This record jump in production in less than a decade led to the largest reduction in hunger on record. For most countries, the best nutrition data available are for children, who are also the segment of society most vulnerable to food scarcity. In Latin America, the share of children who are undernourished dropped from 14 percent in 1980 to 6 percent in 2000.

These gains in eradicating hunger in East Asia and Latin America leave most of the world's hungry concentrated in two regions: the Indian subcontinent and sub-Saharan Africa. In India, with more than a billion people, 53 percent of all children are undernourished. In Bangladesh, the share is 56 percent. And in Pakistan, it is 38 percent. In Africa, the share of children who are undernourished has increased from 26 percent in 1980 to 28 percent today. In Ethiopia, 48 percent of all children are underweight. In Nigeria, the most populous country in Africa, the figure is 39 percent.

Within the Indian subcontinent and sub-Saharan Africa, most of the hungry live in the countryside. The World Bank reports that 72 percent of the world's 1.3 billion poor live in rural areas. Most of them are undernourished; many are sentenced to a short life. These rural poor usually do not live on the productive irrigated plains, but on the semiarid/arid fringes of agriculture or in the upper reaches of watersheds on steeply sloping land that is highly erodible. Eradicating hunger depends on stabilizing these fragile ecosystems.

Recognizing that malnutrition is largely the result of rural poverty, the World Bank is replacing its long-standing agricultural development strategies, which were centered around crop production, with rural development strategies that use a much broader approach. The Bank planners believe that a more systemic approach to eradicating poverty in rural areas—one that embraces agriculture but that also integrates human capital development, the development of infrastructure, and social development into a strategy for rural development—is needed to shrink the number living in poverty. One advantage of encouraging investment in the countryside in both agribusiness and other industries is that it encourages breadwinners to stay in the countryside, keeping families and communities intact. In the absence of such a strategy, rural poverty simply feeds urban poverty.

Demographically, most of the world's poor live in countries where populations continue to grow rapidly, countries where poverty and population growth are reinforcing each other. The Indian subcontinent, for example, is adding 21 million people a year, the equivalent of another Australia. India's population has nearly tripled over the last half-century, growing from 350 million in 1950 to 1 billion in 2000. According to the U.N. projections, India will add 515 million more people by 2050, in effect adding roughly twice the current U.S. population. Pakistan's numbers, which tripled over the last half-century, are now expected to more than double over the next 50 years, going from 156 million to 345 million in 2050. And Bangladesh is projected to add 83 million people during this time, going from 129 million to 212 million. The subcontinent, already the hungriest region on Earth, is thus expected to add another 787 million people by mid-century.

No single factor bears so directly on the prospect of eradicating hunger in this region as population growth. When a farm passes from one generation to the next, it is typically subdivided among the children. With the second generation of rapid population growth and associated land fragmentation now unfolding, farms are shrinking to the point where they can no longer adequately support the people living on them.

Between 1970 and 1990, the number of farms in India with less than 2 hectares (5 acres) of land increased from 49 million to 82 million. Assuming that this trend has continued since then, India may now have 90 million or more families with farms of less than 2 hectares. If each family has six members, then 540 million people—over half India's population—are trapped in a precarious balance with their land.

Whether measuring changes in farm size or in grainland per person, the results of continuing rapid population growth are the same. Pakistan's projected growth from 156 million today to 345 million by 2050 will shrink its grainland per person from 0.08 hectares at present to 0.03 hectares, an area scarcely the size of a tennis court. African countries, with the world's fastest population growth, are facing a similar reduction. For example, as Nigeria's population increases from 111 million today to a projected 244 million in 2050, its per capita grainland, most of it semiarid and unirrigated, will shrink from

0.15 hectares to 0.07 hectares. Nigeria's food prospect, if it stays on this population trajectory, is not promising.

In Bangladesh, average farm size has already fallen below 1 hectare. According to one study, Bangladesh's "strong tradition of bequeathing land in fixed proportions to all male and female heirs has led to increasing landlessness and extreme fragmentation of agricultural holdings." In addition to the millions who are now landless, millions more have plots so small that they are effectively landless.

Further complicating efforts to expand food production are water shortages. Of the nearly 3 billion people to be added to world population in the next 50 years, almost all will be added in countries already facing water shortages, such as India, Pakistan, and many countries in Africa. In India, water tables are falling in large areas as the demands of the current population exceed the sustainable yield of aquifers. For many countries facing water scarcity, trying to eradicate hunger while population continues to grow rapidly is like trying to walk up a down escalator.

Even as the world faces the prospect of adding 80 million people a year over the next two decades, expanding food production is becoming more difficult. In each of the three food systems—croplands, rangelands, and oceanic fisheries—output expanded dramatically during the last half of the twentieth century. Now all that has changed.

Between 1950 and 2000, the world production of grain, the principal product of croplands, expanded from 631 million tons to 1,860 million tons, nearly tripling. In per capita terms, production went from 247 kilograms per person in 1950 to an all-time high of 342 kilograms in 1984, a gain of nearly 40 percent as growth in the grain harvest outstripped that of population. After 1984, production slowed, falling behind population. Production per person declined to 308 kilograms in 2000, a drop of 10 percent from the peak in 1984. (See Figure 1.) This decline is concentrated in the former Soviet Union, where the economy has shrunk by half since 1990, and in Africa where rapid population growth has simply outrun grain production.

Roughly 1.2 billion tons of the world grain harvest are consumed directly as food, with most of the remaining 660 million tons being consumed indirectly in livestock, poultry, and aquacultural products. The share of total grain used for feed varies widely among the "big three" food producers—ranging from a low of 4 percent in India to 27 percent in China and 68 percent in the United States.

Over the last half-century, the world's demand for animal protein has soared. Expanded output of meat from rangelands and of seafood from oceanic fisheries has satisfied most of this demand. World production of beef and mutton increased from 24 million tons in 1950 to 67 million tons in 1999, a near tripling. Most of the growth, however, occurred between 1950 and 1990, when output went up 2.5 percent a year. Since then, beef and mutton production has expanded by only 0.6 percent a year. (See Figure 2.)

An estimated four fifths of the 67 million tons of beef and mutton produced worldwide in 1999, roughly 54 million tons, comes from animals that

Figure 1

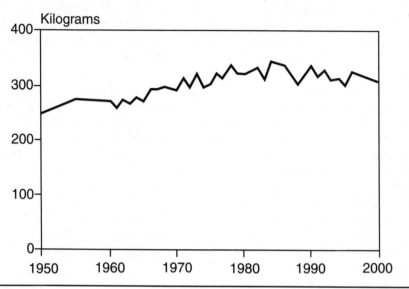

World Grain Production Per Person, 1950–2000

Source: USDA

Figure 2

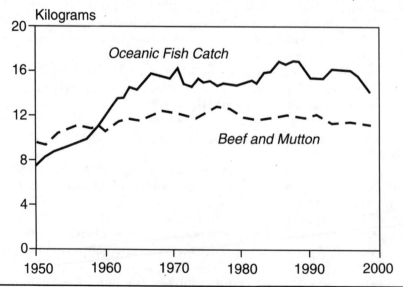

Oceanic Fish Catch and Beef and Mutton Production Per Person, 1950–99

Source: FAO

forage on rangelands. If the grain equivalent of the forage-based output is set at seven kilograms of grain per kilogram of beef or mutton, which is the conversion rate in feedlots, the beef and mutton produced on rangeland are the equivalent of 378 million tons of grain.

The growth in the oceanic fish catch exceeded even that of beef and mutton, increasing from 19 million tons in 1950 to 86 million tons in 1998, the last year for which data are available. This fourfold growth, too, was concentrated in the period from 1950 to 1990, a time during which the annual growth in the oceanic catch—at 3.8 percent—was easily double that of world population. As a result, seafood consumption per person worldwide roughly doubled, climbing from 8 kilograms in 1950 to 16 kilograms in 1990. Since then, it has fallen by some 10 percent. Assuming the fish farm conversion of less than two kilograms of grain for each kilogram of live weight added, then the grain equivalent of the 86-million-ton fish catch in 1998 was 172 million tons of grain.

The new reality is that fishers and ranchers can no longer contribute much to the growth of the world's food supply. For the first time since civilization began, farmers must carry the burden alone.

For a sense of the relative importance of rangelands and oceanic fisheries in the world food economy, compare the grain equivalent of their output with the world grain harvest. With rangelands accounting for the equivalent of 378 million tons of grain and with fisheries at 172 million tons, rangelands contributed 16 percent of the world grain supply and oceanic fisheries 7 percent. (See Table 1.)

Thus rangelands and oceanic fisheries provide the equivalent of nearly one fourth of the world grain supply. With their output no longer expected to expand, all future growth in the food supply must come from the 77 percent of total grain equivalent that is represented by croplands. With little new land to plow, the world's ability to eradicate hunger in the years ahead will depend heavily on how much cropland productivity can be raised. This is also one of the keys to eliminating poverty.

Table 1

Cropland, Rangeland, and Oceanic Fishery Contributions to World Food Supply, Measured in Grain Equivalent, 1999

Source	Quantity of Grain (million tons)	Share of Total (percent)
Grain production from cropland	1,855	77
Grain equivalent of rangeland beef and mutton	378	16
Grain equivalent of oceanic fish catch[1]	172	7
Total	2,405	100

---

[1]Fish production data from 1998.

Source: USDA, *Production, Supply, and Distribution,* electronic database, Washington, DC, updated September 2000; FAO, *Yearbook of Fishery Statistics: Capture Production* (Rome: various years).

Eradicating hunger in the Indian subcontinent and Africa will not be easy. It is difficult to eradicate for the same reasons it exists in the first place—rapid population growth, land hunger, and water scarcity. But there are also new forces that could complicate efforts to eliminate hunger. For example, Bangladesh, a country of 129 million people, has less than one tenth of a hectare of grainland per person—one of the smallest allotments in the world—and is threatened by rising sea level. The World Bank projects that a 1-meter rise in sea level during this century, the upper range of the recent projections by the Intergovernmental Panel on Climate Change, would cost Bangladesh half its riceland. This, combined with the prospect of adding another 83 million people over the next half-century, shows just how difficult it will be for Bangladesh, one of the world's hungriest countries, to feed its people.

Water shortages are forcing grain imports upward in many countries. North Africa and the Middle East is now the world's fastest growing grain import market. In 1999, Iran eclipsed Japan, which until recently was the world's leading importer of wheat. And Egypt, another water-short country, has also edged ahead of Japan.

Many developing countries that are facing acute land and water scarcity will rely on industrialization and labor-intensive industrial exports to finance needed food imports. This brings a need to expand production in exporting countries so they can cover the import needs of the growing number of grain-deficit countries. Over the last-half-century, grain-importing countries, now the overwhelming majority, have become dangerously dependent on the United States for nearly half of their grain imports.

This concentration of dependence applies to each of the big three grains—wheat, rice, and corn. Just five countries—the United States, Canada, France, Australia, and Argentina—account for 88 percent of the world's wheat exports. Thailand, Viet Nam, the United States, and China account for 68 percent of all rice exports. For corn, the concentration is even greater, with the United States alone accounting for 78 percent and Argentina for 12 percent.

With more extreme climate events in prospect if temperatures continue rising, this dependence on a few exporting countries leaves importers vulnerable to the vagaries of weather. If the United States were to experience a summer of severe heat and drought in its agricultural heartland like the summer of 1988, when grain production dropped below domestic consumption for the first time in history, chaos would reign in world grain markets simply because the near-record reserves that cushioned the huge U.S. crop shortfall that year no longer exist.

The risk for the scores of low-income, grain-importing countries is that prices could rise dramatically, impoverishing more people in a shorter period of time than any event in history. The resulting rise in hunger would be concentrated in the cities of the Third World.

POSTSCRIPT

Will the World be Able to Feed Itself in the Foreseeable Future?

Brunel argues that an examination of world agricultural production over a long period of time reveals that it has outpaced population growth. In the last 30 years of the twentieth century, for example, each individual's amount of food (in theory) grew by 20 percent. This equaled 2,700 calories per person per day, more than a person's estimated need. There is a catch, however. Half of the current grain production does not benefit humans. Approximately 20 percent is used to feed cattle, 5 percent is kept for seeds, and 25 percent is lost. In short, ecological factors are not at fault; humans, particularly those in the developed world, are the culprits. Therein lies the solution for Brunel.

Brown considers some but not all of the problem of global hunger to be rooted in human behavior. He asserts that world leaders have not come forward with a comprehensive master plan to address the problem and that they are extremely unlikely to do so in the foreseeable future.

A balanced look at the planet's capacity to feed the UN's projected 2050 population is L. T. Evans's *Feeding the Ten Billion* (Cambridge University Press, 1998). In it, the author takes the reader through the ages, showing how the human race has addressed the agricultural needs of each succeeding billion people. The biggest challenge during the next half century, according to Evans, will be solving two problems: producing enough food for a 67 percent increase in the population and eliminating the chronic undernutrition that afflicts so many people. Solving the first problem requires a focus on the main components of increased food supply.

The second problem brings into play many socioeconomic factors beyond those that are typically associated with agricultural production. Growing enough food worldwide is a necessary but insufficient condition for eliminating problems associated with hunger. Many studies have indicated that the root cause of hunger is poverty. Addressing poverty, therefore, is a prerequisite for ensuring that the world's food supply is distributed such that the challenge of global hunger is met.

Three reports by the UN's Food and Agriculture Organization (FAO) are worth reading: "World Agriculture: Towards 2000" (1988); "The State of Food and Agriculture 2002" (2002); and "World Agriculture: Towards 2015/2030" (2002). The central message of these studies is that the planet will be able to feed a growing population in the foreseeable future if certain conditions are met. Another FAO report, "The State of Food Insecurity in the World 2000" (2000), analyzes the latest data on hunger and discusses approaches to fulfilling the commitments of the World Food Summit of 1996.

ISSUE 8

Is Global Environmental Stress Caused Primarily by Increased Resource Consumption Rather Than Population Growth?

YES: Adil Najam, from "A Developing Countries' Perspective on Population, Environment, and Development," *Population Research and Policy Review* (February 1996)

NO: Lester R. Brown, Gary Gardner, and Brian Halweil, from "Sixteen Impacts of Population Growth," *The Futurist* (February 1999)

ISSUE SUMMARY

YES: Adil Najam, an assistant professor in the Department of International Relations at Boston University, argues that consumption control among developed countries is a major key to avoiding continuing environmental crises.

NO: Lester R. Brown, Gary Gardner, and Brian Halweil, researchers at the Worldwatch Institute, examine 16 dimensions of population growth in the areas of food and agriculture, environment and resources, and economic impacts and quality of life, and they indicate that the effects of such growth are mainly adverse.

In the Woodrow Wilson Center article "Human Population and Environmental Stresses in the Twenty-First Century," *Environmental Change and Security Project Report* (Summer 2000), Richard E. Benedick, a former U.S. State Department official responsible for population and environmental policies, summarizes what he calls the new generation of environmental problems: (1) changing climate, bringing on drought, flooding, and assorted other problems; (2) depletion of the ozone layer, which protects humans, plants, and animals from ultraviolet radiation; (3) loss of biological diversity due to mass extinctions of animal and plant species; (4) spread of arid lands, desertification, and soil erosion on a global scale; (5) pollution of marine and freshwaters, and overfishing; (6) destruction of forests; and (7) worldwide diffusion of hazardous substances.

This issue focuses on a fundamental debate that is currently occurring in both the environmental and population issue areas of the new global agenda: Is the apparent additional stress on the Earth's various systems—such as that reported by Benedick—caused by the arrival of more people on the planet, or is the dramatic increase in per capita consumption, primarily in the developed world, the principal cause of environmental stress and degradation?

In examining this issue, we first make the assumption that both increased consumption and population growth affect the environment in some fashion. Common sense dictates that both play some role, however insignificant. The basic question, then, is whether the impact of one or both factors is substantial enough to damage the environment beyond a certain "normal use" threshold to qualify as the source of environmental "stress."

It is probably useful to break down our analysis into two broad groups of cases—the developed world and the developing sector. The reason for such a division is that different forces are at work in these two sectors. Population growth is almost at a standstill in the industrial sector of the globe. At present growth rates, many wealthy countries will take well over a century to double their populations. On the other hand, most of these developed nations have witnessed huge increases in per-capita resource consumption; in many cases these rates have doubled or tripled in a matter of decades. Given these two characteristics—little or no population growth and large consumption increases— one must assume that any additional environmental stress occurring in the developed world can be attributed to increased resource consumption.

The situation in the developing world is somewhat more complicated. Population growth rates soared for about 30 years. Many developing countries doubled their size during that time, and they continue to add huge numbers each year (despite current declining growth rates because of the built-in momentum). If each individual continued to consume about the same amount of resources during this time, the environmental impact would have doubled during the 30-year period. In this case, more people, not more per-capita consumption, would have made the difference.

On the other hand, evidence suggests that as individuals and countries move up the development ladder, their ability and inclination to use more resources in their daily lives also escalates. Their per-capita resource consumption rises. This phenomenon was quite evident during the demographic transition in the developed world, and it also appears to be currently at work in developing countries that are making strides toward industrialization. Thus, the answer to the question of how much environmental stress in the developing world is attributable to increased resource consumption and how much to population growth requires a more elaborate research strategy.

The two selections for this issue take opposite ends of the argument. Adil Najam supports the position that population growth in the developing world should not be held accountable for global environmental stress. Lester R. Brown, Gary Gardner, and Brian Halweil focus on population growth as the most critical issue of the day.

Adil Najam

 **YES**

A Developing Countries' Perspective on Population, Environment, and Development

In ancient times, people were few but wealthy and without strife. People at present think that five sons are too many, and each son has five sons also and before the death of the grandfather there are already 25 descendants. Therefore people are more and wealth is less; they work hard and receive little. The life of a nation depends upon having enough food, not upon the number of people.

—Han Fei-Tzu (circa 500 BC)

The happiness of a country does not depend, absolutely, upon its poverty or its riches, upon its youth or its age, upon its being thinly or fully inhabited, but upon the rapidity with which it is increasing, upon the degree in which the yearly increase of food approaches to the yearly increase of an unrestricted population.

—Rev. Thomas Robert Malthus (1798)

The causal chain of the deterioration [of the environment] is easily followed to its source. Too many cars, too many factories, too much detergent, too much pesticide, multiplying contrails, inadequate sewage treatment plants, too little water, too much carbon dioxide—all can be traced easily to too many people.

—Paul R. Ehrlich (1968)

The pollution problem is a consequence of population. . . . Freedom to breed will bring ruin to all.

—Garrett Hardin (1968)

From Adil Najam, "A Developing Countries' Perspective on Population, Environment, and Development," *Population Research and Policy Review,* vol. 15, no. 1 (February 1996). Copyright © 1996 by Kluwer Academic Publishers. Reprinted by permission. Notes and references omitted.

Introduction

The perception of population growth as a 'problem' is not new. That the catastrophe predicted by so many has been averted till now does not necessarily disprove the arguments of these Cassandras. However, Pollyannas like Julian Simon (1981) have taken much pleasure in rubbing in this fact and insist that human ingenuity will continue to outpace human propensity for procreation. It is within this context that much of the debate on the subject has been historically framed, with occasional shifts in popular and scholarly sentiments towards one side or the other.

The recent growth of popular interest in environmental issues has generated a renewal of concern about rapid population growth, which is seen as being largely responsible for global trends of environmental degradation. The causal relationship between the two seems intuitively obvious. Yet, it is being contested by a number of critical interests. Although some in the population community may consider such views peripheral to the mainstream debate, the prevalence and persistence of the dissension of the environment-population linkage may be gauged from the fact that in 1987 the World Commission on Environment and Development, reached unanimous agreement on all issues except two-Antarctica and the causal significance of population growth (Shaw 1992). Again, at the 1992 United Nations Conference on Environment and Development (UNCED) population remained the most contentious issue (Najam 1993a).

Most surprising is the reaction of the developing countries. On the one hand, many of them have very high population growth rates and are most immediately vulnerable to its consequences. At the same time, many of them support strong domestic population policies, which have been in place over long periods of time, and are vigorously—and sometimes coercively—enforced. Yet, at the international level, these same states seem hesitant, and sometimes hostile, to the notion of accepting a direct causal link between global environmental degradation and population growth.

This paper attempts to understand why the developing countries of the South are so weary of international population policy in the name of the environment. It is essentially a study of the political behavior of Southern governments. It is argued that the South's response has been, and continues to be, shaped by five inter-related concerns:

1. *Responsibility:* For precedent and principle, the South resists any effort that implies holding population growth (largely in the South) responsible for global environmental stress. Hence, the insistence that 'population growth in the developing countries is a national, not a global problem' (Mahbub-ul-Haq 1976: 124). The South has consistently held that the environmental crisis is of the North's making and has based its demands for assistance/reparations on that 'history of guilt'. In accepting population growth as *the* causal motor of environmental degradation the South loses this perceived leverage.

2. *Efficiency:* From the perspective of international environmental policy, the South insists that if the international concern about population growth stems from its effects on global systems then it would be more efficient to focus on consumption patterns. The argument is that whatever effects population may have, they are only in relation to consumption. As Ramphal (1994) stresses, once consumption is factored in, the 1.2 billion people living in industrialized countries place a pressure equivalent to more than 24 billion living in the developing countries. Thus, it would be more efficient to focus on policies that curb consumption than on those that target procreation.

3. *Efficacy:* From the implementation end, developing countries argue that development is still the best contraceptive. Experience in both North and South shows that 'people in the developed condition do not have too many children' (Keyfitz 1991: 39–40). The South argues, therefore, that if the international community is truly interested in curbing population growth it should spend its dollars on assisting economic and human development rather than simply enhancing contraceptive provision. Although the argument that economic growth will automatically slow population *growth* is contested (e.g., Harrison 1994), the assertion that social *development* will enhance the efficacy of population policies is now widely accepted (e.g., Brower 1994; Chhabra 1994; Harrison 1994; Lutz 1994; Ness 1994). The South builds on this emerging consensus to reinforce and rearticulate its enduring call that development (now more broadly defined) remains the best contraceptive (see Mahbub-ul-Haq 1994; Najam 1993b; Ramphal 1994).

4. *Additionality:* Programatically, the South remains concerned that greater donor assistance for population may translate to a lesser focus on development assistance. Ever since US President Johnson's 1965 statement that 'less than five dollars invested in population control is worth a hundred dollars invested in economic growth', developing countries have doubted the motives behind the West's emphasis on population control. Developing countries want to be sure that in accepting donor priorities (i.e. environment) they would not be asked to forfeit their own (i.e. development). For the South, the end of the cold war means that the geopolitical strategic value of Southern states has diminished, and new claimants to the already shrinking international assistance pie have emerged. This has reinvigorated the urgency of the additionality argument for the South.

5. *Sovereignty:* Politically, the South sees no contradiction between actively pursuing population policies domestically, and resisting them internationally. It sees population as an issue of sovereignty and any interference from the international community as a breach thereof. While recognizing the domestic benefits of slowed population growth and pursuing policies to bring it about, the South is unprepared to hold its policies subservient to external pressure. As Stephen Krasner points out, 'the South has maintained its unity, despite major differences

among individual countries, even in an issue area where the North has offered additional resources' because 'Southern resistance to Northern efforts to develop international population norms is not simply a product of specific national values, of evidence of a concern that antinatalism may be a ploy for subordinating development aid, but is also a reflection of the deep adherence of Third World states to the prerogatives of sovereignty' (Krasner 1985; 276–278).

The rest of this paper will look at how the discussion on population-environment-development policies ignores the South's concerns and, in doing so, alienates the very group of countries that is being required to carry out such policies. . . .

The Population-Environment-Development Nexus

In its most simple articulation, the argument of the new 'green' Cassandras has flowed from two observations: (a) the planet has never had as many people as it has today, and (b) the planet has never seen as much environmental stress on its natural systems as it is experiencing today. The correlation between the two is then extrapolated to imply causality.

In fairness to its proponents, the argument has become far more sophisticated over time. The reigning view is best identified by the Holdren-Ehrlich (1974) identity:

$I = PAT$
(environmental Impact = Population $\times$ Affluence $\times$ Technology)

This identity, while not without serious limitations, is elegant in that it attempts to capture both the number of users and the rate of use of natural systems. However, most scholars who use this, or similar, formulations often end up focusing on the population variable rather than the other two. For example, Nazli Choucri (1991: 100) suggests that 'the population nexus as a whole—the interaction of population, resources, and technological change—must become the focus of global policy'. However, she is quick to add that while population policy alone is by no means sufficient it is nonetheless necessary, implying that it is here that the most emphasis should be invested. Others have made similar arguments on the grounds that population policies will 'help buy time' (Keyfitz 1991; Shaw 1992). Implicit in such arguments seems the belief that changing population patterns is somehow 'easier' than changing patterns of consumption or technology.

From the South's point of view, while the diagnosis suggests that both the number (i.e. population growth) and the rate (i.e. consumption patterns) are at least equally critical motors of causality, the prescription focuses unduly on the first and not enough on the later. For many in the developing world, such a conceptualization adds insult to injury in that the focus on population as the main cause of environment degradation implicitly places the responsibility for such degradation on their doorsteps, even though the 'benefits' have been reaped by those in the North.

Very often, then, the argument becomes merely a more sophisticated re-hash of the more simplistic conception introduced earlier. For example, Nathan Keyfitz (1991: 44, 77) writes:

> In 1950 the world contained 2.5 billion people, and there was little evidence of damage to the biosphere. Now with over 5 billion there is a great deal of evidence with another 2.5 billion and continuance of present trends of production and consumption, disaster faces us. The planet cannot over a long period support that many people; yet an even larger number is threat-ened. . . . Twice as many people cooking with the same wood stoves use up twice as much wood. Twice as many cars of a given kind and given condition of repair put twice as much carbon dioxide into the atmosphere. Twice as many fish eaters require twice as large a catch. *With all else constant,* the requirements are the simplest possible linear function of the number of people. (Emphasis added)

Ceteris, however, is not *paribus.* Keyfitz knows that all else is not constant. He prefaces the above by saying that 'with *given* technology and *given* style of life the requirements from the environment are proportional to the number of peo-ple' (p. 44). However, neither technology nor lifestyle is 'given'. Yet, he chooses (as do most other analysts) to hold consumption constant in arguing for poli-cies that would control the population variable. The implication seems to be that the North's lifestyle as it relates to consumption is accepted as a 'given' be-cause it cannot (or is it, 'should' not?) be changed, but the South's lifestyle as it relates to procreation is not because it can.

The South's Response

Such arguments have the dual implication of holding population growth *re-sponsible* for environmental degradation and touting population control as the most *efficient* opinion for environmental amelioration.

On the first count, the South responds by pointing out, for example, that the average Bangladeshi uses 2 milligrams of CFCs [chlorofluorocarbons] per year in comparison to the average US citizen who uses 2 kilograms per year; as such the environmental impact of an extra Bengali, in CFC terms, is only 1/1000th that of an extra American. On the second, they reason that even if the policy focus is to be only on population and not on consumption, it makes more sense to do so in the North where one averted birth is likely to produce 1000 times the environmental 'benefit' that it would in the South. Further, Southern commentators challenge the assumption that it is somehow 'easier' to reduce population amongst the poor than to curb consumption amongst the rich. If environment is the main concern, they argue, would it be easier to change lifestyles (consumption) of the few who are very rich or the children preferences of the very many who are poor. Arguably, the lifestyle change in-volved in reducing CFC consumption for an individual in USA is no more diffi-cult—in fact, it should be far easier—than changing the children preference demanded from a peasant in Bangladesh. See Mahbub-ul-Haq (1994), Najam (1993b) and Ramphal (1994).

While the question of efficiency relates to the relative importance of the various options to check environmental degradation, the issue of *efficacy* concerns the effectiveness of various means to curb population growth. Since the South nowhere questions the need for population policies per se—and actively pursues them domestically—the efficacy of such policies is crucial for implementation. This brings us to that critical question of why people in poor conditions have high population growth. Environmentalists tend to spend too much effort in arguing *why* population should be controlled, and population experts spend too much time in figuring out *how* it could be controlled, but way too little thought is invested in *why* people have as many children as they do.

Falling mortality rates, old-age security, religion, and the sheer inertia of the demographic momentum are all valid and important factors, but they offer little in way of policy advice. For example, maintaining high mortality is ethically unacceptable, quick changes in social or religious preferences cannot be legislated, nor can changes in the nature of demographic momentum.

For the poorest, the difference between having four children or five is often not the difference between four hungry mouths to feed or five, but that between eight hands to earn with or ten. The rational cost-benefit analysis of childbearing decisions yields very different results where children become earning members before age ten, from where parents have to factor in the escalating costs of an expensive college education before thinking about that extra child. The fundamental, still unbridged, gap between North and South in matters pertaining to population is that what people in the industrialized world see as a problem of 'too many people' is seen by those in the developing countries as the problem of 'too much poverty'. The most vivid exemplars of this persistent chasm remain the following (still relevant) quotes from Paul Ehrlich and Mahmood Mamdani:

> One stinking hot night in Delhi . . . as we crawled through the city [in a taxi], we entered a crowded slum area. The temperature was well over 100, and the air was a haze of dust and smoke. The streets seemed alive with people. People eating, people washing, people sleeping. People visiting, arguing and screaming. People thrusting their hands through the taxi window, begging. People defecating and urinating. People clinging to buses. People herding animals. People, people, people, people. As we moved slowly through the mob, hand horn squawking, the dust, noise, heat, and cooking fires gave the scene a hellish aspect. Would we ever get to our hotel? All three of us were, frankly, frightened . . . since that night I've known the *feel* of overpopulation. (Ehrlich 1968:15)

> The fact is that a hot summer night on Broadway in New York or Picadilly Circus in London would put Ehrlich in the midst of a far larger crowd. Yet such an experience would not spur him to comment with grave concern about 'overpopulation'. On the other

hand, with a little more concern and a little less fear he would have realized that what disturbed him about the crowd in Delhi was not its numbers, but its 'quality'—that is, its poverty. To talk, as Ehrlich does, of 'overpopulation' is to say to people: you are poor because you are too many. . . . People are not poor because they have large families. Quite the contrary: they have large families because they are poor. (Mamdani 1972: 14)

NO

Lester R. Brown, Gary Gardner, and Brian Halweil

Sixteen Impacts of Population Growth

The world's population has doubled during the last half century, climbing from 2.5 billion in 1950 to 5.9 billion in 1998. This unprecedented surge in population, combined with rising individual consumption, is pushing our claims on the planet beyond its natural limits.

The United Nations projects that human population in 2050 will range between 7.7 billion and 11.2 billion people. We use the United Nations' middle-level projection of 9.4 billion (from *World Population Prospects: The 1996 Revision*) to give an idea of the strain this "most likely" outcome would place on ecosystems and governments in the future and of the urgent need to break from the business-as-usual scenario.

Our study looks at 16 dimensions or effects of population growth in order to gain a better perspective on how future population trends are likely to affect human prospects:

Impacts on Food and Agriculture

1. Grain Production

From 1950 to 1984, growth in the world grain harvest easily exceeded that of population. But since then, the growth in the grain harvest has fallen behind that of population, so per-person output has dropped by 7% (0.5% a year), according to the U.S. Department of Agriculture.

The slower growth in the world grain harvest since 1984 is due to the lack of new land and to slower growth in irrigation and fertilizer use because of the diminishing returns of these inputs.

Now that the frontiers of agricultural settlement have disappeared, future growth in grain production must come almost entirely from raising land productivity. Unfortunately, this is becoming more difficult. The challenge for the world's farmers is to reverse this decline at a time when cropland area per person is shrinking, the amount of irrigation water per person is dropping, and the crop yield response to additional fertilizer use is falling.

From Lester R. Brown, Gary Gardner, and Brian Halweil, "Sixteen Impacts of Population Growth," *The Futurist,* vol. 33, no. 2 (February 1999). Copyright © 1999 by The World Future Society. Reprinted by permission of The World Future Society, 7910 Woodmont Avenue, Suite 450, Bethesda, MD 20814.

2. Cropland

Since mid-century, grain area—which serves as a proxy for cropland in general—has increased by some 19%, but global population has grown by 132%. Population growth can degrade farmland, reducing its productivity or even eliminating it from production. As grain area per person falls, more and more nations risk losing the capacity to feed themselves.

The trend is illustrated starkly in the world's four fastest-growing large countries. Having already seen per capita grain area shrink by 40%–50% between 1960 and 1998, Pakistan, Nigeria, Ethiopia, and Iran can expect a further 60%–70% loss by 2050—a conservative projection that assumes no further losses of agricultural land. The result will be four countries with a combined population of more than 1 billion whose grain area per person will be only 300–600 square meters—less than a quarter of the area in 1950.

3. Fresh Water

Spreading water scarcity may be the most underrated resource issue in the world today. Wherever population is growing, the supply of fresh water per person is declining.

Evidence of water stress can be seen as rivers are drained dry and water tables fall. Rivers such as the Nile, the Yellow, and the Colorado have little water left when they reach the sea. Water tables are now falling on every continent, including in major food-producing regions. Aquifers are being depleted in the U.S. southern Great Plains, the North China Plain, and most of India.

The International Water Management Institute projects that a billion people will be living in countries facing absolute water scarcity by 2025. These countries will have to reduce water use in agriculture in order to satisfy residential and industrial water needs. In both China and India, the two countries that together dominate world irrigated agriculture, substantial cutbacks in irrigation water supplies lie ahead.

4. Oceanic Fish Catch

A fivefold growth in the human appetite for seafood since 1950 has pushed the catch of most oceanic fisheries to their sustainable limits or beyond. Marine biologists believe that the oceans cannot sustain an annual catch of much more than 93 million tons, the current take.

As we near the end of the twentieth century, overfishing has become the rule, not the exception. Of the 15 major oceanic fisheries, 11 are in decline. The catch of Atlantic cod—long a dietary mainstay for western Europeans—has fallen by 70% since peaking in 1968. Since 1970, bluefin tuna stocks in the West Atlantic have dropped by 80%.

With the oceans now pushed to their limits, future growth in the demand for seafood can be satisfied only by fish farming. But as the world turns to aquaculture to satisfy its needs, fish begin to compete with livestock and poultry for feedstuffs such as grain, soybean meal, and fish meal.

The next half century is likely to be marked by the disappearance of some species from markets, a decline in the quality of seafood caught, higher prices, and more conflicts among countries over access to fisheries. Each year, the future oceanic catch per person will decline by roughly the amount of population growth, dropping to 9.9 kilograms (22 pounds) per person in 2050, compared with the 1988 peak of 17.2 kilograms (37.8 pounds).

5. Meat Production

When incomes begin to rise in traditional low-income societies, one of the first things people do is diversify their diets, consuming more livestock products.

World meat production since 1950 has increased almost twice as fast as population. Growth in meat production was originally concentrated in western industrial countries and Japan, but over the last two decades it has increased rapidly in East Asia, the Middle East, and Latin America. Beef, pork, and poultry account for the bulk of world consumption.

Of the world grain harvest of 1.87 billion tons in 1998, an estimated 37% will be used to feed livestock and poultry, producing milk and eggs as well as meat, according to the U.S. Department of Agriculture. Grain fed to livestock and poultry is now the principal food reserve in the event of a world food emergency.

Total meat consumption will rise from 211 million tons in 1997 to 513 million tons in 2050, increasing pressures on the supply of grain.

Environment and Resources

6. Natural Recreation Areas

From Buenos Aires to Bangkok, dramatic population growth in the world's major cities—and the sprawl and pollution they bring—threaten natural recreation areas that lie beyond city limits. On every continent, human encroachment has reduced both the size and the quality of natural recreation areas.

In nations where rapid population growth has outstripped the carrying capacity of local resources, protected areas become especially vulnerable. Although in industrial nations these areas are synonymous with camping, hiking, and picnics in the country, in Asia, Africa, and Latin America most national parks, forests, and preserves are inhabited or used for natural resources by local populations.

Migration-driven population growth also endangers natural recreation areas in many industrial nations. Everglades National Park, for example, faces collapse as millions of newcomers move into southern Florida.

Longer waiting lists and higher user fees for fewer secluded spots are likely to be the tip of the iceberg, as population growth threatens to eliminate the diversity of habitats and cultures, in addition to the peace and quiet, that protected areas currently offer.

7. Forests

Global losses of forest area have marched in step with population growth for much of human history, but an estimated 75% of the loss in global forests has occurred in the twentieth century.

In Latin America, ranching is the single largest cause of deforestation. In addition, overgrazing and over-collection of firewood—which are often a function of growing population—are degrading 14% of the world's remaining large areas of virgin forest.

Deforestation created by the demand for forest products tracks closely with rising per capita consumption in recent decades. Global use of paper and paperboard per person has doubled (or nearly tripled) since 1961.

The loss of forest areas leads to a decline of forest services. These include habitat for wildlife; carbon storage, which is a key to regulating climate; and erosion control, provision of water across rainy and dry seasons, and regulation of rainfall.

8. Biodiversity

We live amid the greatest extinction of plant and animal life since the dinosaurs disappeared 65 million years ago, at the end of the Cretaceous period, with species losses at 100 to 1,000 times the natural rate. The principal cause of species extinction is habitat loss, which tends to accelerate with an increase in a country's population density.

A particularly productive but vulnerable habitat is found in coastal areas, home to 60% of the world's population. Coastal wetlands nurture two-thirds of all commercially caught fish, for example. And coral reefs have the second-highest concentration of biodiversity in the world, after tropical rain forests. But human encroachment and pollution are degrading these areas: Roughly half of the world's salt marshes and mangrove swamps have been eliminated or radically altered, and two-thirds of the world's coral reefs have been degraded, 10% of them "beyond recognition." As coastal migration continues—coastal dwellers could account for 75% of world population within 30 years—the pressures on these productive habitats will likely increase.

9. Climate Change

Over the last half century, carbon emissions from fossil-fuel burning expanded at nearly twice the rate of population, boosting atmospheric concentrations of carbon dioxide, the principal greenhouse gas, by 30% over preindustrial levels.

Fossil-fuel use accounts for roughly three-quarters of world carbon emissions. As a result, regional growth in carbon emissions tend to occur where economic activity and related energy use is projected to grow most rapidly. Emissions in China are projected to grow over three times faster than population in the next 50 years due to a booming economy that is heavily reliant on coal and other carbon-rich energy sources.

Emissions from developing countries will nearly quadruple over the next half century, while those from industrial nations will increase by 30%, according to the Intergovernmental Panel on Climate Change and the U.S. Department of Energy. Although annual emissions from industrial countries are currently twice as high as from developing ones, the latter are on target to eclipse the industrial world by 2020.

10. Energy

The global demand for energy grew twice as fast as population over the last 50 years. By 2050, developing countries will be consuming much more energy as their populations increase and become more affluent.

When per capita energy consumption is high, even a low rate of population growth can have significant effects on total energy demand. In the United States, for example, the 75 million people projected to be added to the population by 2050 will boost energy demand to roughly the present energy consumption of Africa and Latin America.

World oil production per person reached a high in 1979 and has since declined by 23%. Estimates of when global oil production will peak range from 2011 to 2025, signaling future price shocks as long as oil remains the world's dominant fuel.

In the next 50 years, the greatest growth in energy demands will come where economic activity is projected to be highest: in Asia, where consumption is expected to grow 361%, though population will grow by just 50%. Energy consumption is also expected to increase in Latin America (by 340%) and Africa (by 326%). In all three regions, local pressures on energy sources, ranging from forests to fossil fuel reserves to waterways, will be significant.

11. Waste

Local and global environmental effects of waste disposal will likely worsen as 3.4 billion people are added to the world's population over the next half century. Prospects for providing access to sanitation are dismal in the near to medium term.

A growing population increases society's disposal headaches—the garbage, sewage, and industrial waste that must be gotten rid of. Even where population is largely stable—the case in many industrialized countries—the flow of waste products into landfills and waterways generally continues to increase. Where high rates of economic and population growth coincide in coming decades, as they will in many developing countries, mountains of waste will likely pose difficult disposal challenges for municipal and national authorities.

Economic Impacts and Quality of Life

12. Jobs

Since 1950, the world's labor force has more than doubled—from 1.2 billion people to 2.7 billion—outstripping the growth in job creation. Over the next half century, the world will need to create more than 1.9 billion jobs in the developing world just to maintain current levels of employment.

While population growth may boost labor demand (through economic activity and demand for goods), it will most definitely boost labor supply. As the balance between the demand and supply of labor is tipped by population growth, wages tend to decrease. And in a situation of labor surplus, the quality of jobs may not improve as fast, for workers will settle for longer hours, fewer benefits, and less control over work activities.

Table 1

The 20 Largest Countries Ranked According to Population Size (In Millions)

1998			2050	
Rank	Country	Population	Country	Population
1	China	1,255	India	1,533
2	India	976	China	1,517
3	United States	274	Pakistan	357
4	Indonesia	207	United States	348
5	Brazil	165	Nigeria	339
6	Pakistan	148	Indonesia	318
7	Russia	147	Brazil	243
8	Japan	126	Bangladesh	218
9	Bangladesh	124	Ethiopia	213
10	Nigeria	122	Iran	170
11	Mexico	96	The Congo	165
12	Germany	82	Mexico	154
13	Vietnam	78	Philippines	131
14	Iran	73	Vietnam	130
15	Philippines	72	Egypt	115
16	Egypt	66	Russia	114
17	Turkey	64	Japan	110
18	Ethiopia	62	Turkey	98
19	Thailand	60	South Africa	91
20	France	59	Tanzania	89

Source: United Nations, *World Population Prospects: The 1996 Revision.*

As the children of today represent the workers of tomorrow, the interaction between population growth and jobs is most acute in nations with young populations. Nations with more than half their population below the age of 25 (e.g., Peru, Mexico, Indonesia, and Zambia) will feel the burden of this labor flood. Employment is the key to obtaining food, housing, health services, and education, in addition to providing self-respect and self-fulfillment.

13. Income

Incomes have risen most rapidly in developing countries where population has slowed the most, including South Korea, Taiwan, China, Indonesia, and Malaysia. African countries, largely ignoring family planning, have been overwhelmed by the sheer numbers of young people who need to be educated and employed.

If the world cannot simultaneously convert the economy to one that is environmentally sustainable and move to a lower population trajectory, economic decline will be hard to avoid.

14. Housing

The ultimate manifestation of population growth outstripping the supply of housing is homelessness. The United Nations estimates that at least 100 million of the world's people—roughly equal to the population of Mexico—have no home; the number tops 1 billion if squatters and others with insecure or temporary accommodations are included.

Unless population growth can be checked worldwide, the ranks of the homeless are likely to swell dramatically.

15. Education

In nations that have increasing child-age populations, the base pressures on the educational system will be severe. In the world's 10 fastest-growing countries, most of which are in Africa and the Middle East, the child-age population will increase an average of 93% over the next 50 years. Africa as a whole will see its school-age population grow by 75% through 2040.

If national education systems begin to stress lifelong learning for a rapidly changing world of the twenty-first century, then extensive provision for adult education will be necessary, affecting even those countries with shrinking child-age populations.

Such a development means that countries which started population-stabilization programs earliest will be in the best position to educate their entire citizenry.

16. Urbanization

Today's cities are growing faster: It took London 130 years to get from 1 million to 8 million inhabitants; Mexico City made this jump in just 30 years. The world's urban population as a whole is growing by just over 1 million people each week. This urban growth is fed by the natural increase of urban populations, by net migration from the countryside, and by villages or towns expanding to the point where they become cities or they are absorbed by the spread of existing cities.

If recent trends continue, 6.5 billion people will live in cities by 2050, more than the world's total population today.

Actions for Slowing Growth

As we look to the future, the challenge for world leaders is to help countries maximize the prospects for achieving sustainability by keeping both birth and death rates low. In a world where both grain output and fish catch per person are falling, a strong case can be made on humanitarian grounds to stabilize world population.

What is needed is an all-out effort to lower fertility, particularly in the high-fertility countries, while there is still time. We see four key steps in doing this:

Assess carrying capacity.   Every national government needs a carefully articulated and adequately supported population policy, one that takes into account the country's carrying capacity at whatever consumption level citizens decide on. Without long-term estimates of available cropland, water for irrigation, and likely yields, governments are simply flying blind into the future, allowing their nations to drift into a world in which population growth and environmental degradation can lead to social disintegration.

Fill the family-planning gap.   This is a high-payoff area. In a world where population pressures are mounting, the inability of 120 million of the world's women to get family-planning services is inexcusable. A stumbling block: At the International Conference on Population and Development in Cairo in 1994, the industrialized countries agreed to pay one-third of the costs for reproductive-health services in developing countries. So far they have failed to do so.

Educate young women.   Educating girls is a key to accelerating the shift to smaller families. In every society for which data are available, the more education women have, the fewer children they have. Closely related to the need for education of young females is the need to provide equal opportunities for women in all phases of national life.

Have just two children.   If we are facing a population emergency, it should be treated as such. It may be time for a campaign to convince couples everywhere to restrict their childbearing to replacement-level fertility.

POSTSCRIPT

Is Global Environmental Stress Caused Primarily by Increased Resource Consumption Rather Than Population Growth?

Brown and his colleagues chose to hold constant the effect of changes in consumption patterns in their analysis of the adverse impacts of population. Their prescription for minimizing such impacts is for each national government to create its own population policy that allows the country to avoid exceeding its carrying capacity "at whatever consumption level citizens decide on." Thus, the article begs the question of the comparative blame of consumption and population growth.

Najam, on the other hand, includes the question of consumption patterns, particularly in the developed world, when examining population effects. He takes to task scholars who choose to focus on population rather than on affluence and technology. He maintains that focusing on population as the principal cause of environmental degradation "implicitly places the responsibility for such degradation on [the] doorsteps" of the developing world.

Najam offers a simple comparison to show where the blame should really lie. The average Bangladeshi uses two milligrams of chlorofluorocarbons (CFCs) per year in comparison to the average American, who uses two kilograms of CFCs per year. Thus, the former's impact is a tiny fraction of an American's. Najam concludes by asserting that it should be far easier to change the consumption lifestyles of the few very rich than the reproductive preferences of the many very poor.

One way to gain some insight into this debate is to examine the Earth's carrying capacity. A work that addresses this question in the context of exploring the planet's ability to feed 10 billion people is Gerhard K. Heileg's "How Many People Can Be Fed on Earth?" in Wolfgang Lutz, ed., *The Future Population of the World: What Can We Assume Today?* 2d ed. (Earthscan, 1996). In it, Heileg concludes that it is much too difficult to answer the question definitively because carrying capacity "is not a natural constant (but) . . . a dynamic equilibrium, essentially determined by human action." And the future choices are far too many to allow a firm prediction. This view is echoed by Joel E. Cohen's *How Many People Can the Earth Support?* (W. W. Norton, 1995).

ISSUE 9

Is the Earth Getting Warmer?

YES: Intergovernmental Panel on Climate Change, from "Climate Change 2001: The Scientific Basis," A Report of Working Group I of the Intergovernmental Panel on Climate Change (2001)

NO: Brian Tucker, from "Science Friction: The Politics of Global Warming," *The National Interest* (Fall 1997)

ISSUE SUMMARY

YES: In the summary of its most recent assessment of climatic change, the Intergovernmental Panel on Climate Change concludes that an increasing set of observations reveals that the world is warming and that much of it is due to human intervention.

NO: Brian Tucker, a senior fellow of the Institute for Public Affairs, argues that there are still too many uncertainties to conclude that global warming has arrived.

At the UN-sponsored Earth Summit in Rio de Janeiro in 1992, a Global Climate Treaty was signed. According to S. Fred Singer, in *Hot Talks, Cold Science: Global Warming's Unfinished Debate* (Independent Institute, 1998), the treaty rested on three basic assumptions: First, global warming has been detected in the climate records of the last 100 years. Second, a substantial warming in the future will produce catastrophic consequences—droughts, floods, storms, a rapid and significant rise in sea level, agricultural collapse, and the spread of tropical disease. And third, the scientific and policy-making communities know (1) which atmospheric concentrations of greenhouse gases are dangerous and which ones are not; (2) that drastic reductions of carbon dioxide (CO_2) emissions as well as energy use in general by industrialized countries will stabilize CO_2 concentrations at close to current levels; and (3) that such economically damaging measures can be justified politically despite there being no significant scientific support for the assertion that global warming is a threat.

Since the Earth Summit, it appears that scientists have opted for placement into one of three camps. The first camp buys into the three assumptions outlined above. In the first Intergovernmental Panel on Climate Change (IPCC)

144

report, published in 1995, 2,500 climate scientists from this first camp announced that the planet was warming due to coal and gas emissions. Scientists in the second camp suggest that while global warming has occurred and continues at the present, the source of such temperature rise cannot be ascertained yet. The conclusions of the Earth Summit were misunderstood by many in the scientific community, the second camp would argue. For these scientists, computer models, the basis of much evidence for the first group, have not yet linked global warming to human activities.

A third group of scientists argues that we cannot be certain that global warming is even taking place, yet alone determine its cause. They present a number of arguments in support of their position. Among them is the contention that presatellite data (pre-1979) showing a century-long pattern of warming is an illusion because satellite data (post-1979) reveal no such warming. Furthermore, when warming *was* present, it did not occur at the same time as a rise in greenhouse gases. Scientists in the third camp are also skeptical of studying global warming in the laboratory. They suggest, moreover, that most of the scientists who have opted for one of the first two camps have done so because of laboratory experiments rather than evidence from the real world.

Despite what appear to be wide differences in scientific thinking about the existence of global warming and its origins, the global community has moved forward with attempts to achieve consensus among the nations of the world for taking appropriate action to curtail human activities thought to affect warming. At a 1997 international meeting in Kyoto, Japan, an agreement was forged for reaching goals established at the 1992 Earth Summit. Thirty-eight industrialized countries, including the United States, agreed to the CO_2 reduction levels outlined in the treaty. However, the U.S. Senate never ratified the treaty, and the George W. Bush administration decided not to support it.

The first of the following selections is from a 2001 report by the Intergovernmental Panel on Climate Change (IPCC), an international body of scientists created by the United Nations to address the issue of global warming. It provides the most recent analysis by a broad scientific community on the twin issues of global warming's existence and causes. The authors of the study predict that the global temperature will likely rise by 1.4–5.8°C by the year 2100. Brian Tucker's position, clarified in the second selection, is that other factors—such as the politics of population, development, and environment—are to blame for the current rush to seek alternatives to oil and gas. Although his article was written prior to the Kyoto conference, his message remains the same.

 **YES**

Summary for Policymakers

This Summary for Policymakers (SPM), which was approved by IPCC [Intergovernmental Panel on Climate Change] member governments in Shanghai in January 2001,[1] describes the current state of understanding of the climate system and provides estimates of its projected future evolution and their uncertainties. Further details can be found in the underlying report, and the appended Source Information provides cross references to the report's chapters.

An increasing body of observations gives a collective picture of a warming world and other changes in the climate system.

Since the release of the Second Assessment Report (SAR[2]), additional data from new studies of current and palaeoclimates, improved analysis of data sets, more rigorous evaluation of their quality, and comparisons among data from different sources have led to greater understanding of climate change.

The global average surface temperature has increased over the 20th century by about 0.6°C.

- The global average surface temperature (the average of near surface air temperature over land, and sea surface temperature) has increased since 1861. Over the 20th century the increase has been 0.6 ± 0.2°C[3,4]. This value is about 0.15°C larger than that estimated by the SAR for the period up to 1994, owing to the relatively high temperatures of the additional years (1995 to 2000) and improved methods of processing the data. These numbers take into account various adjustments, including urban heat island effects. The record shows a great deal of variability; for example, most of the warming occurred during the 20th century, during two periods, 1910 to 1945 and 1976 to 2000.
- Globally, it is very likely[5] that the 1990s was the warmest decade and 1998 the warmest year in the instrumental record, since 1861.
- New analyses of proxy data for the Northern Hemisphere indicate that the increase in temperature in the 20th century is likely[5] to have been the largest of any century during the past 1,000 years. It is also likely[5] that, in the Northern Hemisphere, the 1990s was the warmest decade

and 1998 the warmest year. Because less data are available, less is known about annual averages prior to 1,000 years before present and for conditions prevailing in most of the Southern Hemisphere prior to 1861.

- On average, between 1950 and 1993, night-time daily minimum air temperatures over land increased by about 0.2°C per decade. This is about twice the rate of increase in daytime daily maximum air temperatures (0.1°C per decade). This has lengthened the freeze-free season in many mid- and high latitude regions. The increase in sea surface temperature over this period is about half that of the mean land surface air temperature.

Temperatures have risen during the past four decades in the lowest 8 kilometres of the atmosphere.

- Since the late 1950s (the period of adequate observations from weather balloons), the overall global temperature increases in the lowest 8 kilometres of the atmosphere and in surface temperature have been similar at 0.1°C per decade.
- Since the start of the satellite record in 1979, both satellite and weather balloon measurements show that the global average temperature of the lowest 8 kilometres of the atmosphere has changed by +0.05 ± 0.10°C per decade, but the global average surface temperature has increased significantly by +0.15 ± 0.05°C per decade. The difference in the warming rates is statistically significant. This difference occurs primarily over the tropical and sub-tropical regions.
- The lowest 8 kilometres of the atmosphere and the surface are influenced differently by factors such as stratospheric ozone depletion, atmospheric aerosols, and the El Niño phenomenon. Hence, it is physically plausible to expect that over a short time period (e.g., 20 years) there may be differences in temperature trends. In addition, spatial sampling techniques can also explain some of the differences in trends, but these differences are not fully resolved.

Snow cover and ice extent have decreased.

- Satellite data show that there are very likely[5] to have been decreases of about 10% in the extent of snow cover since the late 1960s, and ground-based observations show that there is very likely[5] to have been a reduction of about two weeks in the annual duration of lake and river ice cover in the mid- and high latitudes of the Northern Hemisphere, over the 20th century.
- There has been a widespread retreat of mountain glaciers in non-polar regions during the 20th century.
- Northern Hemisphere spring and summer sea-ice extent has decreased by about 10 to 15% since the 1950s. It is likely[5] that there has been about a 40% decline in Arctic sea-ice thickness during late summer to early autumn in recent decades and a considerably slower decline in winter sea-ice thickness.

Global average sea level has risen and ocean heat content has increased.

- Tide gauge data show that global average sea level rose between 0.1 and 0.2 metres during the 20th century.
- Global ocean heat content has increased since the late 1950s, the period for which adequate observations of sub-surface ocean temperatures have been available.

Changes have also occurred in other important aspects of climate.

- It is very likely[5] that precipitation has increased by 0.5 to 1% per decade in the 20th century over most mid- and high latitudes of the Northern Hemisphere continents, and it is likely[5] that rainfall has increased by 0.2 to 0.3% per decade over the tropical (10°N to 10°S) land areas. Increases in the tropics are not evident over the past few decades. It is also likely[5] that rainfall has decreased over much of the Northern Hemisphere sub-tropical (10°N to 30°N) land areas during the 20th century by about 0.3% per decade. In contrast to the Northern Hemisphere, no comparable systematic changes have been detected in broad latitudinal averages over the Southern Hemisphere. There are insufficient data to establish trends in precipitation over the oceans.
- In the mid- and high latitudes of the Northern Hemisphere over the latter half of the 20th century, it is likely[5] that there has been a 2 to 4% increase in the frequency of heavy precipitation events. Increases in heavy precipitation events can arise from a number of causes, e.g., changes in atmospheric moisture, thunderstorm activity and large-scale storm activity.
- It is likely[5] that there has been a 2% increase in cloud cover over mid- to high latitude land areas during the 20th century. In most areas the trends relate well to the observed decrease in daily temperature range.
- Since 1950 it is very likely[5] that there has been a reduction in the frequency of extreme low temperatures, with a smaller increase in the frequency of extreme high temperatures.
- Warm episodes of the El Niño-Southern Oscillation (ENSO) phenomenon (which consistently affects regional variations of precipitation and temperature over much of the tropics, sub-tropics and some mid-latitude areas) have been more frequent, persistent and intense since the mid-1970s, compared with the previous 100 years.
- Over the 20th century (1900 to 1995), there were relatively small increases in global land areas experiencing severe drought or severe wetness. In many regions, these changes are dominated by inter-decadal and multi-decadal climate variability, such as the shift in ENSO towards more warm events.
- In some regions, such as parts of Asia and Africa, the frequency and intensity of droughts have been observed to increase in recent decades.

Some important aspects of climate appear not to have changed.

- A few areas of the globe have not warmed in recent decades, mainly over some parts of the Southern Hemisphere oceans and parts of Antarctica.
- No significant trends of Antarctic sea-ice extent are apparent since 1978, the period of reliable satellite measurements.
- Changes globally in tropical and extra-tropical storm intensity and frequency are dominated by inter-decadal to multi-decadal variations, with no significant trends evident over the 20th century. Conflicting analyses make it difficult to draw definitive conclusions about changes in storm activity, especially in the extra-tropics.
- No systematic changes in the frequency of tornadoes, thunder days, or hail events are evident in the limited areas analysed.

Emissions of greenhouse gases and aerosols due to human activities continue to alter the atmosphere in ways that are expected to affect the climate.

Changes in climate occur as a result of both internal variability within the climate system and external factors (both natural and anthropogenic). The influence of external factors on climate can be broadly compared using the concept of radiative forcing.[6] A positive radiative forcing, such as that produced by increasing concentrations of greenhouse gases, tends to warm the surface. A negative radiative forcing, which can arise from an increase in some types of aerosols (microscopic airborne particles) tends to cool the surface. Natural factors, such as changes in solar output or explosive volcanic activity, can also cause radiative forcing. Characterisation of these climate forcing agents and their changes over time is required to understand past climate changes in the context of natural variations and to project what climate changes could lie ahead. . . .

Concentrations of atmospheric greenhouse gases and their radiative forcing have continued to increase as a result of human activities.

- The atmospheric concentration of carbon dioxide (CO_2) has increased by 31% since 1750. The present CO_2 concentration has not been exceeded during the past 420,000 years and likely[5] not during the past 20 million years. The current rate of increase is unprecedented during at least the past 20,000 years.
- About three-quarters of the anthropogenic emissions of CO_2 to the atmosphere during the past 20 years is due to fossil fuel burning. The rest is predominantly due to land-use change, especially deforestation.
- Currently the ocean and the land together are taking up about half of the anthropogenic CO_2 emissions. On land, the uptake of anthropogenic CO_2 very likely[5] exceeded the release of CO_2 by deforestation during the 1990s.
- The rate of increase of atmospheric CO_2 concentration has been about 1.5 ppm[7] (0.4%) per year over the past two decades. During the 1990s the year to year increase varied from 0.9 ppm (0.2%) to 2.8 ppm (0.8%). A large part of this variability is due to the effect of climate variability (e.g., El Niño events) on CO_2 uptake and release by land and oceans.

- The atmospheric concentration of methane (CH_4) has increased by 1060 ppb[7] (151%) since 1750 and continues to increase. The present CH_4 concentration has not been exceeded during the past 420,000 years. The annual growth in CH_4 concentration slowed and became more variable in the 1990s, compared with the 1980s. Slightly more than half of current CH_4 emissions are anthropogenic (e.g., use of fossil fuels, cattle, rice agriculture and landfills). In addition, carbon monoxide (CO) emissions have recently been identified as a cause of increasing CH_4 concentration.

- The atmospheric concentration of nitrous oxide (N_2O) has increased by 46 ppb (17%) since 1750 and continues to increase. The present N_2O concentration has not been exceeded during at least the past thousand years. About a third of current N_2O emissions are anthropogenic (e.g., agricultural soils, cattle feed lots and chemical industry).

- Since 1995, the atmospheric concentrations of many of those halocarbon gases that are both ozone-depleting and greenhouse gases (e.g., $CFCl_3$ and CF_2Cl_2), are either increasing more slowly or decreasing, both in response to reduced emissions under the regulations of the Montreal Protocol and its Amendments. Their substitute compounds (e.g., CHF_2Cl and CF_3CH_2F) and some other synthetic compounds (e.g., perfluorocarbons (PFCs) and sulphur hexafluoride (SF_6)) are also greenhouse gases, and their concentrations are currently increasing.

- The radiative forcing due to increases of the well-mixed greenhouse gases from 1750 to 2000 is estimated to be 2.43 Wm^{-2}: 1.46 Wm^{-2} from CO_2; 0.48 Wm^{-2} from CH_4; 0.34 Wm^{-2} from the halocarbons; and 0.15 Wm^{-2} from N_2O. . . .

- The observed depletion of the stratospheric ozone (O_3) layer from 1979 to 2000 is estimated to have caused a negative radiative forcing (-0.15 Wm^{-2}). Assuming full compliance with current halocarbon regulations, the positive forcing of the halocarbons will be reduced as will the magnitude of the negative forcing from stratospheric ozone depletion as the ozone layer recovers over the 21st century.

- The total amount of O_3 in the troposphere is estimated to have increased by 36% since 1750, due primarily to anthropogenic emissions of several O_3-forming gases. This corresponds to a positive radiative forcing of 0.35 Wm^{-2}. O_3 forcing varies considerably by region and responds much more quickly to changes in emissions than the long-lived greenhouse gases, such as CO_2.

Anthropogenic aerosols are short-lived and mostly produce negative radiative forcing.

- The major sources of anthropogenic aerosols are fossil fuel and biomass burning. These sources are also linked to degradation of air quality and acid deposition.

- Since the SAR, significant progress has been achieved in better characterising the direct radiative roles of different types of aerosols. Direct radiative forcing is estimated to be -0.4 Wm^{-2} for sulphate, -0.2 Wm^{-2} for biomass burning aerosols, -0.1 Wm^{-2} for fossil fuel organic carbon and $+0.2$ Wm^{-2} for fossil fuel black carbon aerosols. There is much less confidence in the ability to quantify the total aerosol direct effect, and its evolution over time, than that for the gases listed above. Aerosols also vary considerably by region and respond quickly to changes in emissions.
- In addition to their direct radiative forcing, aerosols have an indirect radiative forcing through their effects on clouds. There is now more evidence for this indirect effect, which is negative, although of very uncertain magnitude.

Natural factors have made small contributions to radiative forcing over the past century.

- The radiative forcing due to changes in solar irradiance for the period since 1750 is estimated to be about $+0.3$ Wm^{-2}, most of which occurred during the first half of the 20th century. Since the late 1970s, satellite instruments have observed small oscillations due to the 11-year solar cycle. Mechanisms for the amplification of solar effects on climate have been proposed, but currently lack a rigorous theoretical or observational basis.
- Stratospheric aerosols from explosive volcanic eruptions lead to negative forcing, which lasts a few years. Several major eruptions occurred in the periods 1880 to 1920 and 1960 to 1991.
- The combined change in radiative forcing of the two major natural factors (solar variation and volcanic aerosols) is estimated to be negative for the past two, and possibly the past four, decades. . . .

There is new and stronger evidence that most of the warming observed over the last 50 years is attributable to human activities.

The SAR concluded: "The balance of evidence suggests a discernible human influence on global climate". That report also noted that the anthropogenic signal was still emerging from the background of natural climate variability. Since the SAR, progress has been made in reducing uncertainty, particularly with respect to distinguishing and quantifying the magnitude of responses to different external influences. Although many of the sources of uncertainty identified in the SAR still remain to some degree, new evidence and improved understanding support an updated conclusion.

- There is a longer and more closely scrutinised temperature record and new model estimates of variability. The warming over the past 100 years is very unlikely[5] to be due to internal variability alone, as estimated by current models. Reconstructions of climate data for the past 1,000 years also indicate that this warming was unusual and is unlikely[5] to be entirely natural in origin.

- There are new estimates of the climate response to natural and anthropogenic forcing, and new detection techniques have been applied. Detection and attribution studies consistently find evidence for an anthropogenic signal in the climate record of the last 35 to 50 years.

- Simulations of the response to natural forcings alone (i.e., the response to variability in solar irradiance and volcanic eruptions) do not explain the warming in the second half of the 20th century. However, they indicate that natural forcings may have contributed to the observed warming in the first half of the 20th century.

- The warming over the last 50 years due to anthropogenic greenhouse gases can be identified despite uncertainties in forcing due to anthropogenic sulphate aerosol and natural factors (volcanoes and solar irradiance). The anthropogenic sulphate aerosol forcing, while uncertain, is negative over this period and therefore cannot explain the warming. Changes in natural forcing during most of this period are also estimated to be negative and are unlikely[5] to explain the warming.

- Detection and attribution studies comparing model simulated changes with the observed record can now take into account uncertainty in the magnitude of modelled response to external forcing, in particular that due to uncertainty in climate sensitivity.

- Most of these studies find that, over the last 50 years, the estimated rate and magnitude of warming due to increasing concentrations of greenhouse gases alone are comparable with, or larger than, the observed warming. Furthermore, most model estimates that take into account both greenhouse gases and sulphate aerosols are consistent with observations over this period.

- The best agreement between model simulations and observations over the last 140 years has been found when all the above anthropogenic and natural forcing factors are combined. These results show that the forcings included are sufficient to explain the observed changes, but do not exclude the possibility that other forcings may also have contributed.

In the light of new evidence and taking into account the remaining uncertainties, most of the observed warming over the last 50 years is likely[5] to have been due to the increase in greenhouse gas concentrations.

Furthermore, it is very likely[5] that the 20th century warming has contributed significantly to the observed sea level rise, through thermal expansion of sea water and widespread loss of land ice. Within present uncertainties, observations and models are both consistent with a lack of significant acceleration of sea level rise during the 20th century.

Human influences will continue to change atmospheric composition throughout the 21st century.

Models have been used to make projections of atmospheric concentrations of greenhouse gases and aerosols, and hence of future climate, based upon emissions scenarios from the IPCC Special Report on Emission Scenarios (SRES). These scenarios were developed to update the IS92 series, which were used in the SAR and are shown for comparison here in some cases.

Greenhouse gases

- Emissions of CO_2 due to fossil fuel burning are virtually certain[5] to be the dominant influence on the trends in atmospheric CO_2 concentration during the 21st century.

- As the CO_2 concentration of the atmosphere increases, ocean and land will take up a decreasing fraction of anthropogenic CO_2 emissions. The net effect of land and ocean climate feedbacks as indicated by models is to further increase projected atmospheric CO_2 concentrations, by reducing both the ocean and land uptake of CO_2.

- By 2100, carbon cycle models project atmospheric CO_2 concentrations of 540 to 970 ppm for the illustrative SRES scenarios (90 to 250% above the concentration of 280 ppm in the year 1750). These projections include the land and ocean climate feedbacks. Uncertainties, especially about the magnitude of the climate feedback from the terrestrial biosphere, cause a variation of about -10 to $+30\%$ around each scenario. The total range is 490 to 1260 ppm (75 to 350% above the 1750 concentration).

- Changing land use could influence atmospheric CO_2 concentration. Hypothetically, if all of the carbon released by historical land-use changes could be restored to the terrestrial biosphere over the course of the century (e.g., by reforestation), CO_2 concentration would be reduced by 40 to 70 ppm.

- Model calculations of the concentrations of the non-CO_2 greenhouse gases by 2100 vary considerably across the SRES illustrative scenarios, with CH_4 changing by -190 to $+1,970$ ppb (present concentration 1,760 ppb), N_2O changing by $+38$ to $+144$ ppb (present concentration 316 ppb), total tropospheric O_3 changing by -12 to $+62\%$, and a wide range of changes in concentrations of HFCs, PFCs and SF_6, all relative to the year 2000. In some scenarios, total tropospheric O_3 would become as important a radiative forcing agent as CH_4 and, over much of the Northern Hemisphere, would threaten the attainment of current air quality targets.

- Reductions in greenhouse gas emissions and the gases that control their concentration would be necessary to stabilise radiative forcing. For example, for the most important anthropogenic greenhouse gas, carbon cycle models indicate that stabilisation of atmospheric CO_2 concentrations at 450, 650 or 1,000 ppm would require global anthropogenic CO_2 emissions to drop below 1990 levels, within a few decades, about a century, or about two centuries, respectively, and continue to decrease steadily thereafter. Eventually CO_2 emissions would need to decline to a very small fraction of current emissions.

Aerosols

- The SRES scenarios include the possibility of either increases or decreases in anthropogenic aerosols (e.g., sulphate aerosols, biomass aerosols, black and organic carbon aerosols) depending on the extent of fossil fuel use and policies to abate polluting emissions. In addition, natural aerosols (e.g., sea salt, dust and emissions leading to the production of sulphate and carbon aerosols) are projected to increase as a result of changes in climate.

Radiative forcing over the 21st century

- For the SRES illustrative scenarios, relative to the year 2000, the global mean radiative forcing due to greenhouse gases continues to increase through the 21st century, with the fraction due to CO_2 projected to increase from slightly more than half to about three quarters. The change in the direct plus indirect aerosol radiative forcing is projected to be smaller in magnitude than that of CO_2.

Global average temperature and sea level are projected to rise under all IPCC SRES scenarios.

In order to make projections of future climate, models incorporate past, as well as future emissions of greenhouse gases and aerosols. Hence, they include estimates of warming to date and the commitment to future warming from past emissions.

Temperature

- The globally averaged surface temperature is projected to increase by 1.4 to 5.8°C over the period 1990 to 2100. These results are for the full range of 35 SRES scenarios, based on a number of climate models[8,9].
- Temperature increases are projected to be greater than those in the SAR, which were about 1.0 to 3.5°C based on the six IS92 scenarios. The higher projected temperatures and the wider range are due primarily to the lower projected sulphur dioxide emissions in the SRES scenarios relative to the IS92 scenarios.
- The projected rate of warming is much larger than the observed changes during the 20th century and is very likely[5] to be without precedent during at least the last 10,000 years, based on palaeoclimate data.
- By 2100, the range in the surface temperature response across the group of climate models run with a given scenario is comparable to the range obtained from a single model run with the different SRES scenarios.
- On timescales of a few decades, the current observed rate of warming can be used to constrain the projected response to a given emissions scenario despite uncertainty in climate sensitivity. This approach suggests that anthropogenic warming is likely[5] to lie in the range of 0.1 to 0.2°C per decade over the next few decades under the IS92a scenario. . . .
- Based on recent global model simulations, it is very likely[5] that nearly all land areas will warm more rapidly than the global average, particularly those at northern high latitudes in the cold season. Most notable of these is the warming in the northern regions of North America, and

northern and central Asia, which exceeds global mean warming in each model by more than 40%. In contrast, the warming is less than the global mean change in south and southeast Asia in summer and in southern South America in winter.

- Recent trends for surface temperature to become more El Niño-like in the tropical Pacific, with the eastern tropical Pacific warming more than the western tropical Pacific, with a corresponding eastward shift of precipitation, are projected to continue in many models.

Precipitation

- Based on global model simulations and for a wide range of scenarios, global average water vapour concentration and precipitation are projected to increase during the 21st century. By the second half of the 21st century, it is likely[5] that precipitation will have increased over northern mid- to high latitudes and Antarctica in winter. At low latitudes there are both regional increases and decreases over land areas. Larger year to year variations in precipitation are very likely[5] over most areas where an increase in mean precipitation is projected.

Notes

1. Delegations of 99 IPCC member countries participated in the Eighth Session of Working Group I in Shanghai on 17 to 20 January 2001.
2. The IPCC Second Assessment Report is referred to in this Summary for Policymakers as the SAR.
3. Generally temperature trends are rounded to the nearest 0.05°C per unit time, the periods often being limited by data availability.
4. In general, a 5% statistical significance level is used, and a 95% confidence level.
5. In this Summary for Policymakers and in the Technical Summary, the following words have been used where appropriate to indicate judgmental estimates of confidence: *virtually certain* (greater than 99% chance that a result is true); *very likely* (90–99% chance); *likely* (66–90% chance); *medium likelihood* (33–66% chance); *unlikely* (10–33% chance); *very unlikely* (1–10% chance); *exceptionally unlikely* (less than 1% chance). . . .
6. *Radiative forcing* is a measure of the influence a factor has in altering the balance of incoming and outgoing energy in the Earth-atmosphere system, and is an index of the importance of the factor as a potential climate change mechanism. It is expressed in Watts per square metre (Wm^{-2}).
7. ppm (parts per million) or ppb (parts per billion, 1 billion = 1,000 million) is the ratio of the number of greenhouse gas molecules to the total number of molecules of dry air. For example: 300 ppm means 300 molecules of a greenhouse gas per million molecules of dry air.
8. Complex physically based climate models are the main tool for projecting future climate change. In order to explore the full range of scenarios, these are complemented by simple climate models calibrated to yield an equivalent response in temperature and sea level to complex climate models. These projections are obtained using a simple climate model whose climate sensitivity and ocean heat uptake are calibrated to each of seven complex climate models. The climate sensitivity used in the simple model ranges from 1.7 to 4.2°C, which is comparable to the commonly accepted range of 1.5 to 4.5°C.
9. This range does not include uncertainties in the modelling of radiative forcing, e.g. aerosol forcing uncertainties. A small carbon-cycle climate feedback is included.

Brian Tucker  **NO**

Science Friction:
The Politics of Global Warming

As everyone knows, there has been much sound and fury over the past decade about the threat of global warming, a condition held to be the result of "greenhouse" gas emissions caused by human activity. Indeed, the argument over global warming has been the main set piece of the international environmental culture wars for several years, with activist Cassandras and conservative Pollyannas both trying to marshal the authority of science as justification for their views. The ensuing battle has reminded us of a key truth of the sociology of knowledge, that while everyone may be entitled to his own opinions, none is entitled to his own facts.

In Kyoto this coming December [1997] a major UN-sponsored international conference on global warming is scheduled to take place; it is the culmination of several run-up meetings in Berlin, Geneva, and New York that have seen the various players rehearsing for the big event. From these precursor meetings, several things have become clear: that politics and emotions have distorted what is scientifically known about the phenomenon itself; that the issue is far more politically and morally complex than many realize or are prepared to admit; and that governments find themselves besieged by conflicting interests and pressures, both domestically and internationally. In light of all this, the outlook for anything very useful happening at Kyoto is mixed at best. An examination of the real science, the real issues, and the real choices before us shows why.

The Real Science

Many if not most non-expert observers assume that the greenhouse effect necessitates major and relatively rapid reductions in greenhouse gas emissions, particularly carbon dioxide, lest dire climatic changes imperil the biosphere. Typical of much media comment on the issue is the assertion, this one from *The Australian* newspaper, that " . . . some scientists are increasingly worried that greenhouse gases could wreak havoc with the earth's climate in future generations, creating huge floods and droughts from global warming." But those who believe such things do not get their data from the international consortium of climatologists who are continually collating results from recent studies.

Much of the scientific theory about greenhouse gas emissions is well founded, but it does not support the conclusion that calamitous effects from global warming are nigh upon us. The work of the Inter-governmental Panel on Climate Change (IPCC), established in 1988 by the World Meteorological Organization (WMO) and the United Nations Environmental Program, has yielded a worldwide scientific consensus that has become generally, if not universally, accepted as a balanced state-of-knowledge assessment. In essence, its conclusion is that while human activity has caused no significant direct increase of the major greenhouse gas—namely, water vapor—there is no doubt that other such gases are increasing. The most important of these is carbon dioxide, followed by methane and nitrous oxide. Fossil fuel burning is the main reason for the increase of carbon dioxide, and agricultural methods play a major role in the increase of methane and nitrous oxide. It follows that the energy and food requirements resulting from increases in world population together with the demands of the existing population make it inevitable that these gases will continue to accumulate in the foreseeable future. The rate at which this will occur, however, is still unclear.

Although scientific progress has occurred on several fronts, the timing, magnitude, and geographical distribution of climate change due to higher greenhouse gas levels can be estimated only inexactly and in limited detail. Climate scientists know that many important processes are inadequately treated in their models; notable difficulties include specifying the roles of clouds and ocean currents, and imprecise representation of precipitation processes. While sensible and careful attempts are being made to effect improvements, doubts remain as to how accurate our models can become anytime soon at representing the real atmosphere. Hence, the simulation of any but the broadest aspects of climate in a world with higher greenhouse gas levels is an exercise of uncertain validity.

In view of these uncertainties and the cautious way in which the scientists carrying out climate change studies regard their results, it is unfortunate that some on the periphery of this knowledge extrapolate so freely as to predict calamitous consequences even within the next century. There is no scientific justification for such a view. Only one prediction can be hazarded with any reasonable degree of confidence: that when an effective doubling of current greenhouse gases occurs, probably toward the end of [this] century, average near-surface air temperatures will be 2 degrees or 3 degrees C higher than they are at present. This represents the middle range of estimates, with some models predicting one or two degrees higher and others one or two degrees lower. Even this limited prediction depends on a correct treatment of climate processes represented in the models, and we are still not sure of their accuracy extrapolated out as far as eighty or ninety years.

Associated with these air temperature increases are higher near-surface sea temperatures and, with somewhat less confidence, sea level rises of up to half a meter due to thermal expansion of sea water. These changes would be superimposed upon vertical land movements of similar magnitude due to tectonic processes; in some cases, such movements would exacerbate the effects of higher sea levels, in other cases they would mitigate them. Another century on, that is to say around the year 2200, at a time when greenhouse gases may be three times their present concentration, models indicate a total climate change from the present of an additional 50 percent beyond the above-cited figures.

There is scientific consensus on little else. The belief that events such as the January 1995 and summer 1997 floods in Western Europe and the January/February 1996 cold spell in central and eastern parts of the United States were caused by higher greenhouse gas levels is taken seriously by many environmental activists. But they are, strictly speaking, scientifically unjustified. Indeed, all assertions that climate models indicate progressively extreme weather are highly questionable. Suggestions that global warming is increasing the frequency of tropical cyclones or of droughts are unfounded; indeed, some data suggest that a decrease in extreme weather is more likely. Analyses of recent climate model results do reveal a small rise in the frequency of high rainfall events in the presence of twice as much carbon dioxide as currently exists in the atmosphere, but this may be accepting too much detail from the surrogate model's crude spatial resolution and very inexact grasp of rainfall processes.

The Real Issues

Misconstrued science, however, is perhaps the least of the problems associated with the global warming scare. Far more significant is the misconstruction of the underlying politics of the thing.

Casual observers have a very truncated notion of what is at stake in the global warming debate. The typical newspaper reader appears to think that what is at issue is the extent to which the earth's ambient temperature is increasing and what needs to be done to prevent presumed calamity. If the problem were really that simple, we could expect that an objective consensus of qualified experts would suggest the path to prevention and remediation. But even if all the necessary unambiguous scientific facts were available—and as we have seen, this is not so simple a task—the very wording of the main objective of the Framework Convention on Climate Change (FCCC), negotiated at the 1992 Earth Summit at Rio de Janeiro, suggests the key complication. Article 2 of the FCCC commits signatories to achieve " . . . stabilization of greenhouse gas concentrations in the atmosphere at a level that would prevent dangerous anthropogenic interference with the climate system. Such a level should be achieved within a time frame sufficient to allow ecosystems to adapt naturally to climate change, to ensure that food production is not threatened, and to enable economic development to proceed in a sustainable manner."

This is an illuminating formulation for two reasons. First, no empirical definition is given (here or elsewhere in the FCCC) of what constitutes "dangerous" anthropogenic interference with the climate system, or a time frame "sufficient" to allow ecosystems to adapt naturally to climate change. What is dangerous or sufficient in these contexts is not obvious, short of extreme cases. And the entire statement begs the question of the extent to which any measurable change in the climate system has in fact been caused by humans. Not only has this proved difficult to determine, but, thanks to one of the most astonishing scientific episodes of the century, we know something in 1997 that we did not know in 1992: that a slight increase in water vapor, the most plentiful of the greenhouse gases, may come from a constant bombardment of the earth by small ice comets.

Second, and more important for our purposes, the FCCC acknowledges that the greenhouse gas emissions said to cause global warming are to a large extent a consequence of efforts to meet the energy and food requirements of a growing world population. In short, it suggests, obliquely at least, that there are trade-offs to be reckoned. And so there are: The only realistic way to reduce greenhouse gases sharply is to burn less fossil fuel, and the only way to do that without jeopardizing living standards, particularly in the poorest countries, is to rely far more heavily on nuclear-generated electricity. But many of the meliorist-minded political activists who are concerned about the environmental damage caused by modern industrialism are equally concerned about the poverty of the less developed parts of the world, and equally adamant opponents of generating electricity from nuclear power.

We can add to this already vexatious equation the contentious issue of world population control, which is part and parcel of the issue of world poverty. As the failure of the September 1994 World Population Summit at Cairo showed, there is no general international agreement on realistic plans to limit world population increases any more than there is general agreement on a plan to attack poverty in the world's poorer lands. Indeed, this entire melange of issues defines the current standoff between the OECD [Organisation for Economic Co-operation and Development] countries and what used to be called generically the Third World. Many relatively poor countries with high population growth rates have seen population control as a ploy of the advanced countries to deny them their future power, or to fob off responsibility for sharing global prosperity. Representatives of the more advanced countries, with generally more moderate rates of population growth, see such arguments as highly self-defeating and often unjustifiably conspiratorial in inspiration. The global warming argument between rich and poor countries is a variation on the same theme: Efforts by the industrialized powers to impose greenhouse gas emission limits on developing countries are seen by the governments of those countries as an alternative means of reducing their rates of economic growth and hence their national power; while the developing countries' refusal to accept limits on themselves is seen by the developed countries as cynical and self-serving.

The global warming controversy is therefore bound up not only with uncertainties about the objective causes and extent of the phenomenon, but also with impassioned arguments over global development, power, and morality. While most of the relevant government officials appreciate these complexities, many environmentalists have difficulty acknowledging that the Framework Convention objective itself begs the question of what is desirable, let alone what is possible. Having to make painful trade-offs between desirable but irreconcilable objectives is precisely what true believers of all stripes are temperamentally incapable of doing. In this instance, they are extremely reluctant to pose the crucial question: Which is worse, embracing nuclear power with its potential dangers and costs, imposing poverty-inducing restrictions on world energy consumption and agriculture, or accepting the costs of adjustment to climate change?

It is not surprising, therefore, that since the Rio meeting strong emotion has led many governments, including that of the United States, to try to capture the high moral ground of domestic politics by making promises that are, at least for the foreseeable future, manifestly impossible to keep. Indeed, several governments have adopted national interim targets even more ambitious than those of the United States. Germany, for example, aims for a 25 percent reduction of 1990 carbon dioxide emissions by 2005. Of the more than 150 national signatories that have ratified the FCCC and accepted its aim to stabilize greenhouse gas concentrations in the atmosphere, most have adopted national targets that are highly unlikely to be achieved. As Kyoto approaches, these governments find themselves caught in a crisis of wishing, suspended between the short-term political expediency of affirming such goals and the reality of actual limits to achieving them in light of the conflicting political and moral interests at stake.

Why the Alarm?

The scientific evidence thus far suggests that adjusting to what is liable to be rather moderate and very gradual climate change—with some conversion from fossil fuels to nuclear power to provide time for adjustment to a more moderate level and pace of change—is the most sensible answer to the question of trade-offs posed above. But this is not likely to be the conclusion reached in Kyoto. Instead, a combination of international beggar-thy-neighbor politics and a rush of millenarian fear seems to be taking over.

If there is so little evidence to support the notion of calamitous climate change associated with global warming, why has so much alarm been generated? The answer lies in part in the sky-is-falling perspective that seems to characterize the views of environmental activists, and their success at imparting this perspective to the general public through a media industry more interested in selling its wares than it is adept at understanding basic science. Not only is the magnitude of change exaggerated, but environmental activists have also been able to popularize the unwarranted assumption that all change must be deleterious.

Fear is as hardwired into human thought processes as the temptations of utopia, and both have influenced social development throughout human history. No doubt the increased popularity of declinist thought among intellectuals these last few decades owes at least a little something to an awareness of the finite nature of the physical world, and to a keener appreciation that humanity has prospered materially only because we have exploited this finite environment. As a result of these fears, the precautionary principle has become a major plank in many environmental platforms as a means of stopping environmental degradation. This principle states that, when faced with development proposals, regulators should anticipate environmental harm and act to ensure that such harm does not occur.

This sounds innocuous enough; surely no rational person would advocate deliberate and needless environmental degradation. But, as suggested above, there is not just one moral imperative at issue—the environment—but competing moral imperatives, or at least competing desired objectives, including the right to life, food, and comfort, and concern for both international and intergenerational equity. The precautionary principle of environmentalism, however, assumes that any developmental activity has a latent tendency to do more harm than good, and that whatever societal benefit may come from such activity cannot justify further environmental alteration. The key to such an assumption is that the environmental value necessarily takes precedence over all others because if the biosphere is wrecked beyond redemption, then no other value can be served. This key assumption, however, reflects an excessive fear of the future and a highly timorous view of the relationship between humanity and the environment.

The application of this gloomy perspective to the subject of climate change is much encouraged by a tendency to focus on potential damage to systems based on the present climatic state. Yet adaptation has been a hallmark of human resilience throughout history, and opportunities actually to benefit from prospective climate change are consistently ignored. Typically, for example, a broad indication of more rain tends to be regarded as likely to result in more floods rather than fewer droughts, and of less rain as causing more droughts rather than fewer floods. Such misplaced anxiety has caused one senior scientist to bemoan, "We have created the most unnecessarily fearful generation of humankind that ever populated the earth."

Fear of climate change has also been bolstered unwittingly by the development of "scenarios", a term that describes sets of climate change parameters regarded as feasible in the light of emerging model results. Scenarios were first prepared by climate scientists to provide a benchmark for estimating the sensitivity of different sectors of society and the economy to possible climate change. From their inception such scenarios were tagged with strong caveats indicating that they should not be regarded as forecasts, and were openly acknowledged to be based on models widely considered inadequate in many ways. But such warnings and qualifications have often been ignored, partly because many environmentalists feel happiest when they assume the worst, and partly because the explanatory notes issued with early scenarios were sufficiently ambiguous to imply forecasts to the non-expert.

A gradually changing climate may well be the inevitable consequence of human activity, and it may indeed pose serious problems of adaptation. It bears repeating that the evidence does not support the dismissive view that those concerned about global warming are spinning fantasies from thin air. But dramatic descriptions of potential calamity are hyperbole, and the assumption that all change will be for the worse is less a sign of perspicacity than paranoia.

The Real Choices

It is sometimes said that realism is the impact of thinking upon wishing. With world population rising more rapidly than at any previous time in history, and with industrialization spreading fast to formerly impoverished regions, the prospect of reducing world emission of the greenhouse gases necessarily involved in the industrial programs of developing nations and in world food production is, to put it mildly, not good.

It is important to distinguish between greenhouse gas emissions, which are what is put into the atmosphere, and gas concentrations, which are what stay there. An approach to stable emissions will be possible only when world population and associated development expectations have themselves stabilized—probably a century or more hence, or when alternatives to fossil fuel become widely available and commercially competitive. To stabilize concentrations of gases in the atmosphere would require worldwide emission reductions in excess of 50 percent, and that over a time interval during which world population is expected to double.

The situation is not totally irredeemable, however. Fossil fuel combustion accounts for some 75 percent of worldwide carbon dioxide emissions, and substitute means of generating electrical power on a large scale already exist. The most dramatic recent example of a national effort to reduce fossil fuel burning emissions is the French conversion to nuclear energy. France achieved a 40 percent reduction in carbon dioxide emissions during the 1980s, largely by increasing its nuclear-powered electricity generating capacity from about 10,000 to 56,000 megawatts. By 1994, about 78 percent of France's electricity was generated by nuclear power. No other country has undertaken such a massive transfer from fossil fuel to nuclear energy, although several other industrialized countries have moved in this direction. As a result, Sweden, South Korea, Switzerland, and Spain now have some of the lowest per capita rates of carbon dioxide emissions. For the United States, with the highest per capita emissions, carbon dioxide emissions in 1993 would have been 30 percent more than occurred had it not been for 109 nuclear-powered electricity generating stations. For the world as a whole, the amount of carbon dioxide accumulating in the atmosphere would be some 10 percent greater than at present without the existing nuclear production of electricity.

It is clear that policies fostering energy conservation, energy efficiency, and renewable energy alone cannot stabilize worldwide emissions, let alone reduce atmospheric concentrations of greenhouse gases. Nor is the drastic decrease in consumption required for such reductions remotely in prospect, for it would lead to a politically unacceptable reduction in living standards in the economically advanced countries, and to a politically unacceptable limit on development in the less advanced countries. The stabilization of emissions might be achieved, however, within about fifty years if there were an immediate international push to maximize nuclear-powered electricity production. But that, of course, raises concern about cost and safety—and here we come back to another difficult trade-off.

Concerns regarding the safety of nuclear power station operations are not something idly to be dismissed, even if one were satisfied that safe disposal of low, intermediate, and high-level wastes had been achieved. It is not fabricated paranoia to take such problems seriously, nor is it trivial to note that waste disposal problems have still not been solved. But nearly 450 nuclear power reactors in some 30 countries currently supply approximately 17 percent of the world's electricity needs, and nuclear power is today safer and less of an environmental threat than other means of generating electricity, at least as long as there is a willingness to pay the costs of safe reactor operation and waste management procedures. Such costs make nuclear-generated electricity more expensive than the fossil fuel method, but if we seek rapid reductions of carbon dioxide emissions, it is clear that any extra costs involved with nuclear power would be less than those of the only other alternative—namely, the economic and social disruption consequent upon drastic food and energy curtailment.

Political realities being what they are, there is no way that the FCCC objective of reducing emissions—which is likely to be reaffirmed in Kyoto—will be reached within the time frames agreed so far. There is no rapid way either to curtail emissions or to achieve substitution of nuclear-powered electricity generation capacity. The international scientific establishment knows this, having declared two years after Rio that even the "stabilization at present carbon dioxide concentrations . . . obviously is an impossible option." It is even clearer today that the direct economic costs of the magnitude of emission reduction being contemplated as we move toward the Kyoto meeting are enormous. To the extent that there is any political willingness to achieve the Rio targets, it is likely to be soon smothered by realization of the staggering price tag that comes with it.

Perhaps this is just as well since the actual extent of climate change is going to be much less dramatic than most non-experts think. It is important that emotions not lead governments into chasing non-existent solutions to highly exaggerated problems, if only because that will make it harder to achieve less daunting but still necessary solutions to real problems. If dispassionate analysis cannot put an end to the current alarmism, then perhaps the inevitability of failure to meet unrealistic goals will eventually induce a greater sense of realism.

Such a realism would take as its first principle with respect to issues of relative cost that policies aimed at adapting to gradual climate change make more sense than impossible efforts to stop all anthropogenic climate change in its tracks. Put somewhat crudely, it makes more economic and moral sense to build dikes to prevent coastal flooding, to relocate especially vulnerable human habitats, to adjust irrigation schemes for variant rainfall patterns, and to accelerate efforts to replace fossil fuel burning to meet energy requirements than it does to drive world GNP [gross national product] backwards several percentage points per year for half a century or more.

Even such an effort, undertaken over the better part of a century, will be no stroll in the park. It will require, first of all, accelerated research aimed at obtaining improved and unbiased scientific assessments of change, and the most sensible ways to adapt to it. But if one takes a factually based view of the problem, it is clear that the costs for adjustment spread out over time will be vastly lower than any serious effort to meet the Rio targets.

Lost in the Ozone

Now that it is obvious that there is no way to meet the main objective of the Framework Convention on Climate Change, is there any indication of a shift in thinking from a noble but impossible approach to a less dramatic but more rational one? Will we see more sober deliberations in Kyoto? From the evidence of the Conference of Parties to the FCCC convened in April 1995 in Berlin, of a second conference in July 1996 in Geneva, and of the UN/WMO conference in New York in June of [1997], the answer has to be "maybe."

The Berlin meeting was designed to provide a forum for countries to compare and assess their programs to formulate, implement, publish, and regularly update emission reduction measures to mitigate climate change. As expected, representatives of environmental movements also sought to increase and strengthen reduction targets for industrialized countries. A similar line was taken by many poor countries (a more realistic term for what used to be called developing countries), who refused to accept any reduction target applying to themselves.

The impossibility of attaining the 1992 Rio targets was not acknowledged at Berlin, let alone the lunacy of setting still more stringent ones. Little more was decided than to launch a new round of negotiations, termed the Berlin Mandate, aimed at delivering a legally binding instrument on which all delegates could agree at a December 1997 Kyoto conference. Indeed, most participants at Berlin seemed more concerned with making new commitments than with meeting old ones—it was, after all, easier. The real trade-offs were not mentioned, and many strains of hypocrisy were in evidence. Bureaucratic opportunists, having pushed policy formulation well ahead of scientific justification, now saw emerging career prospects from continuing rounds of expensive international negotiation. Environmental opportunists, grasping at any information no matter how selective or exaggerated to foment alarm, appeared completely oblivious to the downstream effects of their extravagant demands. Rich nation opportunists, because their current economic circumstances involve a temporary stabilization of their own gas emissions, saw a means of imposing economic restrictions on competitors not so favorably placed. Poor nation opportunists, while rejecting emission reduction targets for themselves, sought restrictions on rich nations on the grounds that their extravagance caused the problem in the first place.

And so it goes. In Geneva in July 1996, a 1995 scientific update of our understanding of the greenhouse gas-climate change connection, prepared by the Inter-governmental Panel on Climate Change, was received. Its most important component was the report of its Scientific Working Group, because in any logical progression its other two components (estimating impacts and devising strategies to cope) must be based on climate change predictions, which in turn are based on an understanding of how greenhouse gases get into the atmosphere and what happens once they do. The treatment accorded this report was as illuminating as it was discouraging.

Background information and the first draft of documentation were provided by hundreds of scientists who had met to coordinate and articulate the results of their research. An executive group edited the detailed material, prepared summary findings, and published both—which were then sharply criticized from within the ranks of both science and industry. Bias and exaggeration were apparent in the final policymakers' summaries. For example, while there is an implied increase in the frequency of severe droughts and floods in the final wording, there is no "on balance" evidence in the research results to justify this. Also, where there are clear results showing that, for example, tropical cyclone occurrence will remain unchanged, this is interpreted weakly as there not yet being sufficient evidence to show an increase. Careful study of the report by "IPCC watchers" associated with major world industry associations has revealed many examples of last minute changes to the earlier wording, the effect being to "scientifically cleanse" the final report of equivocations and caveats in the background documentation. The World Energy Council has also cited evidence of scientific bias, technical weakness, and political influence in the policymakers' summaries provided by other IPCC working groups.

Whether due to naivete or bias, the credibility of the IPCC has been damaged by these shortcomings. Those responsible for the final wording of the summaries appear to have been unaware that any selected passage that could advance alarmist prejudices would be seized upon by the environmental lobby and widely quoted, whether in context or not. These experiences, in turn, have made prospects for Kyoto even dimmer than they might otherwise be, because the credibility of the core scientific work has now been compromised.

Not all the news is bad, however. From the national reports presented at Geneva, and later in New York, it was evident that few countries would get near their existing and modest target of stabilizing emissions by the year 2000. Developing countries, meanwhile, are still excluded from such targets. Not surprisingly, then, much of the preparatory negotiations for Kyoto involve squabbling over whether any targets should involve "differentiation", that is, considerations of social and economic differences between countries. The United States, currently allied with part of the European Union, opposes differentiation; most other industrialized countries support it. Nevertheless, no country regards such "burden sharing" as other than a means of reducing its own share, while developing countries show little sign of any willingness to make sacrifices.

Still, reading between the lines of these discussions one can detect an increasing recognition of the unrealistic nature of worldwide emission reduction targets, of the still major scientific uncertainties at hand, and of the need to assess the full consequences of a policy before embarking upon it. A small group of countries, including Australia, Iceland, Japan, and Norway, refuses to conform to the pretense of being bound to even more unrealistic emission reduction commitments beyond 2000. Other countries, most notably the United States, while supporting the principle of binding targets, have demanded that they be realistic and verifiable, not compromise national prosperity, and include targets for developing countries.

That is good, of course, but not good enough. It remains politically incorrect, and hence so far impossible, to discuss the issues rationally, with all the attendant assumptions openly scrutinized, and with all the competing values acknowledged and analyzed. From what has been said, it should be clear that the global warming scare is based on four related assumptions: (1) that the greenhouse gas-climate change scientific theory is valid; (2) that any such climate change will necessarily be detrimental to human well-being; (3) that realistically we can achieve a sufficient reduction in greenhouse gas emissions to effect a worldwide stabilization of gas concentrations in the atmosphere; and (4) that the cost to society of such a reduction is less than the cost of adapting to the climate change. It follows that if even one of these assumptions is untrue, then the argument behind a policy of rapid and extensive reduction of greenhouse gas emissions is undermined. As we have seen, the basic scientific theory of global warming is sound, but predicts substantially more moderate effects than those that have captured the popular imagination. Imprecise and uncertain climate change estimates and the complexity of climatic impact assessments make the second assumption indeterminate; while the foregoing arguments have shown the third and fourth to be essentially false. Unless the Kyoto summit acknowledges these basic truths, it looks to be a dreadful affair that will needlessly worsen political tensions between rich and poor, mislead millions of scientific innocents, and misdirect government policies that must deal with the less severe but still real challenge before us.

POSTSCRIPT

Is the Threat of Global Warming Real?

In the early months of the twenty-first century, the U.S. government debated ratifying the Kyoto treaty, while Western European governments sought to influence America to do so. In the summer of 2002, the countries of the world convened in Johannesburg, South Africa, for the 10-year follow-up to the Earth Summit in Rio de Janeiro, where 170 nations originally agreed to voluntarily reduce greenhouse gas emissions.

The conclusions of the IPCC 2001 report are the following: First, the global average surface temperature rose over the twentieth century by about 0.6°C. Second, temperatures have risen during the past four decades in the lowest 8 kilometers of the atmosphere. Third, snow cover and ice extent have decreased. Fourth, global average sea level has risen, and ocean heat has increased. Fifth, changes have occurred in other important aspects of climate. Sixth, concentrations of atmospheric greenhouse gases have continued to increase as a result of human activity. Seventh, there is new and stronger evidence that most of the warming observed over the last 50 years is attributable to human activities. Eighth, human influences will continue to change atmospheric composition throughout the twenty-first century. Ninth, global average temperature and sea level are projected to rise under all IPCC scenarios. And tenth, climate change will persist for many centuries.

Tucker's main thesis is that science "does not support the conclusion that calamitous effects from global warming are nigh upon us." Continuing, he says, "There is no scientific justification for such a view." Tucker raises the stakes by asserting that the global warming controversy is much more than a debate about the causes and extent of the phenomenon. It is also about global development, power, and morality in the struggle between the rich and poor countries, with population "control" a central issue.

Three studies make valuable reading, each of which takes a different position (one at each extreme of the debate and a third one that suggests moderate climate change). S. Fred Singer's *Hot Talk, Cold Science: Global Warming's Unfinished Debate* (Independent Institute, 1998) enhances the author's reputation as one of the leading critics of global warming's reported adverse consequences. At the other extreme, John Houghton, in *Global Warming: The Complete Briefing,* 2d ed. (Cambridge University Press, 1997), accepts global warming as a significant concern and describes how it can be reversed in the future. Finally, S. George Philander, in *Is the Temperature Rising? The Uncertain Science of Global Warming* (Princeton University Press, 1998), concludes that global temperatures will rise 2°C over several decades, creating the prospect of some regional climate changes with major consequences.

ISSUE 10

Is the Threat of a Global Water Shortage Real?

YES: Mark W. Rosegrant, Ximing Cai, and Sarah A. Cline, from "Global Water Outlook to 2025: Averting an Impending Crisis," A Report of the International Food Policy Research Institute and the International Water Management Institute (September 2002)

NO: Bjørn Lomborg, from *The Skeptical Environmentalist: Measuring the Real State of the World* (Cambridge University Press, 2001)

ISSUE SUMMARY

YES: Mark W. Rosegrant, Ximing Cai, and Sarah A. Cline, researchers at the International Food Policy Research Institute, conclude that if current water policies continue, farmers will soon find it difficult to grow sufficient food to meet the world's needs.

NO: Associate professor of statistics Bjørn Lomborg maintains that water is not only plentiful but is a renewable resource that, if properly treated as valuable, should not pose a future problem.

$\mathbf{W}$ater shortages and other water problems are occurring with greater frequency, particularly in large cities. Some observers have speculated that the situation is reminiscent of the fate that befell ancient glorious cities like Rome. Recognition that the supply of water is growing problematic is not new. As early as 1964, the United Nations Environment Programme (UNEP) indicated that close to 1 billion people were at risk from desertification. At the 1992 Earth Summit in Rio de Janeiro, world leaders reaffirmed that desertification was of serious concern.

Moreover, in conference after conference and in study after study, increasing population growth and declining water supplies and quality are being linked together, as is the relationship between the planet's ability to meet its growing food needs and available water. Lester R. Brown, in "Water Deficits Growing in Many Countries: Water Shortages May Cause Food Shortages," *Eco-Economy Update 2002–11* (August 6, 2002), sums up the situation this way: "The world is incurring a vast water deficit. It is largely invisible, historically recent, and growing fast." The World Water Council's study "World Water Actions Report, Third Draft" (October 31, 2002) describes the problem in much the same way: "Water is no longer taken for granted as a plentiful resource that will always be available when we need it." The report continues with the observation that increasing numbers of people in more and more countries are discovering that water is a "limited resource that must be carefully managed for the benefit of people and the environment, in the present and for the future." Some scholars are now arguing that a water shortage is likely to become the twenty-first century's analogue to the oil crisis of the last half of the previous century. The one major difference is that water is not like oil; there is no substitute.

Proclamations of impending water problems abound. Peter H. Gleick, in *The World's Water 1998–1999: The Biennial Report on Freshwater Resources* (Island Press, 1998), reports that the demand for freshwater increased sixfold between 1900 and 1995, twice the rate of population growth. The UN study "United Nations Comprehensive Assessment of Freshwater Resources of the World" (1997) suggests that one-third of the world's population lives in countries that have medium to high water stress. News reports released by the UN Food and Agricultural Organization in conjunction with World Food Day 2002 asserted that water scarcity could result in millions of people having inadequate access to clean water or sufficient food. And the World Meteorological Organization predicts that two out of every three people will live in water-stressed conditions by 2050 if consumption patterns remain the same.

Sandra Postel, in *Pillar of Sand: Can the Irrigation Miracle Last?* (W. W. Norton, 1999), suggests another variant of the water problem. For her, irrigation, the time-tested method of maximizing water usage in the past, may not be feasible as world population marches toward 7 billion. She points to the inadequacy of surface water supplies, increasing depletion of groundwater supplies, the salinization of the land, and the conversion of traditional agricultural land to other uses as reasons why irrigation is unlikely to be a continuing panacea. Yet the 1997 UN study concluded that annual irrigation use would need to increase 30 percent for annual food production to double, which would be necessary for meeting food demands in 2025.

The issue of water quality is also in the news. The World Health Organization reports that in some parts of the world, up to 80 percent of all transmittable diseases are attributable to the consumption of contaminated water. Also, a UNEP-sponsored study, *Global Environment Outlook 2000,* reported that 200 scientists from 50 countries pointed to the shortage of clean water as one of the most pressing global issues.

In the following selection, Mark W. Rosegrant, Ximing Cai, and Sarah A. Cline project that by 2025, water scarcity will result in annual global losses of 350 million metric tons of grain, which is equivalent to approximately the entire U.S. crop. In the second selection, Bjørn Lomborg takes issue with the prevailing wind in the global water debate. His argument can be summed up in his simple statement, "Basically we have sufficient water." Lomborg maintains that water supplies rose during the twentieth century and that we have gained access to more water through technology.

**Mark W. Rosegrant, Ximing Cai,
and Sarah A. Cline**

Global Water Outlook to 2025:
Averting an Impending Crisis

Introduction

Demand for the world's increasingly scarce water supply is rising rapidly, challenging its availability for food production and putting global food security at risk. Agriculture, upon which a burgeoning population depends for food, is competing with industrial, household, and environmental uses for this scarce water supply. Even as demand for water by all users grows, groundwater is being depleted, other water ecosystems are becoming polluted and degraded, and developing new sources of water is getting more costly.

A Thirsty World

Water development underpins food security, people's livelihoods, industrial growth, and environmental sustainability throughout the world. In 1995 the world withdrew 3,906 cubic kilometers (km³) of water for these purposes (Figure 1). By 2025 water withdrawal for most uses (domestic, industrial, and livestock) is projected to increase by at least 50 percent. This will severely limit irrigation water withdrawal, which will increase by only 4 percent, constraining food production in turn.

About 250 million hectares are irrigated worldwide today, nearly five times more than at the beginning of the 20th century. Irrigation has helped boost agricultural yields and outputs and stabilize food production and prices. But growth in population and income will only increase the demand for irrigation water to meet food production requirements (Figure 2). Although the achievements of irrigation have been impressive, in many regions poor irrigation management has markedly lowered groundwater tables, damaged soils, and reduced water quality.

Water is also essential for drinking and household uses and for industrial production. Access to safe drinking water and sanitation is critical to maintain health, particularly for children. But more than 1 billion people across the globe lack enough safe water to meet minimum levels of health and income.

Figure 1

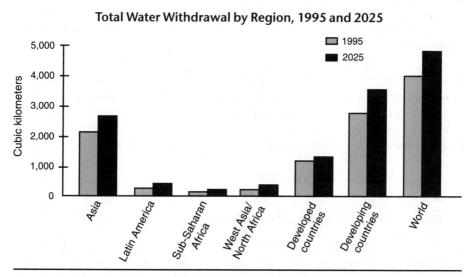

Total Water Withdrawal by Region, 1995 and 2025

Source: Authors' estimates and IMPACT-WATER projections, June 2002.
Note: Projections for 2025 are for the business as usual scenario.

Figure 2

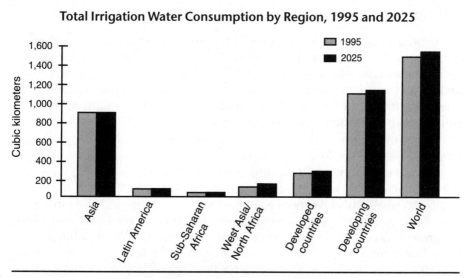

Total Irrigation Water Consumption by Region, 1995 and 2025

Source: Authors' estimates and IMPACT-WATER projections, June 2002.
Note: Projections for 2025 are for the business as usual scenario.

Although the domestic and industrial sectors use far less water than agriculture, the growth in water consumption in these sectors has been rapid. Globally, withdrawals for domestic and industrial uses quadrupled between 1950 and 1995, compared with agricultural uses, for which withdrawals slightly more than doubled.[1]

Water is integrally linked to the health of the environment. Water is vital to the survival of ecosystems and the plants and animals that live in them, and in turn ecosystems help to regulate the quantity and quality of water. Wetlands retain water during high rainfall, release it during dry periods, and purify it of many contaminants. Forests reduce erosion and sedimentation of rivers and recharge groundwater. The importance of reserving water for environmental purposes has only recently been recognized: during the 20th century, more than half of the world's wetlands were lost.[2]

Alternative Futures for Water

The future of water and food is highly uncertain. Some of this uncertainty is due to relatively uncontrollable factors such as weather. But other critical factors can be influenced by the choices made collectively by the world's people. These factors include income and population growth, investment in water infrastructure, allocation of water to various uses, reform in water management, and technological changes in agriculture. Policy decisions—and the actions of billions of individuals—determine these fundamental, long-term drivers of water and food supply and demand.

To show the very different outcomes that policy choices produce, we present three alternative futures for global water.[3] . . .

Business As Usual Scenario

In the business as usual scenario current trends in water and food policy, management, and investment remain as they are. International donors and national governments, complacent about agriculture and irrigation, cut their investments in these sectors. Governments and water users implement institutional and management reforms in a limited and piecemeal fashion. These conditions leave the world ill prepared to meet major challenges to the water and food sectors.

Over the coming decades the area of land devoted to cultivating food crops will grow slowly in most of the world because of urbanization, soil degradation, and slow growth in irrigation investment, and because a high proportion of arable land is already cultivated. Moreover, steady or declining real prices for cereals will make it unprofitable for farmers to expand harvested area. As a result, greater food production will depend primarily on increases in yield. Yet growth in crop yields will also diminish because of falling public investment in agricultural research and rural infrastructure. Moreover, many of the actions that produced yield gains in recent decades, such as increasing the density of crop planting, introducing strains that are more responsive to fertilizer, and improving management practices, cannot easily be repeated.

In the water sector, the management of river basin and irrigation water will become more efficient, but slowly. Governments will continue to transfer management of irrigation systems to farmer organizations and water-user associations. Such transfers will increase water efficiency when they are built upon existing patterns of cooperation and backed by a supportive policy and legal environment. But these conditions are often lacking.

In some regions farmers will adopt more efficient irrigation practices. Economic incentives to induce more efficient water management, however, will still face political opposition from those concerned about the impact of higher water prices on farmers' income and from entrenched interests that benefit from existing systems of allocating water. Water management will also improve slowly in rainfed agriculture as a result of small advances in water harvesting, better on-farm management techniques, and the development of crop varieties with shorter growing seasons.

Public investment in expanding irrigation systems and reservoir storage will decline as the financial, environmental, and social costs of building new irrigation systems escalate and the prices of cereals and other irrigated crops drop. Nevertheless, where benefits outweigh costs, many governments will construct dams, and reservoir water for irrigation will increase moderately.

With slow growth in irrigation from surface water, farmers will expand pumping from groundwater, which is subject to low prices and little regulation. Regions that currently pump groundwater faster than aquifers can recharge, such as the western United States, northern China, northern and western India, Egypt, and West Asia and North Africa, will continue to do so.

The cost of supplying water to domestic and industrial users will rise dramatically. Better delivery and more efficient home water use will lead to some increase in the proportion of households connected to piped water. Many households, however, will remain unconnected. Small price increases for industrial water, improvements in pollution control regulation and enforcement, and new industrial technologies will cut industrial water use intensity (water demand per $1,000 of gross domestic product). Yet industrial water prices will remain relatively low and pollution regulations will often be poorly enforced. Thus, significant potential gains will be lost.

Environmental and other interest groups will press to increase the amount of water allocated to preserving wetlands, diluting pollutants, maintaining riparian flora and other aquatic species, and supporting tourism and recreation. Yet because of competition for water for other uses, the share of water devoted to environmental uses will not increase.

The Water Situation

Almost all users will place heavy demands on the world's water supply under the business as usual scenario. Total global water withdrawals in 2025 are projected to increase by 22 percent above 1995 withdrawals, to 4,772 km^3 (see Figure 1).[4] Projected withdrawals in developing countries will increase 27 percent over the 30-year period, while developed-country withdrawals will increase by 11 percent.[5]

Together, consumption of water for domestic, industrial, and livestock uses—that is, all nonirrigation uses—will increase dramatically, rising by 62 percent from 1995 to 2025. Because of rapid population growth and rising per capita water use, total domestic consumption will increase by 71 percent, of which more than 90 percent will be in developing countries. Conservation and technological improvements will lower per capita domestic water use in developed countries with the highest per capita water consumption.

Industrial water use will grow much faster in developing countries than in developed countries. In 1995 industries in developed countries consumed much more water than industries in the developing world. By 2025, however, developing-world industrial water demand is projected to increase to 121 km^3, 7 km^3 greater than in the developed world. The intensity of industrial water use will decrease worldwide, especially in developing countries (where initial intensity levels are very high), thanks to improvements in water-saving technology and demand policy. Nonetheless, the sheer size of the increase in the world's industrial production will still lead to an increase in total industrial water demand.

Direct water consumption by livestock is very small compared with other sectors. But the rapid increase of livestock production, particularly in developing countries, means that livestock water demand is projected to increase 71 percent between 1995 and 2025. Whereas livestock water demand will increase only 19 percent in the developed world between 1995 and 2025, it is projected to more than double in the developing world, from 22 to 45 km^3.

Although irrigation is by far the largest user of the world's water, use of irrigation water is projected to rise much more slowly than other sectors. For irrigation water, we have computed both potential demand and actual consumption. Potential demand is the demand for irrigation water in the absence of any water supply constraints, whereas actual consumption of irrigation water is the realized water demand, given the limitations of water supply for irrigation. The proportion of potential demand that is realized in actual consumption is the irrigation water supply reliability index (IWSR).[6] An IWSR of 1.0 would mean that all potential demand is being met.

Potential irrigation demand will grow by 12 percent in developing countries, while it will actually decline in developed countries by 1.5 percent. The fastest growth in potential demand for irrigation water will occur in Sub-Saharan Africa, with an increase of 27 percent, and in Latin America, with an increase of 21 percent. Each of these regions has a high percentage increase in irrigated area from a relatively low 1995 level. India is projected to have the highest absolute growth in potential irrigation water demand, 66 km^3 (17 percent), owing to relatively rapid growth in irrigated area from an already high level in 1995. West Asia and North Africa will increase by 18 percent (28 km^3, mainly in Turkey), while China will experience a much smaller increase of 4 percent (12 km^3). In Asia as a region, potential irrigation water demand will increase by 8 percent (100 km^3).

Water scarcity for irrigation will intensify, with actual consumption of irrigation water worldwide projected to grow more slowly than potential consumption, increasing only 4 percent between 1995 and 2025. In developing countries a declining fraction of potential demand will be met over time. The IWSR for developing countries will decline from 0.81 in 1995 to 0.75 in 2025, and in dry river basins the decline will be steeper. For example, in the Haihe River Basin in China, which is an important wheat and maize producer and serves major metropolitan areas, the IWSR is projected to decline from 0.78 to 0.62, and in the Ganges of India, the IWSR will decline from 0.83 to 0.67.

In the developed world, the situation is the reverse: the supply of irrigation water is projected to grow faster than potential demand (although certain basins will face increasing water scarcity). Increases in river basin efficiency will more than offset the very small increase in irrigated area. As a result, after initially declining from 0.87 to 0.85 in 2010, the IWSR will improve to 0.90 in 2025 thanks to slowing growth of domestic and industrial demand (and actual declines in total domestic and industrial water use in the United States and Europe) and more efficient use of irrigation water. . . .

Water Crisis Scenario

A moderate worsening of many of the current trends in water and food policy and in investment could build to a genuine water crisis. In the water crisis scenario, government budget problems worsen. Governments further cut their spending on irrigation systems and accelerate the turnover of irrigation systems to farmers and farmer groups but without the necessary reforms in water rights. Attempts to fund operations and maintenance in the main water system, still operated by public agencies, cause water prices to irrigators to rise. Water users fight price increases, and conflict spills over to local management and cost-sharing arrangements. Spending on the operation and maintenance of secondary and tertiary systems falls dramatically, and deteriorating infrastructure and poor management lead to falling water use efficiency. Likewise, attempts to organize river basin organizations to coordinate water management fail because of inadequate funding and high levels of conflict among water stakeholders within the basin.

National governments and international donors will reduce their investments in crop breeding for rainfed agriculture in developing countries, especially for staple crops such as rice, wheat, maize, other coarse grains, potatoes, cassava, yams, and sweet potatoes. Private agricultural research will fail to fill the investment gap for these commodities. This loss of research funding will lead to further declines in productivity growth in rainfed crop areas, particularly in more marginal areas. In search of improved incomes, people will turn to slash-and-burn agriculture, thereby deforesting the upper watersheds of many basins. Erosion and sediment loads in rivers will rise, in turn causing faster sedimentation of reservoir storage. People will increasingly encroach on wetlands for both land and water, and the integrity and health of aquatic ecosystems will be compromised. The amount of water reserved for environmental purposes will decline as unregulated and illegal withdrawals increase.

The cost of building new dams will soar, discouraging new investment in many proposed dam sites. At other sites indigenous groups and nongovernmental organizations (NGOs) will mount opposition, often violent, over the environmental and human impacts of new dams. These protests and high costs will virtually halt new investment in medium and large dams and storage reservoirs. Net reservoir storage will decline in developing countries and remain constant in developed countries.

In the attempt to get enough water to grow their crops, farmers will extract increasing amounts of groundwater for several years, driving down water tables. But because of the accelerated pumping, after 2010 key aquifers in northern China, northern and northwestern India, and West Asia and North Africa will begin to fail. With declining water tables, farmers will find the cost of extracting water too high, and a big drop in groundwater extraction from these regions will further reduce water availability for all uses.

As in the business as usual scenario, the rapid increase in urban populations will quickly raise demand for domestic water. But governments will lack the funds to extend piped water and sewage disposal to newcomers. Governments will respond by privatizing urban water and sanitation services in a rushed and poorly planned fashion. The new private water and sanitation firms will be undercapitalized and able to do little to connect additional populations to piped water. An increasing number and percentage of the urban population must rely on high-priced water from vendors or spend many hours fetching often-dirty water from standpipes and wells.

The Water Situation

The developing world will pay the highest price for the water crisis scenario. Total worldwide water consumption in 2025 will be 261 km^3 higher than under the business as usual scenario—a 13 percent increase—but much of this water will be wasted, of no benefit to anyone. Virtually all of the increase will go to irrigation, mainly because farmers will use water less efficiently and withdraw more water to compensate for water losses. The supply of irrigation water will be less reliable, except in regions where so much water is diverted from environmental uses to irrigation that it compensates for the lower water use efficiency.

For most regions, per capita demand for domestic water will be significantly lower than under the business as usual scenario, in both rural and urban areas. The result is that people will not have access to the water they need for drinking and sanitation. The total domestic demand under the water crisis scenario will be 162 km^3 in developing countries, 28 percent less than under business as usual; 64 km^3 in developed countries, 7 percent less than under business as usual; and 226 km^3 in the world, 23 percent less than under business as usual.

Demand for industrial water, on the other hand, will increase, owing to failed technological improvements and economic measures. In 2025 the total industrial water demand worldwide will be 80 km^3 higher than under the business as usual scenario—a 33 percent rise—without generating additional industrial production.

With water diverted to make up for less efficient water use, the water crisis scenario will hit environmental uses particularly hard. Compared with business as usual, environmental flows will drop significantly by 2025, with 380 km^3 less environmental flow in the developing world, 80 km^3 less in the developed world, and 460 km^3 less globally. . . .

Sustainable Water Scenario

A sustainable water scenario would dramatically increase the amount of water allocated to environmental uses, connect all urban households to piped water, and achieve higher per capita domestic water consumption, while maintaining food production at the levels described in the business as usual scenario. It would achieve greater social equity and environmental protection through both careful reform in the water sector and sound government action.

Governments and international donors will increase their investments in crop research, technological change, and reform of water management to boost water productivity and the growth of crop yields in rainfed agriculture. Accumulating evidence shows that even drought-prone and high-temperature rainfed environments have the potential for dramatic increases in yield. Breeding strategies will directly target these rainfed areas. Improved policies and increased investment in rural infrastructure will help link remote farmers to markets and reduce the risks of rainfed farming.

To stimulate water conservation and free up agricultural water for environmental, domestic, and industrial uses, the effective price of water to the agricultural sector will be gradually increased. Agricultural water price increases will be implemented through incentive programs that provide farmers income for the water that they save, such as charge-subsidy schemes that pay farmers for reducing water use, and through the establishment, purchase, and trading of water use rights. By 2025 agricultural water prices will be twice as high in developed countries and three times as high in developing countries as in the business as usual scenario. The government will simultaneously transfer water rights and the responsibility for operation and management of irrigation systems to communities and water user associations in many countries and regions. The transfer of rights and systems will be facilitated with an improved legal and institutional environment for preventing and eliminating conflict and with technical and organizational training and support. As a result, farmers will increase their on-farm investments in irrigation and water management technology, and the efficiency of irrigation systems and basin water use will improve significantly.

River basin organizations will be established in many water-scarce basins to allocate mainstream water among stakeholder interests. Higher funding and reduced conflict over water, thanks to better water management, will facilitate effective stakeholder participation in these organizations.

Farmers will be able to make more effective use of rainfall in crop production, thanks to breakthroughs in water harvesting systems and the adoption of advanced farming techniques, like precision agriculture, contour plowing, precision land leveling, and minimum-till and no-till technologies. These technologies will increase the share of rainfall that goes to infiltration and evapotranspiration.

Spurred by the rapidly escalating costs of building new dams and the increasingly apparent environmental and human resettlement costs, developing and developed countries will reassess their reservoir construction plans, with comprehensive analysis of the costs and benefits, including environmental and social effects, of proposed projects. As a result, many planned storage projects will be canceled, but others will proceed with support from civil society groups. Yet new storage capacity will be less necessary because rapid growth in rainfed crop yields will help reduce rates of reservoir sedimentation from erosion due to slash-and-burn cultivation.

Policy toward groundwater extraction will change significantly. Market-based approaches will assign rights to groundwater based on both annual withdrawals and the renewable stock of groundwater. This step will be combined with stricter regulations and better enforcement of these regulations. Groundwater overdrafts will be phased out in countries and regions that previously pumped groundwater unsustainably.

Domestic and industrial water use will also be subject to reforms in pricing and regulation. Water prices for connected households will double, with targeted subsidies for low-income households. Revenues from price increases will be invested to reduce water losses in existing systems and to extend piped water to previously unconnected households. By 2025 all households will be connected. Industries will respond to higher prices, particularly in developing countries, by increasing in-plant recycling of water, which reduces consumption of water.

With strong societal pressure for improved environmental quality, allocations for environmental uses of water will increase. Moreover, the reforms in agricultural and nonagricultural water sectors will reduce pressure on wetlands and other environmental uses of water. Greater investments and better water management will improve the efficiency of water use, leaving more water instream for environmental purposes. All reductions in domestic and urban water use, due to higher water prices, will be allocated to instream environmental uses.

The Water Situation

In the sustainable water scenario the world consumes less water but reaps greater benefits than under business as usual, especially in developing countries. In 2025 total worldwide water consumption is 408 km^3, or 20 percent, lower under the sustainable scenario than under business as usual. This reduction in consumption frees up water for environmental uses. Higher water prices and higher water use efficiency reduces consumption of irrigation water by 296 km^3 compared with business as usual. The reliability of irrigation water supply is reduced slightly in the sustainable scenario compared with business as usual, because this scenario places a high priority on environmental flows. Over time, however, more efficient water use in this scenario counterbalances the transfer of water to the environment and results in an improvement in the reliability of supply of irrigation water by 2025.

This scenario will improve the domestic water supply through universal access to piped water for rural and urban households. Globally, potential domestic water demand under the sustainable water scenario will decrease 9 percent compared with business as usual, owing to higher water prices. However, potential per capita domestic demand for connected households in rural areas will be 12 percent higher than that under business as usual in the developing world, and 5 percent higher in the developed world. This increase is accomplished by expanding universal access to piped water in rural areas even with higher prices for water. And in urban areas, potential per capita water consumption for poor households sharply improves through connection to piped water, while the initially connected households reduce consumption in response to higher prices and improved water-saving technology.

Through technological improvements and effective economic incentives, the sustainable water scenario will reduce industrial water demand. In 2025 total industrial water demand worldwide under the sustainable scenario will be 85 km^3, or 35 percent, lower than under business as usual.

The environment is a major beneficiary of the sustainable water scenario, with large increases in the amount of water reserved for wetlands, instream flows, and other environmental purposes. Compared with the business as usual scenario, the sustainable scenario will also result in an increase in the environmental flow of 850 km^3 in the developing world, 180 km^3 in the developed world, and 1,030 km^3 globally. This is the equivalent of transferring 22 percent of global water withdrawals under business as usual to environmental purposes.

Notes

1. W. J. Cosgrove and F. Rijsberman, *World Water Vision: Making Water Everybody's Business* (London: World Water Council and World Water Vision and Earthscan, 2000); I. A. Shiklomanov, "Electronic Data Provided to the Scenario Development Panel, World Commission on Water for the 21st Century" (State Hydrological Institute, St. Petersburg, Russia, 1999), mimeo.

2. E. Bos and G. Bergkamp, "Water and the Environment," in *Overcoming Water Scarcity and Quality Constraints,* 2020 Focus 9, ed. R. S. Meinzen-Dick and M. W. Rosegrant (Washington, D.C.: International Food Policy Research Institute, 2001).

3. The business as usual, crisis, and sustainable scenarios are compared using average 2025 results generated from 30 hydrologic scenarios. The other scenarios are compared with business as usual based on a single 30-year hydrologic sequence drawn from 1961–90, and results are shown as the average of the years 2021–25.

4. Water demand can be defined and measured in terms of withdrawals and actual consumption. While water withdrawal is the most commonly estimated figure, consumption best captures actual water use, and most of our analysis will utilize this concept.

5. The global projection is broadly consistent with other recent projections to 2025, including the 4,580 km^3 in the medium scenario of J. Alcamo, P. Döll, F. Kaspar, and S. Sieberg, *Global Change and Global Scenarios of Water Use and Availability: An Application of Water GAP 1.0* (Kassel, Germany: Center for Environmental System Research, University of Kassel, 1998), the 4,569 km^3 in the "business-as-usual" scenario of D. Seckler, U. Amarasinghe, D. Molden, S. Rhadika, and R. Barker, *World Water Demand and Supply, 1990 to 2025: Scenarios and Issues,* Research Report Number 19 (Colombo, Sri Lanka: International

Water Management Institute, 1998), and the forecast of 4,966 km^3 (not including reservoir evaporation) of Shiklomanov, "Electronic Data."

6.   Compared with other sectors, the growth of irrigation water potential demand is much lower, with 12 percent growth in potential demand between 1995 and 2025 in developing countries and a slight decline in potential demand in developed countries.

NO

Bjørn Lomborg

Water

There is a resource which we often take for granted but which increasingly has been touted as a harbinger of future trouble. Water.

Ever more people live on Earth and they use ever more water. Our water consumption has almost quadrupled since 1940. The obvious argument runs that "this cannot go on." This has caused government agencies to worry that "a threatening water crisis awaits just around the corner." The UN environmental report *GEO 2000* claims that the water shortage constitutes a "full-scale emergency," where "the world water cycle seems unlikely to be able to cope with the demands that will be made of it in the coming decades. Severe water shortages already hamper development in many parts of the world, and the situation is deteriorating."

The same basic argument is invoked when WWF [World Wildlife Fund] states that "freshwater is essential to human health, agriculture, industry, and natural ecosystems, but is now running scarce in many regions of the world." *Population Reports* states unequivocally that "freshwater is emerging as one of the most critical natural resource issues facing humanity." Environmental discussions are replete with buzz words like "water crisis" and "time bomb: water shortages," and *Time* magazine summarizes the global water outlook with the title "Wells running dry." The UN organizations for meteorology and education simply refer to the problem as "a world running out of water."

The water shortages are also supposed to increase the likelihood of conflicts over the last drops—and scores of articles are written about the coming "water wars." Worldwatch Institute sums up the worries nicely, claiming that "water scarcity may be to the nineties what the oil price shocks were to the seventies—a source of international conflicts and major shifts in national economies."

But these headlines are misleading. True, there may be *regional* and *logistic* problems with water. We will need to get better at using it. But basically we have sufficient water.

How Much Water in the World?

Water is absolutely decisive for human survival, and the Earth is called the Blue Planet precisely because most of it is covered by water: 71 percent of the Earth's surface is covered by water, and the total amount is estimated at the unfathomably large 13.6 billion cubic kilometers. Of all this water, oceans make up 97.2 percent and the polar ice contains 2.15 percent. Unfortunately sea water is too saline for direct human consumption, and while polar ice contains potable water it is hardly within easy reach. Consequently, humans are primarily dependent on the last 0.65 percent water, of which 0.62 percent is groundwater.

Fresh water in the groundwater often takes centuries or millennia to build up—it has been estimated that it would require 150 years to recharge all of the groundwater in the United States totally to a depth of 750 meters if it were all removed. Thus, thoughtlessly exploiting the groundwater could be compared to mining any other non-renewable natural resource. But groundwater is continuously replenished by the constant movement of water through oceans, air, soil, rivers, and lakes in the so-called hydrological cycle. The sun makes water from the oceans evaporate, the wind moves parts of the vapor as clouds over land, where the water is released as rain and snow. The precipitated water then either evaporates again, flows back into the sea through rivers and lakes, or finds its way into the groundwater.

The total amount of precipitation on land is about 113,000 km^3, and taking into account an evaporation of 72,000 km^3 we are left with a net fresh water influx of 41,000 km^3 each year or the equivalent of 30 cm (1 foot) of water across the entire land mass. Since part of this water falls in rather remote areas, such as the basins of the Amazon, the Congo, and the remote North American and Eurasian rivers, a more reasonable, geographically accessible estimate of water is 32,900 km^3. Moreover, a large part of this water comes within short periods of time. In Asia, typically 80 percent of the runoff occurs from May to October, and globally the flood runoff is estimated at about three-quarters of the total runoff. This leaves about 9,000 km^3 to be captured. Dams capture an additional 3,500 km^3 from floods, bringing the total accessible runoff to 12,500 km^3. This is equivalent to about 5,700 liters of water for every single person on Earth *every single day.* For comparison, the average citizen in the EU uses about 566 liters of water per day. This is about 10 percent of the global level of available water and some 5 percent of the available EU water. An American, however, uses about three times as much water, or 1,442 liters every day.

Looking at global water consumption, as seen in Figure 1, it is important to distinguish between water withdrawal and water use. Water withdrawal is the amount of water physically removed, but this concept is less useful in a discussion of limits on the total amount of water, since much of the withdrawn water is later returned to the water cycle. In the EU and the US, about 46 percent of the withdrawn water is used merely as cooling water for power generation and is immediately released for further use downstream. Likewise, most industrial uses return 80–90 percent of the water, and even in irrigation 30–70 percent of the water runs back into lakes and rivers or percolates into aquifers, whence it can be reused. Thus, a more useful measure of water consumption is the amount of water this consumption causes to be irretrievably lost through evaporation or transpiration from plants. This is called water use.

Figure 1

Global, Annual Water Withdrawal and Use, in Thousand km³ and Percentage of Accessible Runoff, 1900–95, and Predictions for 2025

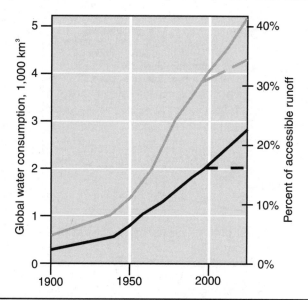

Source: Shiklomanov 2000:22 (high prediction), World Water Council 2000:26 (low prediction).

Over the twentieth century, Earth's water use has grown from about 330 km³ to about 2,100 km³. As can be seen from Figure 1 there is some uncertainty about the future use and withdrawal (mainly depending on the development of irrigation), but until now most predictions have tended to overestimate the actual water consumption by up to 100 percent. Nevertheless, total use is still less than 17 percent of the accessible water and even with the high prediction it will require just 22 percent of the readily accessible, annually renewed water in 2025.

At the same time, we have gained access to more and more water, as indicated in Figure 2. Per person we have gone from using about 1,000 liters per day to almost 2,000 liters over the past 100 years. Particularly, this is due to an approximately 50 percent increase in water use in agriculture, allowing irrigated farms to feed us better and to decrease the number of starving people. Agricultural water usage seems, however, to have stabilized below 2,000 liters per capita, mainly owing to higher efficiency and less water consumption in agriculture since 1980. This pattern is also found in the EU and the US, where consumption has increased dramatically over the twentieth century, but is now leveling off. At the same time, personal consumption (approximated by the municipal withdrawal) has more than quadrupled over the century, reflecting an increase in welfare with more easily accessible water. In developing countries, this is in large part a question of health—avoiding sickness through better access to clean drinking water and sanitation, whereas in developed countries

Figure 2

Global Withdrawal of Water for Agriculture, Industry and Municipal Use, and Total Use, in Liters and Gallons Per Capita Per Day, 1900–95

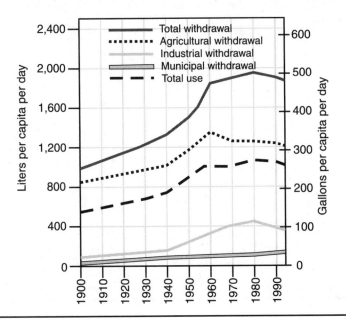

Source: Shiklomanov 2000:24.

higher water use is an indication of an increased number of domestic amenities such as dishwashers and better-looking lawns.

So, if the global use is less than 17 percent of the readily accessible and renewable water and the increased use has brought us more food, less starvation, more health and increased wealth, why do we worry?

The Three Central Problems

There are three decisive problems. First, precipitation is by no means equally distributed all over the globe. This means that not all have equal access to water resources and that some countries have much less accessible water than the global average would seem to indicate. The question is whether water shortages are already severe in some places today. Second, there will be more and more people on Earth. Since precipitation levels will remain more or less constant this will mean fewer water resources for each person. The question is whether we will see more severe shortages in the future. Third, many countries receive a large part of their water resources from rivers; 261 river systems, draining just less than half of the planet's land area, are shared by two or more countries, and at least ten rivers flow through half a dozen or more countries. Most Middle Eastern countries share aquifers. This means that the water question also has an international perspective and—if cooperation breaks down—an international conflict potential.

Beyond these three problems there are two other issues, which are often articulated in connection with the water shortage problem, but which are really conceptually quite separate. One is the worry about water pollution, particularly of potable water. While it is of course important to avoid water pollution in part because pollution restricts the presently available amount of freshwater, it is not related to the problem of water shortage *per se*. . . .

The second issue is about the shortage of *access* to water in the Third World. . . . This problem, while getting smaller, is still a major obstacle for global welfare. In discussing water shortage, reference to the lack of universal access to drinking water and sanitation is often thrown in for good measure, but of course this issue is entirely separate from the question of shortages. First, the cause is *not* lack of water (since human requirements constitute just 50–100 liters a day which any country but Kuwait can deliver, cf. Table 1) but rather a lack of investment in infrastructure. Second, the solution lies not in cutting back on existing consumption but actually in increasing future consumption.

Finally, we should just mention global warming . . . and its connection to water use. Intuitively, we might be tempted to think that a warmer world would mean more evaporation, less water, more problems. But more evaporation also means more precipitation. Essentially, global climate models seem to change *where* water shortages appear (pushing some countries above or below the threshold) but the total changes are small (1–5 percent) and go both ways.

Not Enough Water?

Precipitation is not distributed equally. Some countries such as Iceland have almost 2 million liters of water for each inhabitant every day, whereas Kuwait must make do with just 30 liters. The question, of course, is when does a country not have *enough* water.

It is estimated that a human being needs about 2 liters of water a day, so clearly this is not the restrictive requirement. The most common approach is to use the so-called *water stress index* proposed by the hydrologist Malin Falkenmark. This index tries to establish an approximate minimum level of water per capita to maintain an adequate quality of life in a moderately developed country in an arid zone. This approach has been used by many organizations including the World Bank, in the standard literature on environmental science, and in the water scarcity discussion in *World Resources*. With this index, human beings are assessed to need about 100 liters per day for drinking, household needs and personal hygiene, and an additional 500–2,000 liters for agriculture, industry and energy production. Since water is often most needed in the dry season, the water stress level is then set even higher—if a country has less than 4,660 liters per person available it is expected to experience periodic or regular water stress. Should the accessible runoff drop to less than 2,740 liters the country is said to experience chronic water scarcity. Below 1,370 liters, the country experiences absolute water scarcity, outright shortages and acute scarcity.

Table 1 shows the 15 countries comprising 3.7 percent of humanity in 2000 suffering chronic water scarcity according to the above definition. Many

Table 1

Countries With Chronic Water Scarcity (Below 2,740 Liters Per Capita Per Day) in 2000, 2025, and 2050, Compared to a Number of Other Countries

Available water, liters per capita per day	2000	2025	2050
Kuwait	30	20	17
United Arab Emirates	174	129	116
Libya	275	136	92
Saudi Arabia	325	166	118
Jordan	381	203	145
Singapore	471	401	403
Yemen	665	304	197
Israel	969	738	644
Oman	1,077	448	268
Tunisia	1,147	834	709
Algeria	1,239	827	664
Burundi	1,496	845	616
Egypt	2,343	1,667	1,382
Rwanda	2,642	1,562	1,197
Kenya	2,725	1,647	1,252
Morocco	2,932	2,129	1,798
South Africa	2,959	1,911	1,497
Somalia	3,206	1,562	1,015
Lebanon	3,996	2,971	2,533
Haiti	3,997	2,497	1,783
Burkina Faso	4,202	2,160	1,430
Zimbabwe	4,408	2,830	2,199
Peru	4,416	3,191	2,680
Malawi	4,656	2,508	1,715
Ethiopia	4,849	2,354	1,508
Iran, Islamic Rep.	4,926	2,935	2,211
Nigeria	5,952	3,216	2,265
Eritrea	6,325	3,704	2,735
Lesotho	6,556	3,731	2,665
Togo	7,026	3,750	2,596
Uganda	8,046	4,017	2,725
Niger	8,235	3,975	2,573
Percent people with chronic scarcity	**3.7%**	**8.6%**	**17.8%**
United Kingdom	3,337	3,270	3,315
India	5,670	4,291	3,724
China	6,108	5,266	5,140
Italy	7,994	8,836	10,862
United States	24,420	20,405	19,521
Botswana	24,859	15,624	12,122
Indonesia	33,540	25,902	22,401
Bangladesh	50,293	35,855	29,576
Australia	50,913	40,077	37,930
Russian Federation	84,235	93,724	107,725
Iceland	1,660,502	1,393,635	1,289,976

Source: WRI 1998a.

of these countries probably come as no surprise. But the question is whether we are facing a serious problem.

How does Kuwait actually get by with just 30 liters per day? The point is, it doesn't. Kuwait, Libya and Saudi Arabia all cover a large part of their water demand by exploiting the largest water resource of all—through desalination of sea water. Kuwait in fact covers more than half its total use through desalination. Desalting requires a large amount of energy (through either freezing or evaporating water), but all of these countries also have great energy resources. The price today to desalt sea water is down to 50–80 ¢/m^3 and just 20–35 ¢/m^3 for brackish water, which makes desalted water a more expensive resource than fresh water, but definitely not out of reach.

This shows two things. First, we can have sufficient water, if we can pay for it. Once again, this underscores that *poverty* and not the environment is the primary limitation for solutions to our problems. Second, desalination puts an upper boundary on the degree of water problems in the world. In principle, we could produce the Earth's entire present water consumption with a single desalination facility in the Sahara, powered by solar cells. The total area needed for the solar cells would take up less than 0.3 percent of the Sahara.

Today, desalted water makes up just 0.2 percent of all water or 2.4 percent of municipal water. Making desalination cover the total municipal water withdrawal would cost about 0.5 percent of the global GDP. This would definitely be a waste of resources, since most areas have abundant water supplies and all areas have some access to water, but it underscores the upper boundary of the water problem.

Also, there's a fundamental problem when you only look at the total water resources and yet try to answer whether there are sufficient supplies of water. The trouble is that we do not necessarily know *how* and *how wisely* the water is used. Many countries get by just fine with very limited water resources because these resources are exploited very effectively. Israel is a prime example of efficient water use. It achieves a high degree of efficiency in its agriculture, partly because it uses the very efficient drip irrigation system to green the desert, and partly because it recycles household wastewater for irrigation. Nevertheless, with just 969 liters per person per day, Israel should according to the classification be experiencing absolute water scarcity. Consequently, one of the authors in a background report for the 1997 UN document on water points out that the 2,740 liters water bench-mark is "misguidedly considered by some authorities as a critical minimum amount of water for the survival of a modern society."

Of course, the problem of faulty classification increases, the higher the limit is set. The European Environmental Agency (EEA) in its 1998 assessment somewhat incredibly suggested that countries below 13,690 liters per person per day should be classified as "low availability," making not only more than half the EU low on water but indeed more than 70 percent of the globe. Denmark receives 6,750 liters of fresh water per day and is one of the many countries well below this suggested limit and actually close to EEA's "very low" limit. Nevertheless, national withdrawal is just 11 percent of the available water, and it is estimated that the consumption could be almost doubled without negative environmental consequences. The director of the Danish EPA has stated that, "from the hand of nature, Denmark has access to good and clean groundwater far in excess of what we actually use."

By far the largest part of all water is used for agriculture—globally, agriculture uses 69 percent, compared to 23 percent for industry and 8 percent for households. Consequently, the greatest gains in water use come from cutting down on agricultural use. Many of the countries with low water availability therefore compensate by importing a large amount of their grain. Since a ton of grain uses about 1,000 tons of water, this is in effect a very efficient way of importing water. Israel imports about 87 percent of its grain consumption, Jordan 91 percent, Saudi Arabia 50 percent.

Summing up, more than 96 percent of all nations have at present sufficient water resources. On all continents, water accessibility has *increased* per person, and at the same time an ever higher proportion of people have gained access to clean drinking water and sanitation. While water accessibility has been getting *better* this is not to deny that there are still widespread shortages and limitations of basic services, such as access to clean drinking water, and that local and regional scarcities occur. But these problems are primarily related not to physical water scarcity but to a lack of proper water management and in the end often to lack of money—money to desalt sea water or to increase cereal imports, thereby freeing up domestic water resources.

Will It Get Worse in the Future?

The concerns for the water supply are very much concerns that the current problems will become worse over time. As world population grows, and as precipitation remains constant, there will be less water per person, and using Falkenmark's water stress criterion, there will be more nations experiencing water scarcity. In Figure 3 it is clear that the proportion of people in water stressed nations will increase from 3.7 percent in 2000 to 8.6 percent in 2025 and 17.8 percent in 2050.

It is typically pointed out that although more people by definition means more water stress, such "projections are neither forecasts nor predictions." Indeed, the projections merely mean that if we do not improve our handling of water resources, water will become more scarce. But it is unlikely that we will not become better at utilizing and distributing water. Since agriculture takes up the largest part of water consumption, it is also here that the largest opportunities for improving efficiency are to be found. It is estimated that many irrigation systems waste 60–80 percent of all water. Following the example of Israel, drip irrigation in countries as diverse as India, Jordan, Spain and the US has consistently been shown to cut water use by 30–70 percent while increasing yields by 20–90 percent. Several studies have also indicated that industry almost without additional costs could save anywhere from 30 to 90 percent of its water consumption. Even in domestic distribution there is great potential for water savings. EEA estimates that the leakage rates in Europe vary from 10 percent in Austria and Denmark up to 28 percent in the UK and 33 percent in the Czech Republic.

Figure 3

Share of Humanity With Maximum Water Availability in the Year 2000, 2025, and 2050, Using UN Medium Variant Population Data

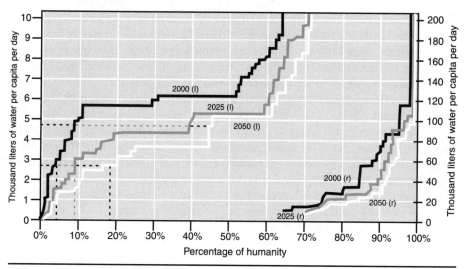

The left side uses the left axis, the right side the right axis.
Source: WRI 1998a.

The problem of water waste occurs because water in many places is not well priced. The great majority of the world's irrigation systems are based on an annual flat rate, and not on charges according to the amount of water consumed. The obvious effect is that participants are not forced to consider whether all in all it pays to use the last liter of water—when you have first paid to be in, water is free. So even if there is only very little private utility from the last liter of water, it is still used because it is free. . . .

This is particularly a problem for the poor countries. The poorest countries use 90 percent of their water for irrigation compared to just 37 percent in the rich countries. Consequently, it will be necessary to redistribute water from agriculture to industry and households, and this will probably involve a minor decline in the potential agricultural production (i.e. a diminished increase in the actual production). The World Bank estimates that this reduction will be very limited and that water redistribution definitely will be profitable for the countries involved. Of course, this will mean increased imports of grain by the most water stressed countries, but a study from the International Water Management Institute indicates that it should be possible to cover these extra imports by extra production in the water abundant countries, particularly the US.

At the same time there are also large advantages to be reaped by focusing on more efficient household water consumption. In Manila 58 percent of all water disappears (lost in distribution or stolen), and in Latin America the figure is about 40 percent. And on average households in the Third World pay only

35 percent of the actual price of water. Naturally, this encourages overconsumption. We know that pricing and metering reduces demand, and that consumers use less water if they have to pay for each unit instead of just paying a flat rate.

Actually, it is likely that more sensible pricing will not only secure future water supplies but also increase the total social efficiency. When agriculture is given cheap or even free water, this often implies a hidden and very large subsidy—in the United States the water subsidy to farmers is estimated to be above 90 percent or $3.5 billion. For the developing countries this figure is even larger: it is estimated that the hidden water subsidy to cities is about $22 billion, and the hidden subsidy to agriculture around $20–25 billion.

Thus, although an increasing population will increase water demands and put extra water stress on almost 20 percent of humanity, it is likely that this scarcity can be solved. Part of the solution will come from higher water prices, which will cut down on inefficient water use. Increased cereal imports will form another part of the solution, freeing up agricultural water to be used in more valuable areas of industry or domestic consumption. Finally, desalting will again constitute a backstop process which can produce virtually unlimited amounts of drinking water given sufficient financial backing.

POSTSCRIPT

Is the Threat of a Global
Water Shortage Real?

The authors of the two selections agree that something must be done—that is, major public policy making must occur—if the future of water is going to be acceptable. They disagree, however, on the urgency of the task and in the level of optimism (or pessimism) that they bring to their analyses of likely success. Rosegrant and his colleagues approach the global water problem from the perspective of its role in food production. They argue that water is one of the main factors that will limit food production in the future. Competition for water comes from many quarters, and farmers find themselves in increasingly competitive situations as they attempt to keep pace with agricultural needs.

Rosegrant and his colleagues' view of past water trends—whether for irrigation or for drinking and household uses—leads them to conclude that the future of water is "highly uncertain" at best. The authors' pessimism stems from the failure of governments to address the issue adequately to date.

Lomborg argues that our ability to find more water has resulted in a global usage rate of 17 percent of the readily accessible and renewable water. The consequence, according to Lomborg, is that the world has "more food, less starvation, more health and increased wealth [so] why do we worry?" He does suggest that there are significant problems, and he identifies three: (1) the unequal distribution of precipitation throughout the globe, (2) increasing global population, and (3) the fact that many countries receive their water through shared river systems. Additionally, water pollution and the shortage of access to water in the developing world are issues, but of a different sort.

Both readings point to a need for aggressive policy action on the part of governments and other actors in the global water regime. This is not to suggest, however, that the world's leaders have been idle. At least 10 major international conferences since 1977 have addressed water issues, resulting in significant action-oriented proposals. For example, the 1992 International Conference on Water and Environment in Dublin, Ireland, established four basic ideas (known as the Dublin Principles). In addition, the 1992 Earth Summit highlighted water as an integral part of the ecosystem. The creation of the World Water Council in 1996 and of the Global Water Partnership the same year are further examples of policy action. But it was the Second World Water Forum, held in The Hague, Netherlands, in March 2000, that was a landmark event in raising global water consciousness. Among the many research reports and documents that emerged from the conference, two stand out: "Vision for Water, Life, and the Environment in the Twenty-First Century" and "Towards Water Security: A Framework for Action."

On the Internet . . .

Worldwide Refugee Information

The Worldwide Refugee Information Web site provides information on the status of refugees and internally displaced persons in 120 countries. Experts from around the world have submitted articles for publication on this site. Search by country or by issue and topic for information.

http://www.refugees.org/world/worldmain.htm

Center for Immigration Studies

The Center for Immigration Studies is a nonpartisan, nonprofit organization that focuses on research and policy analysis of the impacts of immigration on the United States. This site is designed to expand the base of public knowledge about immigration. Included on this site are current articles, statistics, and a brief history of immigration in the United States.

http://www.cis.org

South African Network of Skills Abroad

The South African Network of Skills Abroad links skilled people living abroad who wish to make a contribution to South Africa's economic and social development and connects them with local experts and projects. This site features a brain drain media coverage section that contains a list of and some online links to articles on the subject.

http://sansa.nrf.ac.za

PART 3

The Movement of Peoples

*O*ur ability to travel from one part of the globe to another in a short amount of time has expanded dramatically since the Wright brothers first lifted an airplane off the sand dunes of North Carolina's Outer Banks. As a consequence, the movement of peoples both within and across national boundaries has grown at unprecedented levels.

Some of this movement is voluntary, as individuals willingly leave their homelands in search of better lives. Others are forced to abandon their familiar surroundings in the face of civil war, ethnic cleansing, and the physical deterioration of their land. And, as the events of September 11 have taught us, some people leave their homelands to inflict harm on others.

- Is the Global Community Responding Well to the Plight of Refugees and Displaced Persons?

- Should Immigration to the United States and Other Developed Countries Be Curtailed?

- Is There a Global Brain Drain?

ISSUE 11

Is the Global Community Responding Well to the Plight of Refugees and Displaced Persons?

YES: Roberta Cohen and Francis M. Deng, from "Exodus Within Borders: The Uprooted Who Never Left Home," *Foreign Affairs* (July/August 1998)

NO: David Masci, from "Assisting Refugees," *CQ Researcher* (February 7, 1997)

ISSUE SUMMARY

YES: Roberta Cohen and Francis M. Deng, scholars at the Brookings Institution, contend that the international community should be ready to assist the millions of people who are fleeing persecution and war within their own countries when states fail to protect their own.

NO: David Masci, a staff writer for *CQ Researcher,* argues that there needs to be a reassessment among refugee-aid groups, national governments, the United Nations, and others about whether or not past humanitarian efforts have been appropriate.

One characteristic of the emerging global agenda of the twenty-first century is the host of issues associated with the movement of people, goods, information, ideas, and finances across national boundaries. Some of these issues are solely a product of the global information age, while others have been part of global affairs for some time. Many of the problems in the latter category, although in existence for awhile, have taken on a different meaning and urgency in the new millennium.

One set of issues relates to the forced movement of people, called *displaced persons* (within their homeland) or *refugees* (across national borders). On numerous occasions during the twentieth century, the international community attempted to assist individuals who were trying to flee situations that they deemed intolerable. For example, as early as 1921 the League of Nations, the forerunner to the United Nations, tried to help those who were fleeing the Russian Revolution. Also, during the latter part of World War II, an international relief agency was created to assist people who were displaced by the Western allies as they pushed the Germans eastward away from territory that they had occupied earlier in the war.

It was in the early 1950s, though, with the creation of the United Nations High Commissioner for Refugees (UNHCR), that the UN formally involved itself in the plight of displaced persons. During the cold war, the UN was called upon on several occasions to deal with refugee crises, such as those following U.S. withdrawal from Vietnam and the Soviet invasion of Afghanistan.

With the ending of the cold war in the early 1990s and the subsequent loosening of national borders in the former Soviet bloc of nations and elsewhere, more and more refugee problems emerged—the Kurds in Iraq, Bosnian Muslims in Bosnia and Herzegovina, and Rwandan Hutus in Zaire and Tanzania. Moreover, more people are being displaced within their own countries today because ethnic conflicts have escalated in both number and intensity in the past decade. The refugee situation in central and Western Europe did stabilize in 2001, in that the number of asylum applications submitted across Europe remained around 450,000, as it had been for the previous two years.

But the terrorist attacks of September 11, 2001, promise to affect the refugee situation adversely; thousands are already fleeing areas disrupted by the war on terrorism. The situation promises to become worse as this war intensifies. The American Refugee Committee International estimates that hundreds of thousands of refugees in or near areas of military instability have already offered severe challenges for international agencies.

With the dramatic increase in the number of crises producing refugees and displaced persons as well as the number of refugees and displaced persons themselves, particularly those remaining within their own recognized homelands, a reassessment is beginning to take place within the international community. Should the power structure respond in all cases, no cases, or selected cases? What form should this response take? How can it be more effective than prior efforts? Is the plight of refugees beyond the scope of present-day world organizations?

The two selections for this issue address different aspects of the refugee problem. Roberta Cohen and Francis M. Deng assert that the international community should be ready to act whenever countries are unable to protect their own citizens, no matter what the circumstance. David Masci, on the other hand, argues for a comprehensive review regarding the appropriateness of past efforts by a wide range of international groups and world leaders. He questions whether or not situations have been made worse in the past by international intervention, including the use of soldiers.

**Roberta Cohen and
Francis M. Deng**

 **YES**

Exodus Within Borders:
The Uprooted Who Never Left Home

Tens of millions of people have been forced from their homes during the past decade by armed conflict, internal strife, and systematic violations of human rights, all the while remaining within the borders of their own countries. No continent has been spared. Africa today counts about 10 million internally displaced persons, Europe and Asia some 5 million each, and Latin America up to 2 million. These masses in flight—who, unlike refugees, have not crossed a border—constitute the newest global crisis.

Internal displacement always has severe humanitarian implications. These displaced persons are at the greatest risk of starvation, have the highest rates of preventable disease, and are the most vulnerable to human rights abuses. Internal displacement is a symptom of state dysfunction that poses a threat to political and economic stability at the national and international levels. Both the communities left behind and the towns and villages in which the displaced find refuge are often ravaged. In some cases, so many people flee that whole societies are uprooted. Violence and instability can spread through entire regions, forcing neighboring states to bear the brunt of massive refugee flows. Even countries continents away may have to contend with a wave of desperate refugees.

Today's crisis of internal displacement is no less acute than the refugee crisis that confronted Europe after World War II. Then, humanitarian needs coupled with practical political and economic interests brought about a system of international protection and assistance for those displaced outside their native countries. In 1951, the position of United Nations High Commissioner for Refugees [UNHCR] was created and a U.N. refugee convention adopted. Today, UNHCR has a staff of 5,000, an annual budget of more than $1 billion, and 13.2 million refugees in its care.

But those forced from their homes who remain under their government's jurisdiction are not covered by any international arrangements. Although their numbers now exceed those of refugees, no international institution is specifi-

From Roberta Cohen and Francis M. Deng, "Exodus Within Borders: The Uprooted Who Never Left Home," *Foreign Affairs*, vol. 77, no. 4 (July/August 1998). Copyright © 1998 by The Council on Foreign Relations, Inc. Reprinted by permission.

cally charged with their protection or assistance. The absurdity is that if these people had crossed a border, they would fall under U.N. protection.

A Cold War Legacy

Internal displacement became a subject of international concern in the late 1980s. When the numbers were first compiled in 1982, 1.2 million people were estimated to be displaced in 11 countries. Four years later the total had grown to 14 million. Since the early 1990s, the numbers have fluctuated between 20 million and 25 million in 35 to 40 countries.

The major reason for this dramatic increase was the rise in internal conflicts as the Cold War came to a close. The proxy wars the superpowers fought in the 1980s displaced millions of people who came into full view only as Cold War tensions eased in Angola, Mozambique, Ethiopia, Afghanistan, Cambodia, and El Salvador. The arms the United States and the Soviet Union had supplied to regimes or opposition movements furnished the weaponry for the ethnic and clan warfare that broke out once the superpowers departed. Liberia and Somalia, two countries that plunged into civil war, were among the largest recipients of U.S. military assistance in sub-Saharan Africa during the 1980s. In Europe and Central Asia, the collapse of the Soviet Union lifted the lid on nationalist aspirations and ethnic rivalries that displaced millions more.

Elsewhere, vast disparities in wealth, land ownership, and power have been at the root of conflict. In Rwanda and Burundi, high population density and limited fertile land exacerbated tensions between Hutus and Tutsis. In Colombia, conflict over land has forced hundreds of thousands from their homes. In other cases, struggles between governments and minorities have produced mass displacement. In Sudan, which has the world's largest internally displaced population, the efforts of successive northern governments to impose Islam on the black African south have made four million people homeless. In Turkey, which has the second-largest displaced population, government repression against the Kurdish minority has sent two million people fleeing for their lives.

The 1990s have seen greater willingness on the part of the international community to intervene in these situations, even without the consent of the government concerned. A major motivation has been the desire to forestall international flows of refugees. As the number of refugees has grown, Western governments as well as those in Africa and elsewhere have become less welcoming to those in flight. Their focus has shifted to keeping people in their homelands. The U.N. Security Council justified the international community's precedent-setting intervention on behalf of the Kurds in Iraq in 1991 on the grounds that massive flows of refugees threatened international peace and security. Subsequent Security Council resolutions on Somalia, Bosnia, and Rwanda have also authorized the use of force to facilitate the delivery of relief and, in the latter two cases, to protect internally displaced populations.

Nonetheless, the principles of sovereignty, territorial integrity, and noninterference in the internal affairs of states present formidable obstacles to intervention by international organizations. While some governments, such as Sri

Lanka, invite international assistance, others deliberately bar humanitarian aid, seeing it as strengthening their opponents and undermining their authority. Sudan, for example, has obstructed humanitarian aid, while Turkey has blocked all international assistance to its displaced citizens. Some countries have deliberately starved the displaced while invoking their sovereignty to keep the international community at bay.

Redefining Terms

A response to such conduct requires a broadly recognized standards and arrangements to guide the actions of governments and international humanitarian agencies. The definition of sovereignty should be broadened to include responsibility: a state can claim the prerogatives of sovereignty only so long as it carries out its internationally recognized responsibilities to provide protection and assistance to its citizens. Failure to do so should legitimize the involvement of the international community. States that refuse access to populations at risk could expect calibrated actions ranging from diplomacy to political pressure, sanctions, and, as a last resort, military intervention.

Earlier [in 1998], the representative of the U.N. secretary-general on internally displaced persons outlined guiding principles for the rights of the displaced and the duties and obligations of states and insurgent groups. These principles include a right not to be arbitrarily displaced, access to humanitarian assistance and protection while in flight, and guarantees of reparations upon returning home. While the principles lack legal force, they set standards that should put both governments and insurgent groups on notice and give international and nongovernmental organizations (NGOs) a basis for legitimate action. The United Nations' acknowledgment of these principles and call for their observance would be an important step forward.

A Pick-and-Choose Policy?

During the last ten years, an array of humanitarian, human rights, and development organizations have come forward to provide protection, assistance, and reintegration and development aid to the internally displaced. The efforts of international agencies and the NGOs that work alongside them are being coordinated by the U.N. Office for the Coordination of Humanitarian Affairs. Nonetheless, many internally displaced persons remain neglected because the international response is largely ad hoc. Various agencies pick and choose the situations in which they wish to become involved; no organization has a global or comprehensive mandate to protect the displaced. The result is that the needs of the displaced are met to varying degrees in some countries and not at all in others.

The United Nations' best option is a more targeted approach that takes advantage of existing mandates and capacities. The U.N. emergency relief coordinator could assign principal responsibility in each emergency to one international agency, assisted by a U.N. coordinating mechanism. When a single agency has been made responsible for a group of internally displaced people, greater attention has been paid to their needs.

Since 1992 a representative of the secretary-general has been authorized to monitor displacement worldwide, undertake fact-finding missions, open dialogue with governments, and make proposals for strengthening legal and institutional protection for the internally displaced. He has raised international awareness and mobilized support from governments, foundations, academia, and the legal and NGO communities. The position's effectiveness, however, is limited because it is voluntary and part-time. It has no operational authority and minimal human and material resources at its disposal. The capacity of the representative to act in crises can be strengthened only if the United Nations provides staff and resources commensurate with the task.

To date, the international community has focused on providing food, shelter, and medical supplies. Yet displaced persons have regularly pointed out that security is as great a priority as food. Providing relief to uprooted people while ignoring the fact they are being beaten or raped has led some to call the victims the "well-fed dead." U.N. human rights activities are slowly being integrated into humanitarian programs. A next step would be for the newly appointed High Commissioner for Human Rights, Mary Robinson, to deploy field staff during internal displacement crises to monitor security problems, serve in safe areas and camps, and promote safety during their return home.

Humanitarian assistance and development agencies of all stripes will also have to become more responsible for protecting the people they assist. Many have silently witnessed abuses because of exaggerated fears that confronting a host government will result in expulsion of their personnel and termination of their programs. The effort would be strengthened if governments and insurgent groups were made aware that they were dealing with a united front bound by common human rights and humanitarian standards.

Muscle and Money

In recent years, military interventions have been more successful in preventing starvation than in physically defending people at risk. This failure underscores the need for strategies to prevent genocide and other crimes against humanity that lead to displacement. Early action must be encouraged when there are warning signs, as there were in Rwanda, and forces charged with protecting the displaced must have the equipment, training, and mandates needed to accomplish their task, which was not the case in Bosnia.

Unless accompanied by steps to address the causes of crises, military solutions are only temporary. Humanitarian assistance alone can prolong conflicts. Conflict and internal displacement can be resolved only through a broader commitment to the peaceful management and mediation of disputes.

Since today's conflicts take place mainly within developing countries and may go on for decades, international development and financial institutions cannot afford to wait until they burn themselves out. By getting involved early on, these organizations stand a better chance of influencing the outcome and helping lay a foundation for a transition to peace. Even when societies are still in conflict, they can help stabilize the situation and make reintegration more likely.

In countries devastated by civil wars, up to half the population can be uprooted. Whether in Mozambique, Angola, Afghanistan, Cambodia, or El Salvador, the rehabilitation of areas affected by conflict requires the reintegration of uprooted populations. An expanded role for international development and financial institutions in post-conflict reconstruction could influence the way these societies reintegrate displaced populations. A global reconstruction fund would be an important step toward assuring such transitions. Efforts by international development programs and financial institutions to redress economic inequities could also help prevent future strife.

Conflicts that are allowed to fester can produce mass displacement and leave political and economic scars that damage the economic well-being and political security of neighboring states, regions, and the international system as a whole. The world community cannot let this newest challenge go unchecked.

NO

David Masci

Assisting Refugees

The Issues

On the dusty road to Gisenyi, just inside Rwanda, 200,000 bedraggled people were at last nearing home.

"You could see this river of humanity, with all of their belongings on their heads and their kids straggling along," recalls Samantha Bolton, an aid worker. "Most of them were barefoot and exhausted."

The next day, says Bolton, communications director for Doctors Without Borders USA, another 150,000 people made the trek.

The people returning to Rwanda [November 1996] were Hutus. In 1994, by the hundreds of thousands, they had been chased out of Rwanda by an army of rebel Tutsis, a rival ethnic group. Now the Tutsis were pushing the Hutus back into Rwanda by forcibly closing the refugee camps in Zaire.

Thousands of refugees who had fled to Zaire would never come home. Many children and older people had died of starvation or cholera during the exodus from Rwanda. Others had been killed in the camps by thugs and extremists from their own tribe. Tens of thousands more had died in Hutu-Tutsi fighting that flared up [in 1996] in Zaire.[1]

Throughout the crisis, the United Nations High Commissioner for Refugees (UNHCR) and other aid groups scrambled, often with questionable success, to help. Many of the world's most powerful nations seemed in a quandary about whether to get involved at all, fearful of becoming too entangled in the horrific chaos.

As shocking and tragic as it has been, the situation in Rwanda and Zaire represents nothing more than the planet's latest refugee crisis, sandwiched between the festering upheavals in Bosnia and the next catastrophe yet to appear on the evening news.

Actually, many international experts say, the world is experiencing a level of conflict and population displacement rarely seen in human history. And, they warn, crises are becoming so commonplace that the world community can ignore them only at its own peril. Georgetown University Professor Charles Keeley predicts that multiethnic strife will force many states to collapse in coming decades, leading to still greater chaos.

From David Masci, "Assisting Refugees," *CQ Researcher,* vol. 7, no. 5 (February 7, 1997). Copyright © 1997 by Congressional Quarterly, Inc. Reprinted by permission. Some notes omitted.

The last decade has been especially volatile. While the United States, China, Germany and other global powers have been at peace through most of the post-World War II era, the developing world has seen a steady rise in war and civil strife. With the Cold War coming to an end, human misery and displacement became increasingly common in poorer countries, such as Somalia, Liberia and Burma. Since 1980, the estimated number of refugees in the world has roughly tripled, rising from 5.7 million to 15.3 million in 1996. The number of people displaced within their own countries is even higher, topping 20 million by most estimates.

The refugee increases generally have not been tied to one cause. In the past, a cataclysm on the scale of World War II or the partitioning of India would create millions of refugees for a relatively short period of time. But today, refugee crises are a series of small, constant blips on the world's radar screen, with many crises being precipitated by old ethnic conflicts that flare up periodically.

The growing plight of refugees has generated an unprecedented increase in the size and number of aid groups. The UNHCR, once a small agency that provided legal assistance to refugees seeking asylum in post-World War II Europe, aided more than 27 million people in 1995. In addition, there are now hundreds of non-governmental relief organizations (NGOs)—including CARE, the Red Cross and Doctors Without Borders.

Besides trying to produce food and shelter, UNHCR and other organizations are searching for new and innovative ways to prevent or at least mitigate the number of refugees. They also are focusing more attention on aiding displaced peoples after they have been repatriated, to decrease the chance that they will leave their homes again.

But some observers find fault with the aid efforts, and even question the motives of the aid groups themselves. A number of refugee experts argue that UNHCR may have done more harm than good in Rwanda and Zaire by blindly providing assistance to all Hutus who crossed into Zaire in 1994. The problem, they say, is that among the refugees was a large group of militants who were responsible for the genocidal slaughtering—often by machete—of more than 500,000 Tutsis and moderate Hutus. The Hutu extremists used the refugee camps as de facto military bases and treated the refugees as hostages.

Some cynics among the critics even say that UNHCR and the NGOs see it in their interest to create and perpetuate refugee crises. More starving children with bloated bellies on the evening news means more donations and government funding.

Aid groups call such views ridiculous. But they admit to owing their increased visibility, in no small part, to the news media, even as they fault the media for their short attention span and for ignoring many countries in turmoil. In the final analysis, they acknowledge, the media play the key role in making the global community aware of humanitarian needs.

Citizen awareness is crucial, aid experts say, and not just because it leads to charitable donations. Governments are more inclined to address a problem halfway around the world, they say, if their citizens call for action. Even then, nations in a position to help are often loath to do more than offer money. Many

U.S. lawmakers, among others, are apprehensive about sending soldiers into harm's way to aid refugees or others in need, unless a key American interest is threatened. For example, they say, the Persian Gulf War was really about protecting U.S. access to Saudi Arabian oil, not saving Kuwait from Saddam Hussein's plundering troops.

Aid officials argue that affluent world powers like the United States should, in the interests of humanity and global peace, send troops to protect refugees and aid workers. UNHCR officials say they were rebuffed when they asked the major powers to send soldiers to the refugee camps in Zaire to separate the Hutu extremists from the legitimate refugees. UNHCR officials also say the international community's refusal to help left the agency with no choice but to assist everyone, good and bad, in the camps, in order to prevent innocent refugees from starving.

"If we had not started feeding them immediately, many people would have starved to death," says Soren Jessen-Petersen, director of UNHCR's New York City office.

As refugee advocates, politicians and others debate how to best respond to crises like the one in Zaire, these are some of the questions being asked:

Do the methods used by UNHCR and private relief organizations to help refugees actually make matters worse?

The crisis in the great lakes region of Eastern Africa has rocked the UNHCR to its foundations. Never before has the agency faced such a difficult and chaotic situation on such a vast scale.

In the second half of 1994, an estimated 2 million Rwandans, mainly from the Hutu tribe, fled into Zaire and Tanzania. The flight was triggered by a rebel army of Tutsis, which unseated the country's Hutu-led government.

The coup by the Rwandan Patriotic Front (RPF) was triggered, in turn, by the massacre earlier in 1994 of Tutsis and some moderate Hutus by Hutu extremists. The death toll in what has come to be known as the "Rwandan genocide" is estimated at about 800,000.

When the RPF took control of Rwanda, the Hutus involved in the killings fled. Hundred of thousands of innocent Hutus followed the extremists into exile after being told that they would be targets of Tutsi retribution despite their innocence.

Critics of UNHCR say that it made egregious strategic mistakes at almost every step during the crisis in Zaire, revealing the organization's outmoded, institutional mindset. In particular, they say, UNHCR and the NGOs that support its work immediately moved to feed and care for all the Rwandan refugees streaming across the border, without considering the consequences.

Michael Maren, the author of a recent book highly critical of international aid efforts, contends that UNHCR, however unintentionally, provided aid and comfort to the perpetrators of the genocide when it set up camps for the Hutus. "The Hutu leaders ran the camps and essentially controlled the distribution of aid," he says. In addition, says Maren, a former field worker for the U.S. Agency for International Development, the legitimate refugees were completely controlled by the Hutu killers, giving them enormous leverage with the international community. "The aid groups weren't feeding refugees," he says, "they were feeding hostages."

Instead of setting up camps, Maren says, the UNHCR and the NGOs should have provided services to Hutus in Rwanda. "The agencies should have told them: 'You cross the border, you're on your own,'" he says.

Stephen Stedman, a professor of African Studies at Johns Hopkins University's School of Advanced International Studies, agrees. He argues that if UNHCR had not assisted the Hutu refugees, they ultimately would have returned to Rwanda. This in turn would have denied the Hutu extremists the benefits they derived from the refugee camps.

Stedman also criticizes UNHCR and the NGOs for ignoring their own guidelines. "There are international conventions on what a refugee is, and it is very clear that they cannot be fleeing a country because they committed crimes; cannot be armed; and cannot use their refugee status to reconstruct an armed force," he says.

But others say that the UNHCR made the best of a bad situation. They argue the Hutus so feared for their lives that they would not have returned to Rwanda. If they had stayed, and UNHCR had remained aloof, mass starvation and disease would have resulted.

"What could we have done?" asks Harlan Hale, assistant director for food and logistics at Atlanta-based CARE. "Our humanitarian imperative is to minister to people who are at risk." Those who think that it would have been better in the long run to let some Hutus starve wouldn't feel that way "if it was one of their kids out there," Hale adds.

It is easy to ponder better strategies after the fact, when you don't have to make snap decisions, says Jessen-Petersen. "More than 1 million people fled, and we had to provide life-saving assistance within 72 hours, something that is not very easy when you're in the African bush," he says.

Jessen-Petersen says that it took UNHCR four to six weeks just to stabilize the situation to the point where his agency could begin considering issues beyond providing basic assistance. "Once we got our head above water, we understood that we had a monster on our hands," he says.

UNHCR decided that the only way to tame the monster—the Hutu militants—was to disarm them or drive them from the camps. "The UNHCR knew that the camps contained a lot of killers, and they advocated a police force to separate out the killers from the refugees," says Jeff Drumtra, a policy analyst at the U.S. Committee for Refugees, an advocacy group in Washington. But the appeal fell on deaf ears, according to Drumtra and others. "There was simply no interest in doing this."

UNHCR supporters say that the unwillingness of the United States and other donor nations to commit the forces necessary to separate the extremists left the agency with limited options. "It was a horrible situation, and we faced it alone," says Marie Okabe, a senior UNHCR liaison officer in New York.

Others say that the donor countries' inaction allowed them to blame the UNHCR for anything that went wrong. Instead of taking action, Drumtra says, donor nations assuaged their consciences by pouring money into UNHCR to run the refugee camps in Zaire and Tanzania. UNHCR became "a fig leaf for their own lack of political will," he says.

After it became apparent that no military help was forthcoming, UNHCR officials say they tried to exert more control over the camps by breaking them down into smaller, more manageable units and by distributing food directly to those deemed to be legitimate refugees. But these efforts were stymied by the Hutu militants, who retained a firm hold on the populations in the camps.

Finally, UNHCR hired 1,500 members of the notoriously corrupt Zairian military to provide security in the camps. The intention, Jessen-Peterson says, was to create "a minimum of security" for legitimate refugees inside the camps. But the tiny force proved ineffective against the estimated 50,000–60,000 members of the Hutu militia. "In the end," Jessen-Peterson says, "nothing worked."

Bolton at Doctors Without Borders agrees that UNHCR was in a difficult situation after the refugee camps in Zaire had been established. But she also says the agency made other, big mistakes before the refugee crisis even occurred. "We had two weeks' notice before the refugees crossed the border," she says, but "when they arrived no camps had been set up yet."

Bolton says UNHCR should have used the two-week window to set up many small camps, making it harder for militant Hutus to organize and control the refugees. "They had camps with 250,000 people, which you never do," she says, "because you're always going to have militarization."

But Jessen-Petersen and others say that with so many refugees crossing into Zaire at the same time, it was not logistically possible to break them into small groups while providing food and medical assistance.

Some critics also charge that many of the aid groups that responded to the crisis in Zaire can be criticized for more than shortsightedness. For instance, Maren says, UNHCR and other aid groups perpetuated this and other crises largely out of institutional self-interest. "NGOs don't want to shut down camps because they have a bureaucratic structure built up around these projects," he charges, adding that "there is very little accountability in the aid business, and that leaves lots of room for fat."

This is not true, Hale says. While he admits that groups like his receive more money—in the form of outside donations and UNHCR payments—during well-publicized crises, NGOs do not try to perpetuate disasters in order to improve their balance sheets, nor do they focus only on those situations that catch the public's eye. "It's not a question of helping some refugees and not others," he says, referring to those disasters that receive media attention and those that do not. "We distribute assistance based solely on need."

Should American military personnel be sent overseas to help refugees if it exposes them to danger?

The use of American troops to provide food and other aid to displaced people has become relatively common. In the last few years, U.S. soldiers have been sent to help people in Somalia, Haiti and Bosnia. In November, President Clinton said he would send American troops to Zaire and only reversed that decision after he determined that the situation had improved.

Still, Americans and their leaders are generally wary of committing troops to humanitarian missions in far-off lands. As chairman of the Joint Chiefs of Staff, Gen. Colin L. Powell was an influential advocate for sending American forces rarely, and only when the mission's goals were clear and the use of overwhelming force made the chances of success high.

Much of the reluctance about using force stems from the nation's painful experience in Vietnam, where a small number of American military advisers grew to a half-million combat troops in a matter of years. More recently, Americans watched in horror in 1993 as news reports showed a dead American serviceman in Somalia being dragged through the streets of Mogadishu.

The incident produced what has been dubbed the "Somalia Syndrome," or lawmakers' and the military's fear of sending troops overseas if the possibility exists for even a few casualties. "They really got burned in Somalia, and they don't want to see any more body bags coming home," Bolton says.

Still, many refugee advocates argue that there are times when troops simply must be sent, particularly to protect refugees and aid workers. For instance, it is widely thought that the international community should have sent soldiers to drive Hutus who had participated in the Rwandan genocide out of the refugee camps.

"We had situations where Hutu soldiers just came into hospitals in the camps and shot patients dead," Bolton says. And, Bolton points out, refugees are not the only victims of such violence. Over the last five years, scores of aid workers have been killed in Somalia, Burundi and other hot spots. As recently as Feb. 4, five U.N. workers investigating human rights abuses in Rwanda were ambushed on a country road and shot to death. And in December, six International Red Cross workers in Chechnya were murdered in their beds.

While refugee advocates generally favor multiethnic forces, they argue that some situations demand a U.S. military presence if the intervention is to be effective. CARE's Hale notes that the U.S. has more heavy-lifting capacity, access to intelligence and sheer firepower than any other nation on Earth. Steven Hansch, a senior program officer at the Refugee Policy Group, agrees: "In certain circumstances, our armed forces are the only ones who can do the job."

But, Hansch says, the Somalia syndrome has set "such a conservative threshold of pain" that the nation won't send soldiers anywhere unless the mission is risk free. "We are chickens about this," Hansch says.

According to many refugee advocates, this attitude has soured other nations on working with the U.S. to resolve international problems. "We need to show that we are willing to be a player, that we stand for something," Hale says.

But others argue that the United States military has done more than its share in humanitarian crises, pointing to the developments in Haiti in 1994 and in Bosnia the following year. "We're doing this too routinely, and it's starting to become the norm," says Eugene B. McDaniel, president of the American Defense Institute, a Washington think tank.

Others question sending American soldiers under any circumstances. "Why should it always be the United States that answers the global 911 call?" asks Lionel Rosenblatt, executive director for Refugees International. Rosenblatt and others say that military leaders should be especially wary of sending troops into countries where warring factions are still fighting and civil society has broken down.

In addition, McDaniel says, "Our troops are for defending our country and its national interests," not feeding refugees in a country halfway around the world. This means defending the industrial world's access to oil—as was done during the 1991 Persian Gulf War—but not protecting combatants in Bosnia from each other, McDaniel says. "It's degrading to send our soldiers to those places on those missions," he says.

But Michael Clough, a research associate at the Institute of International Studies at the University of California-Berkeley, calls the argument put forward by McDaniel and others "nonsense." Clough says the United States and other industrialized countries have a clear interest in solving humanitarian and refugee crises around the world. "First, we have an interest in ensuring that places like Africa don't collapse," he says, adding that "by 2025, 30 percent of the world's population will live in Africa, and you just can't write off a part of the world that big."

Another reason rich countries cannot ignore poorer states, Clough says, stems from the fact that large numbers of refugees will eventually reach the shores of European nations and the United States. In addition, he argues, there needs to be stability in the developing world if the global economy is to continue to expand.

But there may be a way to satisfy both McDaniel and Clough. Rosenblatt is among those who propose the establishment of a permanent United Nations rapid deployment force to intervene in humanitarian crises. While the proposal has received some tepid support from the United States and other powers, no action has been taken on the idea, and its prospects are uncertain, at best.

Note

1. By early December [1996] more than 1 million of the 1.5 million Hutu refugees in Zaire had returned to Rwanda. Many of the half-million Hutus who had sought refuge in Tanzania also returned.

POSTSCRIPT

Is the Global Community Responding Well to the Plight of Refugees and Displaced Persons?

Cohen and Deng focus on people who are forced from their homes but who remain within their own countries. Since these displaced people have not crossed national boundaries, no international agreements cover their situation. The authors stress that despite the lack of such arrangements, the international community has become more willing to intervene. These actions have been undertaken in the face of such long-standing international principles as sovereignty, territorial integrity, and noninterference in the internal affairs of states. Nevertheless, the authors advocate UN action because these principles do not free countries from the obligation to treat their citizens humanely.

Masci examines the situation of both refugees and displaced persons. He suggests that some observers have been critical of aid efforts, both of non-governmental relief organizations and of UN agencies. The most damaging attacks suggest that some aid groups find it in their own interest to "create and perpetuate" refugee crises.

Anyone who values human dignity can identify with the plea outlined by Cohen and Deng. Newsmagazines have been replete with vivid accounts of the suffering of those who have been forced to leave their homes. In addition to the obvious kinds of misery that can be captured by a camera, there are more hidden and even insidious negative consequences of such displacement. Entire communities disappear, and ways of life known for centuries die quickly.

Masci goes into greater detail in describing the events surrounding displacement, including what he argues are mistakes made by international interveners. His is a story of episodic events, rather than a description of the broader picture. His examples, however, give one pause as the question of to what extent, if any, the international community has an obligation to assist all who are affected by physical upheaval is addressed.

The United Nations High Commissioner for Refugees publishes a periodic report on its recent accomplishments. The 2000 edition, *The State of the World's Refugees: Fifty Years of Humanitarian Action* (Oxford University Press), spells out some of the newer challenges, including globalization, the changing nature of conflict, the growing complexity of population movements, and the changing nature of humanitarian action. A critical account of humanitarian aid in this area is Michael Maren's *The Road to Hell: The Ravaging Effects of Foreign Aid and International Charity* (Free Press, 1996). A more balanced account is Gil Loescher's *Beyond Charity: International Cooperation and the Global Refugee Crisis* (Oxford University Press, 1993). Also, the U.S. Committee for Refugees, a refugee advocacy group, annually publishes *World Refugee Survey,* which assesses the refugee situation throughout the world.

Several recent books address the question of forced displacement. Michael Dummett's *On Immigration and Refugees* (Routledge, 2001) weaves philosophy and history to address the responsibilities of national governments to refugees and immigrants. *Controlling a New Migration World* by Virginie Guiraudon and Christian Joppke (Routledge, 2001) examines the efforts of contemporary Western countries to control international migration. One major emphasis is on the new linkage between migration and security, a subissue that will only grow as a consequence of September 11.

Anne F. Bayefsky and Joan Fitzpatrick, eds., *Human Rights and Forced Displacement* (Martinus Nijhoff, 2000) addresses the problem of forced displacement and its human rights dimensions in a number of ways. Particularly worth reading is chapter 19, "Turn Back to Look Ahead? Central European Observations on the Future of Regimes Affecting Refugees," in which Boldizsár Nagy identifies three different refugee regime paradigms first articulated by James Hathaway in "A Reconsideration of the Underlying Premise of Refugee Law," *Harvard International Law Journal* (vol. 31, no. 1, 1990). These paradigms are humanitarian, the human rights, and the self-interested control-oriented. The humanitarian model requires countries to make an honest effort to ensure that the welfare of involuntary migrants would be a needs-based concern. The human rights model would address the cause of the flight without suggesting that every involuntary migrant was a refugee. And the self-interest model would allow countries to consider national interests in determining protection decisions.

A well-researched book by one who spent his professional lifetime in the field is *The Price of Indifference: Refugees and Humanitarian Action in the New Century* by Arthur C. Helton (Oxford University Press, 2002). In it, Helton provides a broad picture of recent refugee policy and humanitarian action by focusing on specific cases: Bosnia, Cambodia, East Timor, Haiti, Kosovo, Rwanda, and Somalia. He addresses policy shortcomings as reflected in these crises and proposes major reform.

ISSUE 12

Should Immigration to the United States and Other Developed Countries Be Curtailed?

YES: Mark W. Nowak, from "Immigration and U.S. Population Growth: An Environmental Perspective," *Negative Population Growth Special Report* (December 19, 1997)

NO: John Isbister, from "Are Immigration Controls Ethical?" in Susanne Jonas and Suzie Dod Thomas, eds., *Immigration: A Civil Rights Issue for the Americas* (Scholarly Resources, 1999)

ISSUE SUMMARY

YES: Mark W. Nowak, an environmental writer and a resident fellow of Negative Population Growth, attacks six basic arguments favoring the loosening of U.S. immigration laws, including the argument that immigration contributes little to U.S. population growth.

NO: Professor of economics John Isbister attacks six basic arguments for population controls, among them the contention that open immigration would destroy important American values.

Even before September 11, 2001, if you opened a major American newspaper on any given day, you were likely to find an article addressing some perceived problem relating to immigration, whether the referenced parties were in the United States legally or illegally. This is not surprising; the topic has become one of the most volatile issues of the day, a pattern that has escalated since the law basing immigration on country of origin (and favoring Caucasians) was replaced in 1965 by one that was deemed far less racist. The end result has been a vast opening of American borders to legal and illegal aliens alike, whose countries of origin are typically non-European. The terrorist attacks on the World Trade Center and the Pentagon have reinforced the importance of the issue, as America debates homeland security and other war-on-terrorism issues.

The debate, unlike many within America today, does not fit neatly along conservative and liberal lines. Both groups are somewhat torn on the issue. Conservatives applaud the economic benefits that free-market entrepreneurs receive as a consequence of a larger supply of blue-collar and no-collar workers who will take jobs that are shunned by citizens of longer standing. However, they oppose the increased burden on the American taxpayer, whether an immigrant is participating in the tax system or not. They also take a hard-line stance with regard to vigorously protecting American borders against those who fit the profile of suspected terrorists. Liberals view anti-immigration policies as racist or nativist and welcome the increased diversity that wider openings at the border bring, particularly for those of non-European origin. But they worry about the adverse effects such influxes have on Americans at the bottom of the socioeconomic ladder, particularly when economic times turn rocky. And with respect to the implications of September 11 on the issue of open borders, they worry as much about the denial of basic human rights as they do about the difficulty of keeping terrorists at bay.

John Isbister, in *The Immigration Debate: Remaking America* (Kumarian Press, 1996), outlines specific components of the debate along three lines—demographic, economic, and cultural. A summary of these components will help to set the background for this issue. To Isbister, two questions are at the heart of the demographic debate on immigration: Is the current wave of immigration unusually large, and will it change the ethnic composition of the country? Opponents of immigration believe that the numbers of outsiders entering the United States are unmatched in history and that, if controls were lifted, the resultant influx would be overwhelming. Supporters contend that current levels are actually lower, when measured as a percentage of the resident population, than they were at earlier heights of immigration in American history.

Isbister's economic debate centers around 11 specific questions: Is the quality of the immigrant labor force deteriorating? Are immigrants unusually enterprising? Is immigration desirable because it helps to alleviate labor shortages? Does it reduce American wages? Does it lead to a more unequal distribution of income? Does it increase unemployment? Do immigrants impose a fiscal burden on the rest of Americans? Why do they come? Do they generate enough economic growth to improve the living standards of those before them? Will immigration reduce America's international competitiveness? Will it harm the natural environment and deplete natural resources?

Isbister's cultural debate addresses six issues: Is America of one culture or many cultures? Has the melting pot stopped melting? Can the United States increase its cultural diversity without causing more conflict? How should immigrant applications be prioritized? Could the education and skills of immigrants be upgraded? Has the United States lost control of its borders? The general view of opponents on the cultural dimension is that the typical immigrant of today is so different from that of yesterday that the dominant culture is harmed. Supporters of increased immigration counter that America is not diverse enough and that Anglos dominate too much as a consequence.

In the following selections, Mark W. Nowak attacks six widely accepted reasons for widening U.S. borders, while Isbister counters six fundamental arguments for increasing immigration controls.

Mark W. Nowak

 **YES**

Immigration and U.S. Population Growth: An Environmental Perspective

Controversy over U.S. immigration policy is by no means new to the political landscape. Since 1819, when Congress passed the first significant law regulating immigration into the United States, successive debates over immigration have stirred emotions and polarized perceptions. It is not surprising, then, to find that environmentalists, confronted by the issue with increasing frequency, are by no means in agreement about the relationship between immigration and the environment.

On the one hand are those who argue that immigration, notwithstanding the benefits it provides, is fundamentally a form of population growth. Therefore, say supporters of this position, levels of immigration must be reduced (and fertility held at replacement-level or below) if we are to move toward environmental sustainability in the United States.

Others argue that treating immigration as an environmental issue is a wrong-headed approach to environmental protection. Rather than focusing on immigration, say these proponents, the environment would be better served by addressing issues such as Americans' hyper-consuming lifestyle, which are more to blame for our environmental ills.

It's true that numerous factors—including the high consumption rates of Americans—contribute substantially to environmental degradation, but diminishing or discounting the real role that immigration plays makes little sense. All other factors being equal, the environmental consequences of human activity increase with the growth of the population. This essential relationship is nearly universally recognized—particularly among environmental and population groups—as one of the fundamental bases for providing international population stabilization funding.

If immigration, then, serves as a contributor to U.S. population growth, what is the basis for excluding immigration from any comprehensive analysis of the consequences of U.S. population growth? It's time to look at and evaluate the arguments that are used against viewing immigration as a domestic population issue.

From Mark W. Nowak, "Immigration and U.S. Population Growth: An Environmental Perspective," Negative Population Growth Special Report (December 19, 1997). Copyright © 1997 by Negative Population Growth, Inc. Reprinted by permission. Notes omitted.

Argument: Immigration Contributes Little to U.S. Population Growth

A common assertion about U.S. immigration is that its demographic effect is (and will continue to be) small, implying that any environmental consequence of immigration will be minimal. In reality, the demographic consequences of our current immigration policy will be considerable, as revealed by the population projections prepared by the Census Bureau.

In building its projections, the Census Bureau analyzes four factors—birth rates, death rates, immigration and emigration. Each time the Bureau prepares a substantial update to their projections, they vary their assumptions about each of these components to present a range of possible demographic outcomes.

The Bureau's latest projection, which assumes, essentially, that current demographic trends will continue, projects that the United States will grow from its current 268 million people to 393 million people in 2050, an increase of 125 million people.

Immigration emerges as a prominent component in the calculation: 60% of the population increase in the United States between 1994 and 2050 will be attributable to immigration and the descendants of immigrants.

What this means is that immigration will not be a marginal contributor to future U.S. population growth, but, in fact, the primary one. Further, the absolute numbers in the Census Bureau projection are large: adding 125 million people to the current U.S. population is the equivalent of adding 48 more cities, each the size of Chicago.

Argument: Historic Levels of Immigration Are Higher Than the Current Level

Regardless of immigration's demographic effect, immigration should not be considered solely a demographic issue, say some. It is equally important that we honor our past as an immigrant-receiving nation, and reducing immigration to stabilize the population flies in the face of this tradition. Furthermore, say proponents of this position, the percentage of the U.S. population that is foreign-born is far lower today than it was at the turn of the century, implying that we actually ought to consider increasing immigration levels so that our current policy might be consistent with tradition.

Pointing to the percentage of the population that is foreign-born as a measure of annual immigration makes little sense, however, since each is, at best, an indirect measure of the other. If we are interested in determining the traditional level of immigration into the United States, it makes much more sense to look at annual immigration flows. When we do, we see some surprising facts emerge.

[Figure 1] details legal immigration to the United States between 1821 and 1990, a period that saw more than 61 million people make this country their home. (Annual statistics prior to 1821 are not available since the Federal government didn't begin recording immigration until that year. For the period

1776–1819, however, immigration is estimated at only about 300,000 total.) This chart conveys two extremely important facts about immigration. The first is that no decade appears on the chart that we can point to as "typical." Immigration levels range from a ten-year low of 143,000 (1821–1830) to a ten-year high of 8.8 million (1901–1910). Immigration during the current decade will likely set a new record high. When we observe individual years, the range is even greater: 1823 saw only 6,300 people arrive, while 1.8 million people became legal immigrants in 1991.

Figure 1

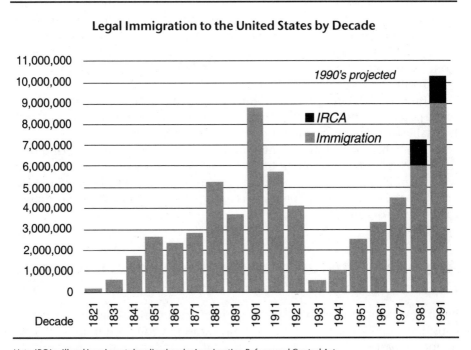

Legal Immigration to the United States by Decade

Note: IRCA = Illegal immigrants legalized under Immigration Reform and Control Act

Sources: US Census Bureau and 1995 Statistical Yearbook of INS (March 97)

In other words, while it is true that the United States has enjoyed a long tradition as an immigrant-receiving nation, there is no such thing as a traditional level of immigration to the United States. A proposal to set a net immigration level of 100,000 annually has as much historical authority as a proposal to cap immigration at 500,000 a year. Consequently, while the simple reminder that we are a nation of immigrants reveals a historic truth, it tells us precious little about what, specifically, our immigration policy should be.

The second fact is that, while the level of immigration into the United States has varied considerably, for the last 65 years absolute levels of immigration have risen steadily and appreciably from one decade to the next. This general trend of rising immigration levels, combined with a relatively steady annual rate of natural increase (births minus deaths), has meant that immigration's share of annual U.S. population growth has been growing larger and larger.

[Figure 2] shows that in 1970, first-generation legal immigrants (annual new arrivals) accounted for 16% of the increase in the U.S. population. By 1996, that figure more than doubled to 36%. These figures represent just the increase contributed by first-generation immigrants. In addition, the Census Bureau has found that first-generation immigrants tend to have higher fertility rates than the native-born. When the demographic contribution of immigrants and their descendants is combined, we find that immigration will account for about two-thirds of future U.S. population growth.

Figure 2

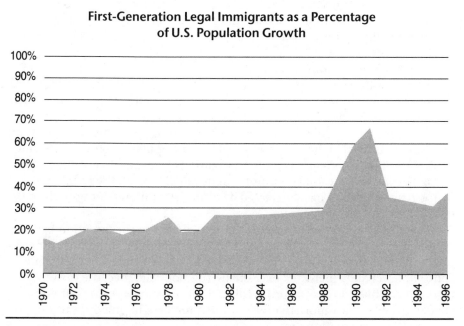

First-Generation Legal Immigrants as a Percentage of U.S. Population Growth

Sources: US Census Bureau and 1995 Statistical Yearbook of INS (March 97)

It's true that the percentage of foreign born in the United States is lower now than it was at the turn of the century, but that is because the U.S. population is so much larger now. The absolute size of the population of foreign born is actually larger today than it was at the turn of the century, even though it represents a smaller percentage of the total population.

In 1910, the United States had 92 million residents. At that time, the nation's 13.5 million first-generation immigrants accounted for a rather significant share of the total population—about 15%. In 1990, when the population had grown to 248 million, the nation's 19.5 million first-generation immigrants constituted a far smaller share—about 8%. But the percentage of the U.S. population that is foreign born has little bearing on the environmental risks of maintaining our current rate of population growth, which as we have seen, will be driven primarily by immigration in the future.

Argument: High Consumption Levels Are More to Blame Than Population Growth

Environmental degradation is a function of more than just sheer population size. The level at which a particular population consumes energy, natural resources and other materials plays a large role in determining total environmental impact, as well. Per capita consumption and waste production rates in the United States are among the highest in the world, leading some to argue that far more benefit would accrue to the environment if we focused our efforts on reducing consumption, rather than working to reduce population growth (and, by extension, immigration). Consumption and population growth, however, are not mutually exclusive issues. Both have a significant consequence, so both must be addressed.

Consider, for example, car and truck ownership by U.S. residents (which includes both the native-born and immigrant population). In 1970, there was one car or truck on the road for every two U.S. residents. Since then, vehicle ownership has increased. By 1994, there was one car or truck on the road for every 1.5 residents. While much of the increase in the number of vehicles on the road stems from higher per capita ownership, population growth accounted for 27% of the rise—not an insignificant figure. This means that had there been no change in the per capita consumption of vehicles at all, the number of cars on the road would still have increased by nearly 26 million—due entirely to the increased size of the population.

The impact of population growth on total energy consumption between 1970 and 1990 is especially dramatic. During this period, numerous conservation and efficiency measures were enacted and, as a result, per capita energy consumption barely increased over the two decades. But, because the U.S. population continued to grow during this period, total energy consumption increased by 36%, with more than 90% of this increase in energy consumption due entirely to population growth.

We can—and should—reduce consumption, but unless we address population growth our net gains will be reduced (or even reversed) by the demands imposed by our growing population.

Argument: Environmental Gains Can Be Made Despite Population Growth

Immigration's strong showing as a contributor to U.S. population growth provides a fundamental basis for including immigration in environmental policy discussions. Some, however, have argued that linking immigration to environmental policy is not necessary because, they say, population stabilization is not a necessary condition for environmental improvement. The basis for this argument is that *every* environmental advance of the last two decades—including improvements in air quality, water quality and habitat restoration, has been made while both immigration rates and total population size were increasing.

The introduction of environmental regulations and the adoption of more efficient technologies have certainly led to some recent gains in environmental quality, but environmental gains are only half the story. Consider some of our environmental losses:

- Annual energy consumption, despite the introduction of efficient technologies, has increased 36% since 1970.
- The United States continues to lose more than a million acres of farmland every year to urban sprawl and erosion.
- More than 90% of all old growth forests in the United States have been cut down.
- About 50% of all wetlands in the United States have been lost. Wetland loss is highest in areas with higher growth rates (and therefore more rapid development): in California, 93% of all wetlands have been destroyed.
- Scientists estimate that since the Declaration of Independence was signed, more than five hundred species in the United States have gone extinct. Nearly 1,100 species are currently listed as threatened or endangered.
- In many parts of the United States, demand for water exceeds supply so that we are now overdrafting our surface waters and "mining" our aquifers. By consuming water faster than the recharge rate, we are destroying an otherwise renewable resource.
- Over the past thirty years, nutrient loads in Chesapeake Bay (including nitrogen and phosphorus from agriculture) have increased as much as 250-fold, while areas of the bay that are oxygen depleted have increased 15-fold. The bay's famous oyster harvest has been decimated and harvests of other species, including crabs, are seriously threatened.
- Emissions of sulfur dioxide, nitrogen oxides and volatile organic compounds have fallen very slowly or not at all since 1980.

The significance of any environmental gains can only truly be measured in the long term. If the population increases, as projected, to 393 million by 2050, will our current environmental victories survive?

Argument: We Should Focus on International Development, Not Immigration Policy

Finally, some contend that immigration is simply too complex to be treated fundamentally as a demographic issue: instead of reducing immigration, which unfairly blames immigrants for problems in the United States, the U.S. should work internationally to reduce the push factors (such as poverty, overpopulation and war) that compel people to leave their own countries and move here.

This argument, which essentially characterizes immigration as an inevitable economic process, identifies incorrectly the basis for most legal immigration. It also fails to recognize that immigration policy and foreign policy goals can be complementary.

"Push" factors explain primarily illegal, not legal, immigration into the United States. Legal immigration, in fact, is driven in part by the desire for economic gain, but a variety of other factors play a role. In 1994, 65% of immigrants arrived solely on the basis of family affiliation—a spouse, parent, child, brother or sister was already living in the United States. About 15% arrived as refugees or asylees (refugees and asylees are able to demonstrate that they would be persecuted if they remained at home), and about 8% arrived on the basis of possessing an identifiable skill. The remaining 12% arrived under one or more of several temporary programs.

In other words, poverty may explain the reason some potential immigrants come to the United States, but no legal immigrant arrives explicitly because he or she is poor—the current immigration law offers no provision to do so. Consequently, dramatically increased foreign assistance might help reduce the flow of illegal immigration (and flight by refugees and asylees), but its consequence for most legal immigration is less certain.

Second, the argument above suggests that trade and foreign aid should serve as a substitute for immigration policy—that rather than setting immigration levels directly, "immigration demand" should be allowed to evolve indirectly as a natural economic consequence of foreign assistance. A far more effective and reasonable strategy is to recognize the value (both environmental and economic) in regulating immigration levels while simultaneously pursuing our foreign aid and trade goals.

Argument: Immigrants Are Being Unfairly Blamed

Some have argued that by identifying immigration's contribution to population growth and advocating a reduction in that contribution, we are unfairly blaming immigrants for America's environmental problems. This argument makes little sense on the face of it. If such an argument were necessarily true, then demographers would be unfairly blaming couples—or their newborn children—for environmental degradation in the United States simply by acknowledging fertility's contribution to population growth. The blame game yields no profit, but does serve to sidetrack a critical policy issue with an unproductive and highly emotional debate.

Our Past—and Possible Futures

Over the past 25 years, the link between population growth and environmental degradation has been so well established that it is hard to find an environmental advocate who does not acknowledge it. Numerous governments have recognized the need for population stabilization, as well, and several international agencies now exist to address population growth directly. In the United States, officials at the highest levels of office have considered whether the United States should adopt a national population policy. In 1972, the recommendations of the Commission on Population Growth and the American Future (also known as the Rockefeller Commission) demonstrated what such a policy might look like: in part, the Commission recommended freezing immigration at its then-current level of about 400,000 a year as part of a national population policy. More recently, the President's Council on Sustainable Development (PCSD) advocated the goal of voluntary population stabilization in the United States, but fell short of recommending a specific immigration level.

Consider, also, the overwhelming national and international consensus on the relationship between population growth and environmental degradation. As recently as 1994 at the International Conference on Population and the Environment, the United States (with the endorsement of numerous U.S. environmental organizations) and dozens of other countries, reaffirmed the goal of stopping population growth as a key element in any environmental protection plan.

Immigration is a highly emotional issue so any debate over immigration policy is likely to be heated. Unfortunately, much of the population and environment community is reluctant even to acknowledge that immigration has a demographic impact, an issue that was long ago resolved by the demographers. Once this first step has been taken, then the environmental debate over immigration can truly begin.

John Isbister  **NO**

Are Immigration Controls Ethical?

Americans typically express their opinions about immigration with certainty and moral outrage. This article argues that many of us are in an ethically fraught position with respect to immigration, however. We have deeply held convictions about the equality of all people. At the same time, though, we use immigration policy to perpetuate a privileged lifestyle at the expense of foreigners. We are not prepared to abandon either this use of immigration policy or the ideology of equality. At the very least, therefore, our moral stance should be one of humility, not outrage.

The moral muddle in which we find ourselves arises in part because of a tendency to confuse our interests with ethics; we persuade ourselves that what is best for us is best, period. It is not.

Most of the current debate about immigration has to do with the national interest or with the interests of various groups inside the U.S. Elsewhere I have explored the interests of Americans with respect to immigration (Isbister, 1996). Briefly, I think that from an economic point of view, some Americans are helped and others are hurt by immigration. On the whole, the former tend to be at the top of the income ladder, the latter at the bottom. Economics is not everything, however, Bill Clinton's 1992 presidential campaign to the contrary notwithstanding. Immigration is transforming the social structure of the U.S., changing it from a white country with several minority groups into a genuinely multicultural country in which no ethnic group will prevail, all ethnic groups will be able to meet each other on a basis of relative equality and the possibilities for mutual enrichment rather than suspicion, and exploitation and racism will be enhanced. Since this social transformation is so important to the U.S., I think that the interest of the country lies in welcoming relatively large numbers of immigrants. The numbers cannot be unlimited, however, since the maintenance of one of the world's highest standards of living requires some restrictions on the inflow of newcomers.

Interests of Americans are central; the national political system responds, it can be argued, to nothing else. Interests are not synonymous with morality, however. By definition, the national interest excludes the interests of foreigners. It would be a curious ethical system that denied moral standing to some people simply because of the accident of their nationality. The ethics of immigration controls require investigation, therefore, separate from the interests that Americans have in immigration. In the end, of course, interests will count more than ethics in the formulation of national immigration policy. We would do well, however, to keep in mind the conflict between the two perspectives.

Equal Moral Worth

The equal moral worth of all people is the principle on which the ethical arguments in this article are grounded. Philosophers have made many attempts to demonstrate equal worth. For the most part, however, those of us who believe in equality treat it as an assumption, an axiom, not something to be proven. The axiom of equal worth is at the core of the two most important statements of political philosophy in U.S. history, the Declaration of Independence and the Gettysburg Address.

In the Declaration of Independence, Thomas Jefferson wrote, "We hold these Truths to be self-evident, that all Men are created equal." He did not argue the point; it was "self-evident." The importance of equality was that equal rights accrue to all people: "They are endowed by their Creator with certain unalienable Rights, that among these are Life, Liberty, and the Pursuit of Happiness."

Neither Jefferson nor his Virginian compatriots conducted their lives in accordance with these "Truths," for they were slave holders. Similarly today, most people do not base their actions on a commitment to the equality of all people. Nevertheless, many of us believe in equality, not because we have reasoned it out and considered the arguments pro and con, but because it is "self-evident."

The radical equality of the Declaration of Independence was restricted sharply by the more conservative Constitution of 1789, a document that among other things protected slavery. As Gary Wills has argued, however, Abraham Lincoln's Gettysburg Address of 1863 had the effect of subverting the Constitution by restoring equality as the central American value (Wills, 1992). It began, in words now as familiar to Americans as those of the Declaration, and more familiar than any in the Constitution, "Fourscore and seven years ago, our fathers brought forth upon this continent a new nation, conceived in liberty and dedicated to the proposition that all men are created equal."

Neither document says "all Americans are created equal." Jefferson and Lincoln may or may not have implied "people" by their use of the word "men," but in our current reading, we do. These most formative of American documents assume the equality of all human beings.

People are equally valuable and therefore have equal rights. It does not follow from this that they have unlimited rights. Rights often conflict one with another; when they do, the liberal state (that is, the state based on the presumption of equal worth) is justified in restricting some rights, in order to protect others. Whenever it restricts rights, however, the state must be able to give morally justifiable reasons why it has done so—else it forfeits its claim to liberalism and descends into despotism.

An Argument for Open Borders

Immigration controls restrict free movement by establishing groups that have unequal rights. Among the people who wish to live in the U.S., some favored ones are allowed to, while others are not. Can one successfully argue that, while immigration controls infringe on some people's liberties, they are justified because they protect more important rights and liberties? Or does their allocation of unequal rights to people who are of equal worth make them morally impermissible?

Freedom of movement is a facet of the "Liberty" that the Declaration of Independence takes to be an inherent right of equal human beings. Countries that systematically restricted the movement of their people are rightly criticized. In recent years, the clearest example of the morally unjustified restriction of internal movement was in South Africa, which enforced its apartheid system with pass laws.

As a mental exercise, one could ask how a law passed by the residents of the city of New York, restricting the permanent entry of Americans who were not city residents, would be judged. Leaving aside the fact that it would be unconstitutional, would it be morally justified? The people of New York could offer some good reasons for the law. New York is already crowded and cannot tolerate further population growth, they might argue. The sanitation system is close to breaking down, the schools are crowded, the welfare system is bankrupt, the homeless shelters are inadequate, and the unemployment rate is rising.

These sorts of arguments would not prove convincing to most Americans, who would find the restriction on personal freedom too onerous. Every day, people migrate into (and out of) New York for compelling reasons. They move in order to accept a job or to look for a job or because their job has been relocated to New York. They could not have the same job in Boston or Chicago because New York is unique (as also Boston and Chicago are). They move to New York to be with their family or to care for a friend or for one of any number of other reasons. The decision to migrate to New York is seldom taken lightly; people have good reasons. The interests that New Yorkers may have in restricting entry, while perhaps meritorious, are not of sufficient weight to permit such massive violations of the rights and interests of outsiders. Morally, New York could not justify an immigration policy of its own.

If this argument is accepted, how can one accept immigration restrictions in the U.S.? What makes the U.S. fundamentally different from New York? Nothing much, except sovereign power.

Reasoning by analogy, it is difficult to find an ethical justification for the United States to restrict entry across its borders. In fact, it is more difficult, since the people of the U.S. are privileged, *vis-à-vis* the rest of the world, in a way that the residents of New York are not, in comparison to other Americans. New Yorkers could argue plausibly that among U.S. cities, their city is not so special, that people denied entry into it could find comparable amenities in other cities. The U.S. occupies a unique position in the world, however, or at least the long lines of potential immigrants would so indicate. The great majority of immigrants and potential immigrants hope to enjoy a significantly higher standard of living in the U.S. than they experienced in their home countries. Immigration controls on the U.S. border therefore restrict access to privilege.

It is the protection of privilege that is so damaging, ethically, to the country's immigration laws. It makes U.S. border controls even less justifiable than New York's would be. The purpose and effect of U.S. immigration controls are to maintain a state of inequality in the world between the haves (the Americans) and the have-nots (the foreigners, especially the potential immigrants). Americans maintain immigration laws because they fear that unrestricted entry would lead to a major influx of people, that the newcomers would compete for scarce resources and jobs in the U.S. and that they would drive down the standard of living of residents. No doubt it is in the interest of the privileged to protect their privileges, but it cannot be ethical if that protection has the effect of further disadvantaging the unprivileged.

The ethical case against immigration controls is based, therefore, not just on the fact that they convey unequal rights to morally equal people, but that they do so in a particularly damaging way, so as to protect advantage and deepen disadvantage. To understand the importance of this, one can consider the argument that some types of unequal treatment are morally justified. For example, a system of preferential hiring in which race is taken into account treats different groups of people unequally, but it may be fair if it is designed to benefit people who have been exploited or to dismantle a system of racial injustice (Wasserstrom, 1986). Unequal treatment is clearly unjust, however, when it is used to perpetuate rather than break down a system of privilege and disadvantage. This is just what U.S. immigration controls do. They violate the equal worth and equal rights of people in an egregious way, by sheltering already advantaged Americans at the expense of relatively disadvantaged potential immigrants.

Arguments for Immigration Controls

This section considers six ethical arguments that are made in favor of immigration controls, in ascending order of persuasiveness. They are not the only arguments made in the academic literature and the political discourse, but they seem to me to be the strongest. In the end, I find that the first five are rationalizations used to perpetuate a structure of advantage and disadvantage in the world while allowing Americans to avoid this truth, and that their moral logic fails. The sixth argument, however, has merit and leaves us with a conflict of rights.

> *1. "Immigration is theft." It is argued that the U.S., with all its wealth, is the property of Americans, to treat as they wish. Immigrants want to share in the country's wealth, but they have no right to it unless Americans willingly offer it to them. People have the legal and moral right to protect their property against theft. They are under no obligation to give it away.*

Is the country the private property of Americans? Can one say that Americans own the U.S., just as an individual owns a house? Or would a better analogy be to a public beach? Whoever arrives first on a secluded beach would perhaps like to claim title and keep others from access to it, but the first-comer has no right to do so, because the beach is common property. International law gives Americans the legal right to treat the country as their private house, but I think that morally the U.S. has more of the characteristics of a public beach.

The U.S. is both a land area and the society that has been built on it. As to the land itself, Americans are in a weak position to claim exclusive ownership of it, since they are such recent arrivals themselves. Most Americans are only a few generations removed from immigration. If they were to take seriously the claim of prior ownership, they would have to return the land to the descendants of the Native Americans, whom their forbears pushed aside and slaughtered. They are not prepared to do this. They cannot successfully claim, therefore, that a new wave of entrants has fewer rights of access to the land than their immigrant predecessors did.

The U.S. is more than the land, however; it is the improvements as well. Today the U.S. is far different from the place the settlers wrested from the Natives. It is now a rich country—and it is because of its standard of living, not its land mass, that it is a Mecca for immigrants. Can Americans claim that they created the wealth they enjoy and that they therefore have the sole right to decide whether and which foreigners have access to it? I think not.

For the most part, Americans now living did not create the wealth; they inherited it. At the time they were born, the U.S. was already one of the world's richest countries and its wealth was growing rapidly. If anything, the collective actions of Americans over the last quarter century have diminished their inheritance. By living beyond their means, both personally and publicly, Americans have actually lowered the standard of living of a good many people. They are living off their inheritance.

In other words, Americans now living found a lovely beach; they did not build or buy a new house. Since they are not responsible for their good fortune, they do not have the moral right to deny other people the opportunity to share in it.

Not so, Americans might reply. We have a right to our inheritance. Our ancestors built this country not just for their own pleasure, but for the well-being of their children. If one could ask them who has a claim to the riches they built up, they would reply that it is we, their descendants.

This argument is not silly, but it is not transparently valid either. Inheritance presents ethical difficulties. Ordinary, small inheritances are certainly morally justifiable, since they represent a simple expression of family love. Just as a mother cares for her baby, so does she want to pass some of her gains on to her children after her death. To want to do so is not wrong; indeed, to fail to do so would be thought by most people to mark a lack of parental love. It is when inheritance is so substantial that it perpetuates a system of privilege and disadvantage that it lacks moral justification, since it violates the norms of equal access, equal opportunity, and equal treatment, the norms that follow from the axiom of equal worth. Americans would reject as immoral the rigid class system of medieval Europe or the caste system of India, in which one's prospects in life were completely determined by the accident of birth. That sort of inheritance is immoral. It is just the sort of inheritance that Americans as a whole have received from their ancestors, an inheritance that gives them enormously greater privilege than most of the rest of humankind. It does not give them the right to exclude others from their country.

One cannot, therefore, defend immigration restrictions as protection against theft. Immigrants are not burglars.

2. "Open immigration would destroy important American values." *The second argument for border controls is that unrestricted immigration might destroy institutions or values in the U.S. that are of transcendent importance. Open immigration might destroy something that is vital not only to Americans, but also to foreigners and potential immigrants.*

What might such institutions or values be? Political scientist Frederick Whelan argues that liberal values—equal moral worth, democracy, equal access, etc.—are scarce and precious in the world and need protection. It is a mistake to use liberal principles to argue for open access if open access would diminish liberal values and institutions in the world (Whelan, 1988).

The concern is one of absorptive capacity and the assimilative powers of the U.S. A massive influx of foreigners who were unfamiliar with and uncommitted to American political values might use the democratic procedures of the country to destroy the institutions that support those procedures. This might happen before they had the opportunity to adopt as their own the principles of freedom and self-government. Under such circumstances, Whelan and others have argued that the state is morally justified in restricting immigration.

A closely related argument is that members of a community have the right to protect their values and culture by excluding outsiders if necessary (Walzer, 1983).

This argument perhaps has theoretical merit. However, no evidence exists that it relates to a real danger in the U.S. Native groups have often opposed the political positions taken by immigrants, regarding them as inimical to their interests, but this is simply evidence that the immigrants have adapted to the political norms of the country and have used the political process to their own benefit. Disagreement over political goals is not the same as rejection of basic political philosophy. During the entire period before the 20th century, when the U.S. maintained open borders, the principal threat to a liberal political system came not from recent immigrants, but from the slave holders who were descendants of the first white settlers.

3. "We have better ways of helping the world's disadvantaged." *The third argument in favor of immigration controls is that while Americans may be morally obliged to redress the imbalance of privilege in the world, they can do so more effectively through policies to improve standards of living in poor countries than they can by allowing a few residents of those countries to immigrate. Immigration focuses the benefits of U.S. resources upon just a few people. Better to take action to improve the entire economies of poor countries, actions that will affect far more people. The policy tools available to the U.S. government are in-*

numerable, including foreign aid, loans and technical assistance, and trade, in-
vestment, and fiscal and monetary policies designed to benefit poor countries.
In addition, individuals can contribute best to world justice by making personal
charitable contributions.

An effective rejoinder to this argument is that while rich countries and
their citizens may have a duty to assist in the economic development of poor
countries, fulfillment of this duty does not relieve them of the obligation to allow immigration. The two policies are not alternatives; they can be thought of
rather as complements with different sorts of effects. Foreign economic aid,
while never enough and often ineffective, is normally designed to effect small
improvements in the lives of large numbers of people. Immigration enables a
relatively small number of people, the immigrants, to make a radical change in
their living situation. One would like to find a policy that combines the merits
of both, that in a short period of time lifts the economic disadvantage of a large
fraction of the world's people. No such policy exists. In its absence, therefore,
the residents of the world's richer countries have a duty to do what they can on
all fronts. The case for open borders is not refuted just because other ways exist
for Americans to redress injustice.

4. *"Americans have a special obligation to their fellow citizens." The fourth*
argument for immigration controls is that Americans have greater obligations
to their fellow citizens than to foreigners. If there is a reasonable possibility that
immigration may hurt some U.S. citizens, therefore, Americans are justified in
restricting entry, even though restrictions may violate the rights of foreigners.

Consider a parallel case that is clearer. A middle-class couple has enough
savings to provide either a basic education for a poor orphan or supplementary
educational enrichment for their own child. The orphan has the greater disadvantage and the greater need, but few would criticize the parents for deciding
that their priority was to provide the best possible education for their own
child. Can Americans not make a similar case relating to their fellow citizens?

Perhaps they can. Perhaps one can posit a series of widening circles, outside each of which our obligations diminish. Our greatest obligation is to our
family, then perhaps to our friends, then to our neighbors, then perhaps to our
ethnic group, then to our fellow citizens, and finally to all people. No doubt a
great many people think this way. To test oneself, one might think back to the
spring and summer of 1994 when hundreds of thousands of innocent Rwandans were murdered and ask whether that event occasioned as much moral
anger in oneself as did a smaller, more local violation of rights, such as the bur-
glary of one's home. Most Americans had some sympathy for the Rwandans,
but not as much as they would have, had they been neighbors.

People think this way, but that does not mean that this kind of thinking is
morally justified. Before accepting this view as the proper basis for national immigration policy, one should understand how devastating it is to the proposition with which this inquiry began: that all people are morally equal and as a
consequence have equal rights. The fourth argument denies this "Truth" of the

Declaration of Independence. If the argument is accepted, Americans have greater moral standing than foreigners and more rights, at least as seen through the eyes of Americans.

I think we must concede that kin have greater standing than nonrelatives in the moral considerations of most people and that this is natural and good. In all likelihood, it is an evolutionary necessity. Further departures from the norm of equal moral worth seem to me, however, to be unjustified. The failure to recognize the full humanity of people who are different from oneself—people who have different ethnic backgrounds, languages, religions, citizenships, and so forth—is at the root of much of the warfare and suffering in the word today. We may feel a more natural connection to people who are like us in some ways, but if we act on that feeling, we are in danger of creating terribly unjust situations as, for example, in Rwanda and Bosnia.

I do not mean to disparage patriotism. Love of one's country can be an affirming, healthy sentiment. Yet an ethical patriot recognizes that his or her patriotism does not diminish the rights of others. It is critical that patriotism and other kinds of group identity be grounded in the understanding that everyone else has equal rights and that one's own group is not justified in reducing the rights of other groups in any way.

Although it is tempting, therefore, the fourth argument for immigration controls fails. If we develop public policy on the presumption of unequal worth, we are heading down a path toward conflict and selfishness, not moral clarity.

5. "Americans are not obliged to be heroic." *The fifth argument for immigration controls is that although people may have an obligation to come to the aid of other people in distress, that obligation holds only if it can be done without undue sacrifice. The idea is sometimes called the "the principle of mutual aid" or "the cutoff for heroism." To incur or risk great sacrifice on behalf of someone else is commendable because it is heroic, but because it is heroic it is not required. Ordinary people have an obligation to take ordinary care for their fellow humans, but not to be saints (Wolf, 1982). Andrew Shacknove describes the principle this way:*

"In its classic form, the principle of mutual aid envisions a passer—by who encounters a drowning child. If the passer-by can easily save the child, she is morally obligated to do so, even if the child is a complete stranger. Failure to save the child under such conditions would be universally condemned as callousness. If, to the contrary, the passer-by could hope to rescue the child only at great personal risk, no one would chastise her for not acting. The passer-by is not expected to sacrifice life or limb because of a chance encounter with a stranger. Doing so is heroic, but heroism by its nature is voluntary; it exceeds the limits of obligation" (Shacknove, 1988).

The analogy to immigration policy is obvious. If the cutoff for heroism is a valid principle and if it takes precedence over other moral principles, if we are required to help our fellow human beings only when doing so is not very difficult or dangerous, then we may be rescued from the obligation to open our borders to all comers. For in the long run at least, open borders would probably entail a real sacrifice for many Americans. Immigrants come to the U.S. partly because of its high standard of living; they would likely continue to come until they had driven down wages and driven up unemployment to such an extent that immigration was no longer so attractive.

Perhaps we can say, then, that morality does not require such a heroic sacrifice. Americans are obliged to do what they can to alleviate suffering in the world, as long as they do not have to pay a significant price.

The argument is respectable, but it is too comforting to the rich and privileged to stand unmodified. It is a justification for continued injustice. If it is applicable to Americans, why is it not equally applicable to anyone in a position of privilege? Could it not be used by any reactionary to justify keeping his status and income: by the King of France before the Revolution, by southern slave holders before the Civil War, by South African whites before the transfer of power? Any readjustment of a privilege—disadvantage relationship calls for sacrifice on the part of the privileged. If the principle of the cutoff for heroism is dominant, if it holds greater weight than the principle of the equal moral worth of all people, then the privileged are morally justified in maintaining their privilege.

Attractive though the fifth objection is, therefore, it fails as an ethical justification for immigration controls. The cutoff for heroism may be a valid principle when a passer-by is considering saving a drowning child, because in that case the two people are not connected by a relationship of privilege and subordination. When it is used as an argument for protecting privilege, however, it does not have greater moral weight than the principle of equal worth or even the Golden Rule.

6. *"Immigration controls protect the disadvantaged." The sixth and strongest argument against open borders is that immigration controls may not protect the privileged at all, but rather the unprivileged. The discussion so far has tacitly assumed that all Americans are the same, but they are not. Some are privileged while some are severely disadvantaged: the poor, the unskilled, and many members of non-Anglo ethnic groups.*

It must be conceded that the research conducted so far on the impact of immigration on the prospects of disadvantaged Americans is not definitive. Nevertheless, it is likely that an increased and unending supply of low-wage labor from Third World countries would keep the earnings of unskilled workers low and profits high, thereby increasing the gap between the poor and the rich in the U.S. If so, it is argued, immigration should be curtailed.

Sometimes this argument is mixed with either the fourth or the fifth argument. It should not be, because those arguments can be refuted, as we have seen. The sixth argument should not depend on the assertion that Americans are more worthy than foreigners; they are not. Neither should it depend on the assertion that immigration should require no sacrifice from Americans; sacrifice may be morally required.

Can the sixth argument be refuted successfully? One way of attempting to do so would be to say that immigration policy need not bear the burden of reducing the inequities among Americans. Many other policy tools are available, one might argue, among them welfare, education, social insurance programs, the minimum wage, job training, etc. In the conservative political atmosphere of the mid-1990s, these policies are being de-emphasized—but if the country had a commitment to reducing poverty, it could reverse those trends. If it did, it could promote social justice at home while still allowing increased immigration. Put differently, the obligation to rectify the imbalances of privilege in the world rests on the shoulders of the privileged in the rich countries, not on everyone in those countries. Morality requires them to transfer resources to their less fortunate brethren at home *and* to allow immigrants from the Third World to enter their country.

This refutation is valid, but only up to a point. Certainly rich and middle-class Americans cannot ethically argue that the welfare of their poor fellow citizens requires them to discriminate against poor foreigners when they are actually reducing their help to their own people. Still, it is likely that, even if the U.S. made a massive good-faith effort to reduce domestic poverty, it could never be successful if completely open immigration were permitted. Whatever advances were made in the welfare of the U.S. poor would just be swallowed up by new immigrants from poor countries who were seeking to take advantage of American generosity.

A second possible refutation is to assert that the needs of foreigners are more pressing than the needs of even disadvantaged Americans. The relative gap between the rich and the poor in the U.S. is substantial and shameful, but it is far from the largest in the world; the gap in many Third World countries, including the largest senders of immigrants to the U.S., is greater. The standard of living of most poor Americans is higher than the *average* standard of living in most poor countries and certainly higher than the standard of living of most immigrants. Many Mexicans, for example, enter the U.S. to earn wages that, while low by American standards, are several multiples of what they could earn at home. In the hierarchy of advantage, therefore, poor Americans occupy an intermediate position, much worse off than other Americans, but still better off than many immigrants. Therefore, one might maintain, it is morally permissible to harm poor Americans, if this is the price that must be paid to improve the lot of poor immigrants.

This refutation is not persuasive. Poor Americans are genuinely needy and unfairly impoverished. It simply cannot be right to take conscious, public action to worsen their plight. Morality obliges us to protect the welfare not just of the most disadvantaged people in the world, but of all who suffer disadvantage.

The sixth argument in favor of border controls is therefore valid. So, however, is the argument in the previous section for open borders. We are left with an ethical conflict. Disadvantaged foreigners have a moral right to enter the world's most privileged country, in an attempt to improve their position. On the other hand, poor Americans have the right of protection against so much competition from low-income newcomers that their own circumstances deteriorate. The conflict cannot be resolved completely; in the end, the best solution may be a compromise.

Although the conflict of rights cannot be resolved completely, it can be resolved partially. Recall the point made above that as long as the U.S. is a rich country, with a majority of its people very well off or at least comfortable, it is not ethical to depend upon immigration restrictions alone to improve the position of poor Americans. Such a policy would put the burden of sacrifice on even poorer foreigners, rather than on comfortable Americans, the truly privileged among equally worthy human beings.

The sixth argument is therefore a valid justification for immigration restrictions only (1) if the restrictions are accompanied by a major national commitment to improve the quality of life of the U.S. poor and (2) if, in the absence of restrictions, the flow of immigrants would be too great to allow that program to be successful.

Conclusion

An ethically defensible immigration policy, based on the principle of equal moral worth, would be just one component of an integrated program in which rich and middle-class Americans fulfilled their obligations to those less fortunate. The policy would harm no one who was in an already disadvantaged position, would help as many disadvantaged people as possible, and would require sacrifice on the part of the privileged.

The program would consist of at least the following elements. First, privileged Americans (the majority in the U.S.) would greatly increase their commitment to improving the well-being of the American poor, by transferring such a significant portion of their resources to them that they bore a real sacrifice. Second, the U.S. would increase its commitment to foreign aid and other ways of improving the standard of living in poor countries.

Third, the U.S. would raise the flow of immigrants, in order to increase the number of foreigners who had a chance of participating in the advantages of American life. Two reasons exist for thinking that immigration could be increased substantially without hurting the domestic poor: (1) the fact that empirical research can detect little effect of current immigration upon the welfare of the U.S. poor and (2) the increased commitment to the poor postulated in the previous paragraph.

Fourth, the U.S. would maintain some quantitative limit on immigration, or at least the stand-by authority to impose a limit if the flow of immigration became too great to be absorbed without sacrifice by the U.S. poor.

Fifth, if the U.S. had to restrict the number of immigrants, and therefore had to choose among applicants, it would give first priority to the neediest, the refugees who are increasing in number around the world and who have no secure place to live, even if this meant cutting back on immigrants who were admitted for reasons of employment or family reunification.

In the end, the case for completely open borders can probably not be sustained; some limits on immigration may be required. Note, however, how different this argument is from the usual discourse. The usual view is that American immigration policy should maximize the interests of American residents. The argument I am making is that privileged Americans have a moral obligation to redress the balance of privilege in the world, through immigration policies and other policies. How exactly to do this is complex, because the structure of privilege and disadvantage in the world is not straightforward. Any adjustment of that relationship in the direction of greater equity, however, will require sacrifice by those who currently have the advantage, not the enhancement of their own self-interest.

This conclusion is troubling because it is so completely at variance not only with U.S. immigration policy, but also with every country's policy and with the opinions of almost every American. Faced with such overwhelming rejection of one's reasoning, a prudent person should consider the possibility of being in error. Of course, I may be. I think, however, that what has been uncovered is a fundamental dilemma. Americans are not about to admit so many immigrants that they bear a significant cost. For those who are comfortable simply with the pursuit of self-interest, this will not cause any sleepless nights. Those Americans who are genuinely altruistic, however, those who try to think out their positions from an ethical perspective and who believe in the equal worth of all people, those most admirable of Americans need to come to terms with the fact that they are among the world's privileged (their differential location on that scale in the U.S. notwithstanding), that immigration controls protect their privilege, and that they are not going to abandon that protection.

It is a dilemma familiar to thoughtful people, if not necessarily in the context of immigration. Who has not, at least momentarily, compared his or her life to Mother Teresa's and concluded that it is wanting in courage and ethics? Can those of us who are not willing to abandon our worldly goods and devote our lives completely to the service of the poor claim to live moral lives?

This investigation into the ethics of immigration policy has concluded that, if completely open borders are not required, Americans should still welcome immigrants to their country in such large numbers that they sacrifice and that their position of privilege in the world is reduced. We can be certain that this conclusion is not going to be translated into policy. Almost no American would stand for it. We are up against the uncomfortable realization, therefore, that our actions to maintain our standard of living through our immigration policy are inherently immoral. If this understanding does not change our behavior, it may at least clarify where we stand.

POSTSCRIPT

Should Immigration to the United States and Other Developed Countries Be Curtailed?

Nowak attacks six basic arguments in favor of immigration: (1) immigration contributes little to U.S. population growth; (2) historic levels of immigration are higher than the current level; (3) high consumption levels are more to blame for environmental degradation than population growth; (4) environmental gains can be made despite population growth; (5) the United States should focus on international development, not immigration policy; and (6) immigrants are being unfairly blamed for America's environmental problems.

With respect to population projections, Nowak asserts that the government's own figures reveal that 60 percent, or 75 million, of the projected population growth of 125 million in the United States between now and 2050 will be attributed to immigration, making it more than simply a marginal factor. The second argument really has no bearing, says Nowak, for many reasons. With respect to the third argument about consumption, Nowak argues that reducing per capita consumption will mean little if population numbers are allowed to rise dramatically. The fourth argument cannot be answered in the short run because America has witnessed environmental losses as well. The argument that the United States should focus on international development, asserts Nowak, ignores the fact that immigration policy and foreign policy goals can be complementary. Finally, Nowak simply dismisses the last argument about unfair treatment.

Isbister's arguments flow from the concept of "equal moral worth of all people," the core of the Declaration of Independence and the Gettysburg Address. Thus, he attacks what he terms "six ethical arguments" in favor of immigration controls, made in ascending order of importance. The bottom line is that he finds the first five to be "rationalizations used to perpetuate a structure of advantage and disadvantage in the world while allowing Americans to avoid this truth . . . [thus] their moral logic fails." Isbister admits that the sixth argument has some validity, leaving him with a conflict of rights. The five arguments that Isbister deems without merit are the following: (1) "immigration is theft" because U.S. wealth "is the property of Americans, to treat as they wish"; (2) "open immigration would destroy important American values"; (3) "we have better ways of handling the world's disadvantaged"; (4) "Americans have a special obligation to their fellow citizens"; and (5) "Americans are not obliged to be heroic."

Isbister finds the first argument untenable for one basic reason: Americans found "a lovely beach" through no good deeds of their own, so they have no right to deny it to others. The author counters the second argument about the threat to values by asserting that there is no real evidence of such danger. In response to the third argument, Isbister maintains that fulfillment of the duty to assist in economic development "does not relieve [Americans] of the obligation to allow immigration." For the fourth argument, the author suggests that "if we develop public policy on the presumption of unequal worth, we are heading down a path toward conflict and selfishness, not moral clarity." To the fifth argument, Isbister says, "It fails as an ethical justification for immigration controls." To the sixth argument—"immigration controls protect the disadvantaged"—the author counters that disadvantaged non-Americans "have a moral right to enter the world's most privileged country . . . [while] poor Americans have the right of protection against so much competition from low-income newcomers that their own circumstances deteriorate."

Nowak provides a succinct alternative view to each of the pro-immigration positions that he outlines. In short, if one views America's role in the global arena as one of simply advancing the interests of its own citizens, then Nowak's viewpoint makes sense. Isbister sets up his own set of arguments to attack. For him, humanitarian obligations do not stop at a nation's shores. The key question in both cases is whether America's own interests and the interests of the larger global society are one and the same, somewhat similar, or rather different. If one believes that America has an obligation primarily to its own citizens, then Nowak's assessment makes sense. If one holds that America's humanitarian obligations go far beyond its own borders, then Isbister provides sufficient intellectual justification for open borders.

Isbister develops his arguments further in *The Immigration Debate: Remaking America* (Kumarian Press, 1996). Kenneth K. Lee, in *Huddled Masses, Muddled Laws: Why Contemporary Immigration Policy Fails to Reflect Public Opinion* (Praeger, 1998), examines immigration policy making from a historical perspective. Two books that examine immigration beyond the American case are W. T. S. Gould and A. M. Findlay, eds., *Population Migration and the Changing World Order* (John Wiley, 1994) and Christian Joppke's *Immigration and the Nation-State: The United States, Germany, and Great Britain* (Oxford University Press, 1999). Also, *The No-Nonsense Guide to International Migration* by Peter Stalker (Verso, 2001) describes why people migrate, how destinations are chosen, and the economic consequences of such movement.

An important government document is the report of the U.S. Commission on Immigration Reform entitled *Becoming an American: Immigration and Immigrant Policy* (U.S. Government Printing Office, 1997). A useful source for other material is the Center for Immigration Studies. One of its post–September 11 studies, *Safety in (Lower) Numbers: Immigration and Homeland Security,* by Mark Krikorian (October 2002), suggests that a new consensus has emerged on the need for tighter immigration enforcement. As the author relates, "Gone are the days when *The Wall Street Journal* repeatedly called for a constitutional amendment that would say 'There shall be open borders.'"

ISSUE 13

Is There a Global Brain Drain?

YES: William J. Carrington, from "International Migration and the 'Brain Drain,'" *The Journal of Social, Political and Economic Studies* (Summer 1999)

NO: Jean M. Johnson and Mark C. Regets, from "International Mobility of Scientists and Engineers to the United States—Brain Drain or Brain Circulation?" National Science Foundation, Division of Science Resources Studies Issue Brief (November 10, 1998)

ISSUE SUMMARY

YES: William J. Carrington, an economist at the Bureau of Labor Statistics, details what he considers an extensive global brain drain, or the migration of the more qualified citizens of the developing world to the richer countries.

NO: Jean M. Johnson and Mark C. Regets, senior analysts in the Division of Science Resources Studies at the National Science Foundation, contend that roughly half of all foreign doctoral recipients leave the United States immediately upon completion of their studies, while others leave some years later, creating a brain circulation rather than a total drain.

One characteristic of globalization is the ease with which individuals can move from one country to another. The visit may be a short one, lasting a month or so. Or it may involve studying abroad, where the length of time is a function of the period needed to complete educational requirements. Finally, it may be an even longer move, approaching permanency, designed to find better employment opportunities. The current issue focuses on this last type of international movement.

The term *brain drain* does not imply the movement of less fortunate individuals within poor societies to wealthier countries in search of jobs, no matter how meager. Developing nations are far less concerned about their unemployable or underemployable residents emigrating to places that offer a potentially better life. To these countries, each emigrant means one less job to find, one less mouth to feed, one less potential health care recipient to provide for, and one less fertile individual to contribute to future population growth. This less fortunate group is viewed as net consumers in their societies rather than effective producers who will make important contributions on a daily basis.

One can acquire a better sense of the nature of the issue by examining the problem from an intracountry perspective. That is, consider the problem of brain drain *within* the United States itself. America is replete with communities that are dying because their young people, particularly those with greater education or sophisticated employment skills, are leaving for greener pastures. The local area is typically a blue-collar town, the industrial base of which is primarily associated with the earlier industrial era but which now faces competition from abroad or from some more modern and, thus, more desirable location in the Sun Belt. Without employment opportunities, the young go off to college and rarely return home, opting instead for jobs in more alluring locations. The result is that the community where the individuals grew up never benefits from the time and effort spent nurturing these people from birth to adulthood. In short, it never receives a return on its initial investment. It does not benefit from the assistance that its "best and brightest" could give in helping to solve local community problems by staying home or returning there after completing their education.

Many scholars suggest that this situation also exists at the international level. Developing countries invest heavily in preparing their citizens for productive lives within their homelands, the argument goes, only to witness their more qualified people leaving for the developed world, where they can pursue careers or the education that will give them a career in their newly adopted home. There is wide agreement among scholars and policymakers about the number and places of origin of individuals who travel to countries of the developed world in pursuit of education and training. There is also agreement that the original country suffers in proportion to the number of individuals who do not return home.

Brain drain is not a new phenomenon. The United Nations Development Programme's *Human Development Report 2001* says that in the 1960s, only 16 percent of all Koreans who studied in the United States returned home. But two recent influences may change the brain drain picture. First, emerging new technologies like the Internet mean that individuals do not need to relocate to the developed world to take commercial advantage of developed market economies. And second, countries of origin as well as receiving countries are developing programs to encourage and help individuals to return home.

The basic issue is twofold. First, to what extent is the brain drain really simply a "brain circulation," where citizens *do* return home after a short period of time to take their places as contributors to their society? And second, what, if anything, should be done by the developed world to slow or stop one-way movement?

The two selections for this issue focus on the first question. William J. Carrington explores the nature and extent of the brain drain to the wealthy sector of the globe. Jean M. Johnson and Mark C. Regets, on the other hand, conclude that society is witnessing a movement away from permanent emigration (brain drain) toward a more temporary visit (brain circulation).

William J. Carrington

 **YES**

International Migration and the "Brain Drain"

P erhaps the oldest question in economics is why some countries are rich while others are poor. Economic theory has emphasized that differences in the educational levels of the population are an important part of the answer and that improved schooling opportunities should raise incomes in developing countries. Yet, while there is little doubt that highly educated workers in many developing countries are scarce, it is also true that many scientists, engineers, physicians, and other professionals from developing countries work in Canada, the United States, and Western Europe. This phenomenon, often referred to as the "brain drain," was noticed as early as the 1960s and has been a contentious issue in the North-South debate ever since. One important implication of the brain drain is that investment in education in a developing country may not lead to faster economic growth if a large number of its highly educated people leave the country. Also, efforts to reduce specific skill shortages through improved educational opportunities may be largely futile unless measures are taken to offset existing incentives for highly educated people to emigrate.

But how extensive is the brain drain? Which countries and regions are especially affected? Do highly educated professionals from developing countries living abroad represent a sizable proportion of the pool of skilled workers in their countries of origin or too small a number to worry about? Unfortunately, attempts to answer these important questions quickly come up against a formidable barrier: there is no uniform system of statistics on the number and characteristics of international migrants. Also, source countries typically do not keep track of emigrants' characteristics; and, although some receiving countries do, their definitions of immigration differ. Thus, it is difficult to measure precisely the flow and levels of education of immigrants. Further, it has only recently become possible to measure the stock of educated workers in each source country—the pool from which brainpower is drained.

The Brain Drain From the OECD to the U.S.A.

Despite the lack of systematic data about international migrants, estimates of the stock of migrants by educational level in member countries of the Organization

From William J. Carrington, "International Migration and the 'Brain Drain,'" *The Journal of Social, Political and Economic Studies,* vol. 24, no. 2 (Summer 1999). Copyright © 1999 by The Council for Social and Economic Studies. Reprinted by permission. Notes omitted.

for Economic Cooperation and Development (OECD) can be constructed using a variety of data sources. The resulting estimates are less than perfect in many respects, but they significantly improve our knowledge of the magnitude of the brain drain. The study on which we based this article (Carrington and Detragiache, 1998) covers migration from sixty-one developing countries accounting for about 70 percent of the total population of developing countries. Because of the lack of data, we have not attempted to estimate the extent of either the brain drain from the former Soviet Union and Eastern Europe, even though casual evidence suggests that it is substantial, or migratory flows among developing countries.

In our study we followed a two-step procedure: first, estimates of the brain drain to the United States were constructed using 1990 U.S. census data and other sources of information. Then, these estimates were used—together with data on migrants to OECD countries other than the United States drawn from the OECD's Continuous Reporting System on Migration—to estimate the extent of the brain drain to all OECD countries. While the resulting estimates should be reasonably precise for migration to the United States (which accounts for 54.3 percent of the total migration from the developing countries in our sample to all OECD countries), they are much more tentative for the brain drain to all OECD countries.

The U.S. census reports whether individuals polled are foreign born and, if they are, their country of origin and the number of years of schooling received is also reported for each individual. After individuals under 25 years of age are eliminated to ensure compatibility with the data on educational attainment described below, all foreign-born individuals in the census are put into one of three broad educational categories: primary (0 to 8 years of schooling), secondary (9 to 12 years of schooling), and tertiary (more than 12 years of schooling). A further adjustment involves subtracting from the group of foreign-born individuals with a tertiary education all graduate students in U.S. universities, using data from the Institute of International Education. This procedure yields, for each developing country in the sample, the number of migrants in the United States in each of the three educational categories. To assess the extent of the brain drain from each country considered, these estimates must be compared with the number of individuals in each educational group who remain in their home country. Doing this requires a breakdown by educational category of the population of each developing country in the sample, for which we rely on a data set recently assembled by Robert Barro and Jong-Wha Lee (Barro and Lee, 1993), which provides the best estimates available to date of educational attainment for individuals more than 25 years of age in a large sample of countries.

Source Countries

The first striking feature of the U.S. migration data is that immigration flows of individuals with no more than a primary education are quite small, both in absolute terms and relative to other educational groupings (about 500,000 individuals out of a total of 7 million immigrants). Foreign-born individuals with little or no education, however, may be undercounted by the census if they are in the country illegally or do not speak English. The largest group of immigrants into the United States (about 3.7 million) consists of individuals with secondary education from other North American countries (understood here to include Central American and Caribbean countries), primarily Mexico. Perhaps surprisingly, the second largest group (almost 1.5 million individuals) consists of highly educated migrants from Asia and the Pacific. Total immigration from South America and, especially, Africa is quite small. It is noteworthy, however, that immigrants from Africa consist primarily of highly educated individuals (about 95,000 of the 128,000 African migrants).

Among the countries in Asia and the Pacific, the biggest source is the Philippines, with 730,000 migrants. Of these, the great majority have a tertiary education. The second largest stock of migrants is from China (400,000), which is split almost equally between the secondary and tertiary educational groups. Both India and Korea have seen more than 300,000 people migrate to the United States. It is striking that more than 75 percent of Indian immigrants have a tertiary education, compared with only 53 percent of Korean immigrants.

The biggest migratory flows from Africa to the United States are from Egypt, Ghana, and South Africa, with more than 60 percent of immigrants from those three countries having a tertiary education. Migration of Africans with only a primary education is almost nil. The picture is quite different for the migratory flows from the Western Hemisphere: Mexico is by far the largest sending country (2.7 million), with the large majority of its migrants (2.0 million) having a secondary education and fewer than 13 percent having a tertiary education. This pattern is also observed for the smaller countries of Central America, but not for the two Caribbean countries for which we have information, for which migrants with a tertiary education are a more substantial percentage of the total (42 percent for Jamaica and 46 percent for Trinidad and Tobago). Finally, migration from South America to the United States is relatively small in absolute numbers, with immigrants split almost equally between the secondary and the tertiary educational groups.

In each sending country, how do the numbers of emigrants compare with the size of the population with a given educational attainment? For most countries, people with a tertiary education have the highest migration rate, with the exceptions of the Central American countries, Ecuador, and Thailand (in Thailand, people with a secondary education and those with a tertiary one have approximately the same migration rates). Thus, migrants to the United States tend to be better educated than the average person in their home (that is, the sending) country, and the proportion of very highly educated people who migrate is particularly high. Also, migration from Central America seems to follow a somewhat different pattern than migration from other developing countries, in that the highest migration rate is for persons with a secondary education, rather than those with a tertiary education. The brain drain to the United States from many Central American and Caribbean countries is substantial: for persons with a tertiary education, immigration rates for virtually all these countries are above 10 percent, and some appear to be 50 percent or even higher. In South

America, the country with by far the largest brain drain is Guyana, from which more than 70 percent of individuals with a tertiary education have moved to the United States; for the rest of the region, the immigration rates for this educational group are much lower. The Islamic Republic of Iran has had a substantial drain of highly educated individuals (more than 15 percent) and so has Taiwan Province of China (8–9 percent).

Problems of Methodology

To construct estimates of the brain drain from developing countries to OECD countries, we have relied on the OECD Continuous Reporting System on Migration. Unfortunately, unlike the U.S. census, this data source does not report the years of schooling that migrants have received. For lack of any practical alternatives, we have assumed that the distribution of immigrants by educational category from each source country is the same for the United States as for other OECD countries. Although this is the only feasible approach, which often produces numbers that are consistent with anecdotal evidence, there are some instances in which it yields implausible results, particularly for countries with low rates of immigration to the United States but high rates to one or more of the other OECD countries. Immigrants to the United States from such countries are likely to be better educated than immigrants to other OECD countries, who thus may be more representative of the source country's population.

A second problem with the data for OECD countries other than the United States lies in the different criteria for classifying individuals as immigrants. Although Australia, Canada, and the United States define an immigrant as a person who was born abroad to noncitizens, most European countries define immigrant status based on the ethnicity or immigration status of the parent. A third difficulty with the OECD data is that they did not permit us to exclude immigrants under the age of 25. Finally, the OECD records immigrants from only the top 5 or 10 countries from which they come to each OECD country. Thus, for example, the OECD figures for Canada would include specific information on the numbers of immigrants from China and Mexico, but not those from Jamaica and El Salvador. This is a problem when emigration flows are significant for the source country but small for the receiving country. Thus, particularly for small countries, our estimates of immigration to OECD countries other than the United States may be seriously understated.

If, as a rule of thumb, we consider estimates to be unreliable when migrants to the United States account for less than one-third of the total of immigrants to all OECD member countries, then all estimates for immigration from the Asian and Pacific countries are reliable with the exceptions of those for Malaysia and Sri Lanka. Turkey is also an exception. Among the remaining countries, the extent of the brain drain to all OECD members is substantial—and it increases significantly compared with the U.S. data for the Islamic Republic of Iran, Korea, and, to a lesser extent, the Philippines. For the Islamic Republic of Iran, the fraction of the population with a tertiary education living in OECD countries is around 25 percent; for Korea, 15 percent; and for the

Philippines, about 10 percent. For Pakistan, the migration rate of individuals with a tertiary education is more than 7 percent, while for India it is about 2.7 percent; these figures, however, fail to take into account the sizable flow of professionals from the Indian subcontinent to Bahrain, Kuwait, Oman, Qatar, and the United Arab Emirates and therefore neglect an important component of the brain drain from the relevant source countries. The migration rate of highly educated individuals from China is about 3 percent.

For Africa, the estimates are unreliable for Algeria, Senegal, and Tunisia, from which migrants go mainly to France. For most other countries in the sample, however, migration to OECD countries other than the United States is quite small, so the results derived for the United States remain essentially valid. There are, however, some exceptions: for Ghana, the migration rate of highly educated individuals is a dramatic 26 percent; for South Africa, it is more than 8 percent; for Egypt, the brain drain includes 2.5 percent of such individuals emigrating to the United States and another 5 percent emigrating to other OECD countries. For countries in the Western Hemisphere, the bulk of migration is to the United States, and inclusion of flows to the rest of the OECD makes little difference. The only exception is Jamaica, which has a considerable stock of migrants living in the United Kingdom. The drain from Jamaica's population with secondary education is 33 percent, while that from its population with tertiary education is more than 77 percent.

Our estimates show that there is an overall tendency for migration rates to be higher for highly educated individuals. With the important exceptions of Central America and Mexico, the highest migration rates are for individuals with a tertiary education. A number of countries especially small countries in Africa, the Caribbean, and Central America lost more than 30 percent of this group to migration. We have also found sizable brain drain from Iran, Korea, the Philippines, and Taiwan Province of China. These numbers suggest that in several developing countries the outflow of highly educated individuals is a phenomenon that policymakers cannot ignore.

Why the Brain Drain?

More research, especially empirical studies, is needed to evaluate the impact of the brain drain on source economies and on worldwide welfare, as well as the reasons for such migration. In regard to the latter subject, immigration policies in OECD countries tend to favor better educated people, which may explain why the educational composition of total migration is skewed toward the better educated but cannot explain why so many skilled workers are willing to leave developing countries.

Wage differentials may be part of the explanation, but this raises the question of what accounts for such differentials. Differences in the quality of life, educational opportunities for children, and job security may also play a role, as may the desire to interact with a broader group of similarly skilled colleagues. Another important issue is the extent to which the benefits of education acquired by citizens of developing countries are externalities that individuals cannot be expected to take into account when making their private decisions. If such externalities are substantial, as is emphasized by the "new growth theory," then policies to curb the brain drain may be warranted.

Our research also indicates several ways in which estimates of the brain drain could be improved using existing data. The first would be to use census information for other large immigrant-receiving countries, such as Australia, Canada, France, and Germany. Together with the United States, these four countries account for about 93 percent of total migratory flows to OECD countries, so the resulting figures would be a very good approximation of the total. Another promising direction for future research would be to try to obtain, from census data or other sources, more detailed information about the occupational categories of highly skilled migrants, in order to assess whether the brain drain from a given country is especially marked for particular professional groups. This type of analysis could be useful for evaluating the problems that policy programs—such as health sector reform, financial liberalization, or civil service reform—may encounter in developing countries.

Jean M. Johnson and Mark C. Regets  **NO**

Brain Drain or Brain Circulation?

Foreign-born scientists and engineers (S&Es) contribute significantly to the brain power of the United States. Considering the U.S. labor force with doctoral degrees in S&E fields, immigrants are 29 percent of those conducting R&D.

Several decades ago, the emigration of such highly skilled personnel to the United States was considered one-way mobility, a permanent brain drain depriving the countries of origin of the "best and the brightest." More recently, however, the mobility of highly talented workers is referred to as "brain circulation," since a cycle of study and work abroad may be followed by a return to the home country to take advantage of high-level opportunities. What do the data tell us about foreign-born S&E personnel in the United States? Are we seeing brain drain or brain circulation? This issue brief discusses student flows into U.S. higher education, the stay rates of foreign doctoral recipients, and their short- and long-term employment in U.S. industry, universities, and Government.

U.S. Higher Education and Foreign S&E Graduate Students

The large foreign component of U.S. human intellectual capital is linked to the ability of U.S. higher education to attract, support, and retain foreign S&E graduate students. Foreign students, particularly those from Asia, represent a large fraction of enrollment and degrees in S&E fields in U.S. graduate institutions. In 1995, of the 420,000 graduate students in S&E programs, roughly 100,000 were foreign students, mainly from a dozen countries of origin. In 1995, at the doctoral level, foreign students (including those with permanent and temporary visas) earned 39 percent of the natural science degrees, 50 percent of the mathematics and computer sciences degrees, and 58 percent of the engineering degrees. Students from China, India, South Korea, and Taiwan accounted for over half of these S&E doctorates.

Financial support available from academic research activities appears to be a major factor associated with attracting foreign students to U.S. doctoral programs. More than 75 percent of the 10,000 foreign doctoral recipients at U.S. universities in 1996 reported their universities as the primary source of support

From Jean M. Johnson and Mark C. Regets, "International Mobility of Scientists and Engineers to the United States—Brain Drain or Brain Circulation?" National Science Foundation, Division of Science Resources Studies Issue Brief (November 10, 1998). Washington, DC: U.S. Government Printing Office, 1998. Notes omitted.

for their graduate training. Of those who did so, the majority reported that their primary support came in the form of research assistantships. Financial resources for research assistantships are provided to universities by Federal Government agencies, industry, and other non-Federal sources in the form of research grants. At the same time that academic research expenditures have been growing, the number of foreign doctoral students supported by university S&E departments has also been increasing. From 1985–96, academic research expenditures increased from $13 to $21 billion in constant (1992) dollars. During the same period, the number of foreign doctoral students primarily supported as research assistants more than tripled—from 2,000 in 1985 to 7,600 in 1996.

Between 1988 and 1996, foreign students from major Asian and European countries, Canada, and Mexico earned over 55,000 U.S. S&E doctoral degrees.

During this period, about 63 percent of these doctoral recipients planned to remain in the United States after completion of their studies, and about 39 percent had firm plans to do so. The proportion of foreign students who remain in the United States, referred to as the "stay rate," differs widely by country. In the last decade, approximately half of the foreign doctoral recipients from China and India have sought and received firm opportunities for further study and employment in the United States. In contrast, only 23 percent of the doctoral recipients from South Korea and 28 percent from Taiwan accepted firm offers to remain in the United States.

What Do Foreign S&E Doctoral Recipients Who Stay in the United States Do?

Foreign S&E doctoral recipients remaining in the United States do so mainly by entering postdoctoral study. Of the 55,000 foreign students from these major countries of origin who earned S&E doctoral degrees between 1988 and 1996, about 22 percent (12,000) stayed on for postdoctoral study, and 17 percent (9,000) accepted employment in the United States.

Firm employment offers to foreign doctoral recipients are strongly geared toward research and development, mainly within business or industry. The decision to remain for a postdoctoral appointment is, not surprisingly, greatest in fields where post-doctorates are a common career path. Over 50 percent of all foreign students earning doctoral degrees in the biological sciences remained in the United States for postdoctoral experiences; only 5 percent were offered jobs at universities or in industry. In contrast, in computer sciences, only 7 percent remained for postdoctoral experiences, while over 38 percent accepted employment.

Do Foreign S&E Doctoral Recipients Who Plan to Stay Actually Join the U.S. Labor Force?

A recent study of foreign doctoral recipients working and earning wages in the United States (Finn, 1997) shows that about 47 percent of the foreign students on temporary student visas who earned doctorates in 1990 and 1991 were working in the United States in 1995. The majority of the 1990–91 foreign doctoral recipients from India (79 percent) and China (88 percent) were still working in the United States in 1995. In contrast, only 11 percent of South Koreans who completed S&E doctorates from U.S. universities in 1990–91 were working in the United States in 1995.

Do Foreign S&E Doctorates Stay in the United States in the Long Term?

The same study looked at foreign doctoral recipients from 1970–72. Finn estimated that 47 percent were working in the United States in 1995, and that the stay rate for that group was around 50 percent during the 25 years leading up to 1995. There is no evidence of significant net return migration of these scientists and engineers after 10 or 20 years of work experience in the United States. The fairly constant stay rates indicate that any tendency of the 1970–72 cohorts to leave the United States after gaining work experience here has been largely offset by others from the same cohort returning to the United States after going abroad. Remaining in the United States does not represent a complete brain drain on their home country. Choi has shown extensive networking by Asian-born faculty and researchers working in the United States to advise, disseminate information, and assist in building their home-country S&E infrastructure. This is particularly true for the foreign-born faculty in S&E departments. In 1993, foreign-born faculty in U.S. higher education represented 37 percent of the engineering professors and over a quarter of the mathematics and computer science teachers.

Conclusions

Data on mobility and stay rates of foreign-born S&Es working in the United States support the notion of brain circulation for some countries (Taiwan and South Korea) and somewhat more brain drain for other countries (China and India). In the aggregate, roughly half of all foreign doctoral recipients leave the United States immediately after completing their graduate education, and others leave after some years of teaching or industrial experience in the United States. In addition, some of those who remain in the United States network with home-country scientists. More research is needed, however, on the activities of foreign doctoral recipients who return to their home countries. For example, we need to know more about their contributions to their home countries' S&E infrastructure, including research, teaching, and science administration. Also, we need to be able to identify patterns of circulation and lengths of stay that are beneficial to the United States, the countries of origin, and the diffusion of S&E knowledge in the world. Information presented in this issue brief on foreign doctoral recipients and their planned stay rates comes mainly from the forthcoming NSF report: *Statistical Profiles of Foreign Doctoral Recipients in Science and Engineering:*

Plans to Locate in the United States. Data for this report were collected in the Survey of Earned Doctorates (SED) conducted by the National Opinion Research Center (NORC) for NSF and four other Federal agencies. Information on foreign-born scientists and engineers in the U.S. labor force is from the NSF Division of Science Resources Studies (SRS), SESTAT data system, available on the SRS World Wide Web site (http://www.nsf.gov/sbe/srs/).

POSTSCRIPT

Is There a Global Brain Drain?

The perceived brain drain from one country to another is typically, although not exclusively, from developing countries to developed countries. In the minority of cases in which the flow is from one developed country to another, it is likely to be the consequence of the immigrant's inability to maximize his or her employment potential in the home country.

Carrington's article is based on his earlier analysis of migration from 61 developing countries to the United States. His conclusion is that education of the immigrant plays an important role in the immigration decision. The largest group of immigrants in his study came with a secondary education from other North American countries; the second-largest group consisted of highly educated immigrants from Asia and the Pacific. Among this latter group, the number of immigrants from the Philippines is the largest, with a majority having a tertiary education. The same can be said of a majority of immigrants from Africa. Carrington's basic conclusion is that many countries are experiencing a sizable brain drain.

Carrington is quick to point out a number of methodological problems (not of his making) associated with the study. The principal limitation is the large number of unreliable estimates or missing data. Much work needs to be done before definitive statements about the brain drain in the author's sample can be made with confidence.

Johnson and Regets's report begins with the observation that "foreign-born scientists and engineers . . . contribute significantly to the brain power of the United States." The authors caution, however, that one ought not to conclude that the situation qualifies as a brain drain or a one-way ticket from the countries of origin, as was the case several decades ago. Rather, they have coined the phrase "brain circulation" to characterize the current situation, which they suggest features many foreign-born scientists and engineers' returning home after finishing their education or engaging in a cycle of work abroad. Their general conclusion is that about half of all foreign doctoral students leave the United States after obtaining their degrees. The percentage is not uniform among the countries of origin, however. And some of those who stay in the United States network with their counterparts back home, although the exact percentage is unknown.

The United Nations Development Programme's *Human Development Report 2001: Making New Technologies Work for Human Development* (Oxford University Press, 2001) addresses the issue of brain drain from a number of developing countries, paying special attention to India. The report estimates that India loses $2 billion a year because of the immigration of computer professionals to the United States alone. In addition, about 100,000 Indians emigrate each year. The report also alludes to efforts to convince nationals to return home, including programs in South Korea and Taiwan designed to help citizens find jobs to entice them to return home. Other Asian countries are considering punitive measures to address the brain drain problem. One such measure is the imposition of a flat "exit tax" to be paid by the employer or company when a visa is issued. Another idea is a loan system whereby each student in tertiary education receives a loan (study subsidy) from the government that must be repaid if the student leaves the country.

Even host countries have gotten into the act. The National Institutes of Health (NIH) has begun a new program to pay foreign-born scientists who were trained at the NIH to return home. The NIH has also set aside $1 million per year to pay part of the salaries of developing world scientists who return home.

The International Organization for Migration (IOM), a nongovernmental organization based in Switzerland, has created a program to encourage African nationals to return home. Although it is currently a modest operation, it represents a beginning.

The National Science Foundation continues to study the issue, releasing its findings periodically. One such analysis is *Statistical Profiles of Foreign Doctoral Recipients in Science and Engineering: Plans to Stay in the United States* by Jean M. Johnson (November 1998). Hyaeweol Choi examines America's Asian scientific community in *An International Scientific Community: Asian Scholars in the United States* (Praeger, 1995). Xiaonan Cao presents a thoughtful article about the subject in "Debating 'Brain Drain' in the Context of Globalisation," *Compare* (vol. 26, no. 3, 1996). Finally, W. T. Gould and Anne M. Findlay, eds., *Population Migration and the Changing World Order* (John Wiley, 1994) cites several examples of brain drain, with special emphasis on that from Ireland.

Globalization: Threat or Opportunity?

This Web site contains the article "Globalization: Threat or Opportunity?" by the staff of the International Monetary Fund (IMF). This article discusses such aspects of globalization as current trends, positive and negative outcomes, and the role of institutions and organizations. "The Challenge of Globalization in Africa," by IMF acting managing director Stanley Fischer, and "Factors Driving Global Economic Integration," by Michael Mussa, IMF's director of research, are also included on this site.

http://www.imf.org/external/np/exr/ib/2000/041200.htm

Center for Democracy & Technology

The Center for Democracy & Technology (CDT) works to preserve and enhance free expression on the Internet as well as to protect privacy. Democratic values are promoted through public policy advocacy, online grassroots organizing, and public education campaigns. Explore this site to learn about the CDT's proposals for achieving its goals as well as its thoughts on other Internet issues, such as digital authentication, bandwidth, terrorism, and access to government information.

http://www.cdt.org

Popular Culture Association/American Culture Association

The Popular Culture Association/American Culture Association Web site, a member of the H-NET Humanities OnLine initiative, fosters intellectual discussion of popular culture. View the discussion logs by month or by specific topic. Also included on this site is a list of links to sites on specific aspects of American culture.

http://www2.h-net.msu.edu/~pcaaca/

PART 4

The Flow of Information and Ideas

*T*he global information network and the technology that drives it are creating a global communication village. However, the devices that allow people to connect with the world also allow the world to connect with them. Fiber-optic lines carry information, products, ideas, values, and cultures across the globe. The impact has yet to be calculated, but we know that it is both uplifting and unsettling.

This section examines the impact of international communication networks and the actors who manage them on people around the globe.

- Is Globalization a Positive Development for the World Community?

- Will the Digital/Computer World Lead to Greater Individual Freedom?

- Is the Globalization of American Culture a Positive Development?

ISSUE 14

Is Globalization a Positive Development for the World Community?

YES: Thomas L. Friedman, from "DOScapital," *Foreign Policy* (Fall 1999)

NO: Christian E. Weller, Robert E. Scott, and Adam S. Hersh, from "The Unremarkable Record of Liberalized Trade," Economic Policy Institute Briefing Paper (October 2001)

ISSUE SUMMARY

YES: Journalist Thomas L. Friedman argues that globalization is built around three balances, each of which makes the system more individualized and democratic by empowering individuals on the world stage.

NO: Economists Christian E. Weller, Robert E. Scott, and Adam S. Hersh contend that the data on economic liberalization and globalization clearly show that it has failed to alleviate poverty or foster greater development. They call for significant reforms to correct the situation.

Globalization is a phenomenon and a revolution. It is sweeping the world with increasing speed and changing the global landscape into something new and different. Yet, like all such trends, its meaning, development, and impact puzzle many. We talk about globalization and experience its effects, but few of us really understand the forces that are at work in the global political economy.

When people use their cell phones, log onto the Internet, view events from around the world on live television, and experience varying cultures in their own backyards, they begin to believe that this process of globalization is a good thing that will bring a variety of new and sophisticated changes to people's lives. Many aspects of this technological revolution bring fun, ease, and sophistication to people's daily lives. Yet the anti–World Trade Organization (WTO) protests in Seattle, Washington, in 1999 and Washington, D.C., in 2000 are graphic illustrations of the fact that not everyone believes that globalization is a good thing. Many Americans who have felt left out of the global economic boom as well as Latin Americans, Africans, and Asians who feel that their job skills and abilities are being exploited by multinational corporations (MNCs) in a global division of labor believe that this system does not meet their needs. Local cultures that believe that Walmart and McDonald's bring cultural change and harm rather than inexpensive products and convenience criticize the process. In this way, globalization, like all revolutionary forces, polarizes people, alters the fabric of their lives, and creates rifts within and between people.

Many in the West, along with the prominent and elite—among MNCs, educators, and policymakers—seem to have embraced globalization. They argue that it helps to streamline economic systems, disciplines labor and management, brings forth new technologies and ideas, and fuels economic growth. They point to the relative prosperity of many Western countries and argue that this is proof of globalization's positive effects. They see little of the problems that critics identify. In fact, those who recognize some structural problems in the system argue that despite these issues, globalization is like an inevitable tide of history, unfortunate for some but unyielding and impossible to change. Any problems that are created by this trend, they say, can be solved.

Many poor and middle-class workers, as well as hundreds of millions of people across the developing world, view globalization as an economic and cultural wave that tears at the fabric of centuries-old societies. They see jobs emerging and disappearing in a matter of months, people moving across the landscape in record numbers, elites amassing huge fortunes while local cultures and traditions are swept away, and local youth being seduced by promises of American material wealth and distanced from their own cultural roots. These critics look past the allure of globalization and focus on the disquieting impact of rapid and system-wide change.

The irony of such a far-ranging and rapid historical process such as globalization is that both proponents and critics may be right. The realities of globalization are both intriguing and alarming. As technology and the global infrastructure expand, ideas, methods, and services are developed and disseminated to greater and greater numbers of people. As a result, societies and values are altered, some for the better and others for the worse.

In the selections that follow, the authors explore the positive and negative impacts of globalization and come to different conclusions. Thomas L. Friedman argues that globalization has an empowering effect. He surmises that individuals will gain greater control over their own destinies and that life will become more democratic, with greater freedom of choice as a result. Christian E. Weller, Robert E. Scott, and Adam S. Hersh examine a wealth of data and ascertain that poverty has increased, development has not been widespread, and the inequities between the rich and the poor within and among countries are increasing. This, they argue, refutes the notion that globalization supports global economic development.

Thomas L. Friedman

 **YES**

DOScapital

$\mathbf{I}$f there can be a statute of limitations on crimes, then surely there must be a statute of limitations on foreign-policy cliches. With that in mind, I hereby declare the "post-Cold War world" over.

For the last ten years, we have talked about this "post-Cold War world." That is, we have defined the world by what it wasn't because we didn't know what it was. But a new international system has now clearly replaced the Cold War: globalization. That's right, globalization—the integration of markets, finance, and technologies in a way that is shrinking the world from a size medium to a size small and enabling each of us to reach around the world farther, faster, and cheaper than ever before. It's not just an economic trend, and it's not just some fad. Like all previous international systems, it is directly or indirectly shaping the domestic politics, economic policies, and foreign relations of virtually every country.

As an international system, the Cold War had its own structure of power: the balance between the United States and the USSR, including their respective allies. The Cold War had its own rules: In foreign affairs, neither superpower would encroach on the other's core sphere of influence, while in economics, underdeveloped countries would focus on nurturing their own national industries, developing countries on export-led growth, communist countries on autarky [economic self-sufficiency], and Western economies on regulated trade. The Cold War had its own dominant ideas: the clash between communism and capitalism, as well as detente, nonalignment, and perestroika. The Cold War had its own demographic trends: The movement of peoples from East to West was largely frozen by the Iron Curtain; the movement from South to North was a more steady flow. The Cold War had its own defining technologies: Nuclear weapons and the Second Industrial Revolution were dominant, but for many developing countries, the hammer and sickle were still relevant tools. Finally, the Cold War had its own defining anxiety: nuclear annihilation. When taken all together, this Cold War system didn't shape everything, but it shaped many things.

Today's globalization system has some very different attributes, rules, incentives, and characteristics, but it is equally influential. The Cold War sys-

From Thomas L. Friedman, "DOScapital," *Foreign Policy,* no. 116 (Fall 1999). Copyright © 1999 by The Carnegie Endowment for International Peace. Reprinted by permission of *Foreign Policy;* permission conveyed via The Copyright Clearance Center, Inc.

tem was characterized by one overarching feature: division. The world was chopped up, and both threats and opportunities tended to grow out of whom you were divided from. Appropriately, that Cold War system was symbolized by a single image: the Wall. The globalization system also has one overarching characteristic: integration. Today, both the threats and opportunities facing a country increasingly grow from whom it is connected to. This system is also captured by a single symbol: the World Wide Web. So in the broadest sense, we have gone from a system built around walls to a system increasingly built around networks.

Once a country makes the leap into the system of globalization, its elite begin to internalize this perspective of integration and try to locate themselves within a global context. I was visiting Amman, Jordan, in the summer of 1998 when I met my friend, Rami Khouri, the country's leading political columnist, for coffee at the Hotel Inter-Continental. We sat down, and I asked him what was new. The first thing he said to me was "Jordan was just added to CNN's worldwide weather highlights." What Rami was saying was that it is important for Jordan to know that those institutions that think globally believe it is now worth knowing what the weather is like in Amman. It makes Jordanians feel more important and holds out the hope that they will profit by having more tourists or global investors visiting. The day after seeing Rami I happened to interview Jacob Frenkel, governor of the Bank of Israel and a University of Chicago-trained economist. He remarked to me: "Before, when we talked about macroeconomics, we started by looking at the local markets, local financial system, and the interrelationship between them, and then, as an afterthought, we looked at the international economy. There was a feeling that what we do is primarily our own business and then there are some outlets where we will sell abroad. Now, we reverse the perspective. Let's not ask what markets we should export to after having decided what to produce; rather, let's first study the global framework within which we operate and then decide what to produce. It changes your whole perspective."

Integration has been driven in large part by globalization's defining technologies: computerization, miniaturization, digitization, satellite communications, fiber optics, and the Internet. And that integration, in turn, has led to many other differences between the Cold War and globalization systems.

Unlike the Cold War system, globalization has its own dominant culture, which is why integration tends to be homogenizing. In previous eras, cultural homogenization happened on a regional scale—the Romanization of Western Europe and the Mediterranean world, the Islamization of Central Asia, the Middle East, North Africa, and Spain by the Arabs, or the Russification of Eastern and Central Europe, and parts of Eurasia, under the Soviets. Culturally speaking, globalization is largely the spread (for better and for worse) of Americanization from Big Macs and iMacs to Mickey Mouse.

Whereas the defining measurement of the Cold War was weight, particularly the throw-weight of missiles, the defining measurement of the globalization system is speed—the speed of commerce, travel, communication, and innovation. The Cold War was about Einstein's mass-energy equation, $e = mc^2$.

Globalization is about Moore's Law, which states that the performance power of microprocessors will double every 18 months. The defining document of the Cold War system was "the treaty." The defining document of the globalization system is "the deal." If the defining anxiety of the Cold War was fear of annihilation from an enemy you knew all too well in a world struggle that was fixed and stable, the defining anxiety in globalization is fear of rapid change from an enemy you cannot see, touch, or feel—a sense that your job, community, or workplace can be changed at any moment by anonymous economic and technological forces that are anything but stable.

If the defining economists of the Cold War system were Karl Marx and John Maynard Keynes, each of whom wanted to tame capitalism, the defining economists of the globalization system are Joseph Schumpeter and Intel chairman Andy Grove, who prefer to unleash capitalism. Schumpeter, a former Austrian minister of finance and Harvard University professor, expressed the view in his classic work *Capitalism, Socialism, and Democracy* (1942) that the essence of capitalism is the process of "creative destruction"—the perpetual cycle of destroying old and less efficient products or services and replacing them with new, more efficient ones. Grove took Schumpeter's insight that only the paranoid survive for the title of his book about life in Silicon Valley and made it in many ways the business model of globalization capitalism. Grove helped popularize the view that dramatic, industry-transforming innovations are taking place today faster and faster. Thanks to these technological breakthroughs, the speed at which your latest invention can be made obsolete or turned into a commodity is now lightning quick. Therefore, only the paranoid will survive— only those who constantly look over their shoulders to see who is creating something new that could destroy them and then do what they must to stay one step ahead. There will be fewer and fewer walls to protect us.

If the Cold War were a sport, it would be sumo wrestling, says Johns Hopkins University professor Michael Mandelbaum. "It would be two big fat guys in a ring, with all sorts of posturing and rituals and stomping of feet, but actually very little contact until the end of the match, when there is a brief moment of shoving and the loser gets pushed out of the ring, but nobody gets killed." By contrast, if globalization were a sport, it would be the 100-meter dash, over and over and over. No matter how many times you win, you have to race again the next day. And if you lose by just one one-hundredth of a second, it can be as if you lost by an hour.

Last, and most important, globalization has its own defining structure of power, which is much more complex than the Cold War structure. The Cold War system was built exclusively around nation-states, and it was balanced at the center by two superpowers. The globalization system, by contrast, is built around three balances, which overlap and affect one another.

The first is the traditional balance between nation-states. In the globalization system, this balance still matters. It can still explain a lot of the news you read on the front page of the paper, be it the containment of Iraq in the Middle East or the expansion of NATO against Russia in Central Europe.

The second critical balance is between nation-states and global markets. These global markets are made up of millions of investors moving money around the world with the click of a mouse. I call them the "Electronic herd." They gather in key global financial centers, such as Frankfurt, Hong Kong, London, and New York—the "supermarkets." The United States can destroy you by dropping bombs and the supermarkets can destroy you by downgrading your bonds. Who ousted President Suharto in Indonesia? It was not another superpower, it was the supermarkets.

The third balance in the globalization system—the one that is really the newest of all—is the balance between individuals and nation-states. Because globalization has brought down many of the walls that limited the movement and reach of people, and because it has simultaneously wired the world into networks, it gives more direct power to individuals than at any time in history. So we have today not only a superpower, not only supermarkets, but also super-empowered individuals. Some of these super-empowered individuals are quite angry, some of them quite constructive but all are now able to act directly on the world stage without the traditional mediation of governments or even corporations.

Jody Williams won the Nobel Peace Prize in 1997 for her contribution to the International Campaign to Ban Landmines. She managed to build an international coalition in favor of a landmine ban without much government help and in the face of opposition from the major powers. What did she say was her secret weapon for organizing 1,000 different human rights and arms control groups on six continents? "E-mail."

By contrast, Ramzi Ahmed Yousef, the mastermind of the February 26, 1993, World Trade Center bombing in New York, is the quintessential "super-empowered angry man." Think about him for a minute. What was his program? What was his ideology? After all, he tried to blow up two of the tallest buildings in America. Did he want an Islamic state in Brooklyn? Did he want a Palestinian state in New Jersey? No. He just wanted to blow up two of the tallest buildings in America. He told the Federal District Court in Manhattan that his goal was to set off an explosion that would cause one World Trade Center tower to fall onto the other and kill 250,000 civilians. Yousef's message was that he had no message, other than to rip up the message coming from the all-powerful America to his society. Globalization (and Americanization) had gotten in his face and, at the same time, had empowered him as an individual to do something about it. A big part of the U.S. government's conspiracy case against Yousef (besides trying to blow up the World Trade Center in 1993, he planned to blow up a dozen American airliners in Asia in January 1995) relied on files found in the off-white Toshiba laptop computer that Philippine police say Yousef abandoned as he fled his Manila apartment in January 1995, shortly before his arrest. When investigators got hold of Yousef's laptop and broke into its files, they found flight schedules, projected detonation times, and sample identification documents bearing photographs of some of his co-conspirators. I loved that—Ramzi Yousef kept all his plots on the C drive of his Toshiba laptop! One should have no illusions, though. The super-empowered angry men are out there, and they present the most immediate threat today to the United States and the stability of the new globalization system. It's not because Ramzi Yousef can ever be a super-power. It's because in today's world, so many people can be Ramzi Yousef.

So, we are no longer in some messy, incoherent "post-Cold War world." We are in a new international system, defined by globalization, with its own moving parts and characteristics. We are still a long way from fully understanding how this system is going to work. Indeed, if this were the Cold War, the year would be about 1946. That is, we understand as much about how this new system is going to work as we understood about how the Cold War would work in the year Churchill gave his "Iron Curtain" speech.

Nevertheless, it's time we recognize that there is a new system emerging, start trying to analyze events within it, and give it its own name. I will start the bidding. I propose that we call it "DOScapital."

NO

Christian E. Weller, Robert E. Scott, and Adam S. Hersh

The Unremarkable Record of Liberalized Trade

Recently, a growing number of policy makers have touted the potential for global economic integration to combat poverty and economic inequity in the world today. On September 24, 2001, for instance, U.S. Trade Representative Robert Zoellick (2001), arguing for new "fast track" trade promotion authority, cited a World Bank study claiming that globalization "reduces poverty because integrated economies tend to grow faster and this growth is usually widely diffused" (World Bank 2001a, 1). Yet the empirical evidence suggests that reductions in poverty and income inequality remain elusive in most parts of the world, and, moreover, that greater integration of deregulated trade and capital flows over the last two decades has likely undermined efforts to raise living standards for the world's poor.

In 1980, median income in the richest 10% of countries was 77 times greater than in the poorest 10%; by 1999, that gap had grown to 122 times. Inequality has also increased within many countries. Over the same period, any gains in poverty reduction have been relatively small and geographically isolated. The number of poor people rose from 1987 to 1998, and the share of poor people increased in many countries—in 1998 close to half the population were considered poor in many parts of the world. In 1980, the world's poorest 10%, or 400 million people, lived on 72 cents a day or less. The same number of people had 79 cents (nominally) per day in 1990 and 78 cents in 1999.

While many social, political, and economic factors contribute to poverty, the evidence shows that unregulated capital and trade flows contribute to rising inequality and impede progress in poverty reduction. Trade liberalization leads to more import competition and to a growing use of the threat to move production to lower-wage locales, thereby depressing wages. Deregulated international capital flows have led to rapid increases in short-term capital flows and more frequent economic crises, while simultaneously limiting the ability of governments to cope with crises. Economic upheavals disproportionately harm the poor, and thus contribute to the lack of success in poverty reduction and to rising income inequality.

The world's poor may stand to gain from global integration, but not under the unregulated version currently promoted by the World Bank and others. The lesson of the past 20 years is clear: it is time for a different approach to global integration, whereby living standards of the world's poor are raised rather than jeopardized.

Deregulated Global Trade and Capital Markets As the Culprit

Over the past decades international capital mobility has grown as capital controls were reduced or eliminated virtually everywhere. Consequently, capital flows to developing countries have grown rapidly, from $1.9 billion in 1980 to $120.3 billion in 1997, the last year before the global financial crisis, or by more than 6,000%. Even in 1998, in the wake of the financial crisis, capital flows remained remarkably high at $56 billion. A substantial share of these capital flows (e.g., 36% in 1997) consisted of short-term portfolio investments (IMF 2001b).

Faster capital mobility in a relatively deregulated environment leads to rising inequality, both within countries and between countries, and to less poverty reduction or even increasing poverty. . . .

The burdens of financial crisis are disproportionately borne by a country's poor. Since higher-income earners have better access to insurance mechanisms that protect them from the fallout of a crisis (including capital flight), macro-economic crises lead to a more unequal income distribution within countries (Lustig 2000). Thus, economic crises increase the need for well-functioning social safety nets. Yet unfettered capital flows limit governments' abilities to design policies to help the poor when they need it most—in the middle of a crisis. The International Monetary Fund often opposes increased government expenditures to assist the poor during economic crises, and investors withdraw their funds following increased government expenditures (Blecker 1999).

Finally, developing countries are prone to experience more severe economic crises with greater frequency than are developed economies (Lustig 2000; Lindgren, Garcia, and Saal 1996), leading to greater inequality between countries.

Trade liberalization—the complement to deregulated capital markets in the global deregulation agenda—also plays a significant role in raising inequality and limiting efforts at poverty reduction. By inducing rapid structural change and shifting employment within industrializing countries, trade liberalization leads to falling real wages and declining working conditions and living standards (Bannister and Thugge 2001; Scott et al. 1997; Scott 2001a; Scott 2001b; Mishel et al. 2001).

Trade liberalization also gives teeth to employers' threats to close plants or to relocate or outsource production abroad—where labor regulations are less stringent and more difficult to enforce—and undermines workers' attempts to organize and bargain for improved wages and working conditions (Bronfenbrenner 1997, 2000). This trend fuels a race to the bottom in which national governments vie for needed investment by bidding down the cost to employers (and livings standards) of working people.

The connection between rapid trade liberalization and inequality appears to be universal, indicating downward wage pressures and rising inequality following trade liberalization in industrializing and industrialized economies (USTDRC 2000). A report by UNCTAD (1997) found that trade liberalization in Latin America led to widening wage gaps, falling real wages for unskilled workers (often more than 90% of the labor force in developing countries), and rising unemployment.

Rising Inequality Is Common Within Many Countries

Defenders of the current regime of global deregulation, including the World Bank, acknowledge that inequality has increased within countries. But in its most recent and rather comprehensive document on globalization and poverty (World Bank 2001a), the Bank raised two issues that supposedly mute the fact of rising intra-country inequality. First, data for China dwarfs observations for all other countries, thereby suggesting that rising inequality in globalizing countries does not exist outside of China (World Bank 2001a, 47). However, data for other countries show that growing inequality is indeed a widespread trend. Second, the World Bank also claimed that rising inequality is not a result of increasing poverty, which thus makes it presumably less troubling (World Bank 2001a, 48). While this claim may hold true in China, it does not describe the trend in many other parts of the world.

There is a broad consensus that income inequality has risen in industrialized countries since 1980. The World Bank reports that there was a "serious . . . increase in within-country inequality in industrialized countries reversing the trend of [the period 1950–80]" (World Bank 2001a, 46). Similarly, Gottschalk and Smeeding (1997, 636) found that "almost all industrial economies experienced some increase in wage inequality among prime-aged males" in the 1980s and early 1990s. Further, data from the Luxembourg Income Study (LIS 2001) show that, among 24 countries, 18 experienced increasing income inequality, five (Denmark, Luxembourg, the Netherlands, Spain, and Switzerland) experienced declining inequality, and one (France) saw no change.

Income inequality is also rising in industrializing countries. There has been an unambiguous rise in inequality in Latin America in the 1980s and 1990s (Lustig and Deutsch 1998; IADB 1999; UNCTAD 1997; ECLAC 1997). Other areas also saw inequality rise in the 1980s and 1990s (Faux and Mishel 2000; Ravallion and Chen 1997). Deininger and Squire (1996) found rising inequality in East Asia, Eastern Europe, and Central Asia since 1981, and growing polarization in South Asia. Only sub-Saharan Africa shows a trend toward more income equality since the 1980s.

While a widening gap between the rich and the poor within countries is not universal, it appears to have occurred at least in the majority of countries, and is affecting the income of the majority of people around the globe, contrary to claims by the World Bank that rising inequality within countries has been rare.

Poverty Remains a Large and Widespread Problem

The World Bank tries to divert attention from rising inequality by emphasizing its analyses of poverty reduction. It argues that "the long [term] trends of rising global inequality and rising numbers of people in absolute poverty have been halted and perhaps even reversed" due to greater globalization (World Bank 2001a, 49). However, the purported success in poverty reduction is elusive: the number of poor people is on the rise, relative poverty shares remain high in many parts of the world, and poverty shares are rising in many regions.

In assessing global poverty trends, the World Bank relies on a study that highlights the World Bank's *Global Poverty Monitoring* database and provides an overview of poverty trends from 1987 to 1998 (Chen and Ravallion 2001). The authors themselves, though, conclude that "[i]n the aggregate, and for some large regions, all . . . measures suggest that the 1990s did not see much progress against consumption poverty in the developing world" (Chen and Ravallion 2001, 18). Also, the IMF (2000, Part IV, p. 1) reports that "[p]rogress in raising real incomes and alleviating poverty has been disappointingly slow in many developing countries."

The assessment of poverty trends by the World Bank suffers from several problems. First, measuring poverty is a difficult undertaking that can easily lead to errors. Different measures of poverty exist. The World Bank's *Global Poverty Monitoring* database, for example, uses an international poverty line of $1.08 per day in 1993 dollars based on purchasing power parity (PPP) exchange rates (Chen and Ravallion 2001; World Bank 2001b). But absolute poverty lines such as this one ignore regional or country-by-country differences.

The evidence shows that the use of an international poverty line tends to understate the share of people living in poverty, compared to other poverty measures. For example, a method using individual national poverty lines finds an additional 14% of the population to be considered poor compared to a method using the international poverty line (World Bank 2001b). An alternative to both the national and international poverty line methods is to use a relative poverty line based on mean consumption or income levels in each country. Using such a relative poverty line instead of the international poverty line shows on average an additional 8% of the population to be considered poor (Chen and Ravallion 2001).

Second, poverty lines are often inadequate to measure the true hardships people are facing in meeting the basic necessities of life. For instance, a recent U.S. study showed that 29% of working families did not earn enough to afford basic necessities, suggesting that a better approach to understanding poverty may lie in measuring household budgets rather than simple poverty lines (Boushey et al. 2001).

The third problem with the Bank's poverty assessment is that even the poverty reduction gains it does find are small and geographically isolated. In

Table 1

Share of People Living Below Relative Poverty Lines

	1987	1990	1993	1996	1998
East Asia	33.01%	33.69%	29.82%	19.03%	19.56%
East Asia, excluding China	45.06	38.68	30.76	23.16	24.55
Eastern Europe and Central Asia	7.54	16.19	25.34	26.08	25.60
Latin America and Caribbean	50.20	51.48	51.08	51.95	51.35
Middle East and North Africa	18.93	14.49	13.62	11.40	10.76
South Asia	45.20	44.21	42.52	42.49	40.20
Sub-Saharan Africa	51.09	52.05	54.01	52.80	50.49
Share of world:					
Living under $1.08/day	28.31%	28.95%	28.15%	24.53%	23.96%
Living under relative poverty lines	36.31	37.41	36.73	32.79	32.08
Maximum daily consumption of world's poorest 400 million (nominal)	$0.79	$0.79	$0.56	$0.84	$0.75

Notes: The drop in 1993 reflects sharp decreases in per capita GDP in Nigeria, Ethiopia, Myanmar, and the Democratic Republic of Congo that, combined, made up 58% of the sample population in 1993. Calculations for the world's poorest 400 million are based on average nominal per capita GDP.

Sources: Chen and Ravallion (2001); IMF (2001a, 2001b); and authors' calculations.

1998, the share of the population living in poverty in industrializing countries was 32%, under a relative poverty line. Although that percentage was down from 36% in 1987, the actual number of people living in poverty increased from 1.5 billion to 1.6 billion. In 1998, the share of the population in poverty remained very high in some regions: over 40% in South Asia and over 50% in sub-Saharan Africa and Latin America (Table 1). Since 1987, the share of the poor has stayed relatively constant in sub-Saharan Africa and Latin America but more than tripled in Eastern Europe and Central Asia.

Another way to look at the global trends in poverty is to consider the incomes of an absolute number of poor people. Take, for instance, the poorest 10% of the population in 1980, consisting of about 400 million people, based on average per capita GDP. The poorest 400 million lived on a nominal $0.72 a day in 1980, $0.79 a day in 1990, $0.84 in 1996, and $0.78 in 1999 (Table 1). In other words, the income of the world's poorest did not even keep up with inflation. Clearly, the economic burden worsened for a large number of people in the 1990s.

Fourth, since the data do not extend beyond 1998, the full impact of the crises in Asia, Latin America, and Russia is not included, making it likely that future revisions will show less progress in poverty reduction. Lustig (2000) argues that frequent macroeconomic crises are the single most important cause of rapid increases in poverty in Latin America. Consequently, future revisions to the poverty trends in the late 1990s could show smaller average reductions or larger increases in the crisis-stricken areas. In fact, revisions to past data already show less success in poverty reduction than previously assumed. Chen and Ravallion (2001), for example, show that the reduction of people living below the poverty line between 1987 and 1993 was not four percentage points, as estimated in 1997 (Ravallion and Chen 1997), but less than one percentage point.

Finally, the World Bank's conclusion that the lot of the poor has improved during the era of increasing trade and capital flow liberalization relies substantially on data from China and India, but the experiences of both countries are anomalies. In reality, the facts in these countries undermine the case for a connection between greater deregulation of capital and trade flows and falling poverty and inequality. While in China the percentage who are poor has fallen, there has been a rapid rise in inequality (World Bank 2001a). Most notably, inequality between rural and urban areas and provinces with urban centers and those without grew from 1985 to 1995. Also, a large number of China's workers labor under abhorrent, and possibly worsening, slave or prison labor conditions (USTDRC 2000; U.S. Department of State 2000, 2001). This situation not only means that many workers are left out of China's economic growth, it also makes China an unappealing development model for the rest of the world. Thus, improvements in China are not universally shared and leave many workers behind, often in deplorable conditions. . . .

Conclusion

Criticism of the unregulated globalization agenda has been met with policy makers' renewed adherence to the doctrine that greater global deregulation of trade and capital flows helps to improve inequality between countries, to raise equality within countries, and to accelerate poverty reduction. But income distribution between countries worsened in the 1980s, and its apparent improvement (or leveling off) in the 1990s is the result solely of rising per capita income in China, where the enormous population tends to distort world averages. Within-country income inequality is also growing and is a widespread trend in countries with both advanced and developing economies. Success in reducing poverty has been limited. The number of poor people has risen, and the share of poor people has grown in many areas, especially in Eastern Europe and Central Asia. And the share of poor people remained high at 40–50% in Latin America, sub-Saharan Africa, and South Asia.

The promises of more equal income distribution and reduced poverty around the globe have failed to materialize under the current form of unregulated globalization. Thus, it is time for multinational institutions and other international policy makers to develop a different set of strategies and programs to provide real benefits to the poor.

References

Bannister, G. J., and Kamau Thugge. 2001. "International Trade and Poverty Alleviation." Working Paper. Washington, D.C.: International Monetary Fund.

Blecker, Robert. 1999. *Taming Global Finance.* Washington, D.C.: Economic Policy Institute.

Boushey, Heather, Chauna Brocht, Bethney Gundersen, and Jared Bernstein. 2001. *Hardships in America: The Real Story of Working Families.* Washington, D.C.: Economic Policy Institute.

Bronfenbrenner, Kate. 1997. "The Effects of Plant Closings and the Threat of Plant Closings on Worker Rights to Organize." Supplement to *Plant Closings and Workers' Rights: A Report to the Council of Ministers by the Secretariat of the Commission for Labor Cooperation.* Lanham, Md.: Bernam Press.

Bronfenbrenner, Kate. 2000. "Uneasy Terrain: The Impact of Capital Mobility on Workers, Wages, and Union Organizing." Washington, D.C.: U.S. Trade Deficit Review Commission. http://www.ustdrc.gov/research/research.html.

Chen, S., and M. Ravallion. 2001. "How Did the World's Poorest Fare in the 1990s?: Methodology." *Global Poverty Monitoring Database.* Washington D.C.: World Bank. http://www.worldbank.org/research/povmonitor/method.htm.

Deininger, Klaus, and Lyn Squire. 1996. "A New Data Set Measuring Income Inequality." *World Bank Economic Review* 10(3).

Economic Council on Latin America and the Caribbean (ECLAC). 1997. "The Equity Gap: Latin America, the Caribbean, and the Social Summit." LG/G, 1954. (CONF86/3). Santiago, Chile, March.

Faux, Jeff, and Lawrence Mishel. 2000. "Inequality and the Global Economy." In Will Hutton and Anthony Giddens, eds., *On the Edge: Living With Global Capitalism.* London, U.K.: Jonathan Cape.

Gottschalk, Peter, and Timothy M. Smeeding. 1997. "Cross-National Comparisons of Earnings and Income Inequality." *Journal of Economic Literature* 35(2): 633–87.

Inter-American Development Bank (IADB). 1999. "Facing Up to Inequality in Latin America." *Economic and Social Progress in Latin America, 1998–1999 Report.* Washington, D.C.: IADB.

International Monetary Fund (IMF). 2000. *World Economic Outlook.* May. Washington, D.C.: IMF.

International Monetary Fund (IMF). 2001b. *World Economic Outlook.* May. Washington, D.C.: IMF.

Lindgren, C., G. Garcia, and M. Saal. 1996. *Bank Soundness and Macroeconomic Policy.* Washington, D.C.: IMF.

Lustig, N. 2000. "Crises and the Poor: Socially Responsible Macroeconomics." Sustainable Development Department Technical Paper Series No. POV-108. Washington, D.C.: IADB.

Lustig, N., and R. Deutsch. 1998. "The Inter-American Development Bank and Poverty Reduction: An Overview." No. POV-101-R. Washington, D.C.: IADB.

Luxembourg Income Study (LIS). 2001. "LIS Key Figures—Income Inequality Measures." http://lisweb.ceps.lu/keyfigures/ineqtable.htm.

Mishel, Lawrence, Jared Bernstein, and John Schmitt. 2001. *The State of Working America 2000/2001.* Ithaca, N.Y.: Cornell University Press.

Ravallion, M., and S. Chen. 1997. "What Can New Survey Data Tell Us About Recent Changes in Distribution and Poverty?" *World Bank Economic Review* 11(2).

Scott, Robert E., Thea Lee, and John Schmitt. 1997. "Trading Away Good Jobs: An Examination of Employment and Wages in the U.S., 1979–94." Briefing Paper. Washington, D.C.: Economic Policy Institute.

Scott, Robert E. 2001a. "Our Kind of Trade: Alternatives to Neoliberalism That Can Unite Workers in the North and South." In *Workers in the Global Economy—Project Papers and Workshop Report.* Ithaca, N.Y.: Cornell University School of Industrial Relations.

Scott, Robert E. 2001b. "NAFTA's Hidden Costs." *NAFTA at Seven: Its Impact on Workers in All Three Nations.* Washington, D.C.: Economic Policy Institute.

United Nations Conference on Trade and Development (UNCTAD). 1997. *Trade and Development Report.* Geneva, Switzerland: UNCTAD.

U.S. Department of State. 2000. *1999 Country Reports on Human Rights Practices.* Washington, D.C.: U.S. Department of State.

U.S. Department of State. 2001. *2000 Country Reports on Human Rights Practices.* Washington, D.C.: U.S. Department of State.

U.S. Trade Deficit Review Commission (USTDRC). 2000. *The U.S. Trade Deficit: Causes, Consequences and Recommendations for Action.* Washington, D.C.: USTDRC.

World Bank. 2001a. "Draft Policy Research Report: Globalization, Growth and Poverty: Facts, Fears and an Agenda for Action." Washington, D.C.: World Bank.

World Bank. 2001b. *Global Poverty Monitoring Database.* http://www.worldbank.org/research/povmonitor/.

Zoellick, R. B. 2001. "American Trade Leadership: What Is at Stake?" Remarks at the Institute for International Economics, Washington, D.C., September 24.

POSTSCRIPT

Is Globalization a Positive Development for the World Community?

It is hard to argue that this kind of revolution is all positive or all negative. Many will find the allure of technological growth and expansion too much to resist. They will adopt values and ethics that seem compatible with a materialistic Western culture. And they will embrace speed over substance, technical expertise over knowledge, and wealth over fulfillment.

Others will reject this revolution. They will find its promotion of materialism and Western cultural values abhorrent and against their own sense of humanity and being. They will seek enrichment in tradition and values rooted in their cultural pasts. This resistance will take many forms. It will be political and social, and it will involve actions ranging from protests and voting to division and violence.

Trying to determine whether a force as dominant and all encompassing as globalization is positive or negative is like determining whether the environment is harsh or beautiful. It is both. One can say that in the short term, globalization will be destabilizing for many millions of people because the changes that it brings will cause some fundamental shifts in beliefs, values, and ideas. Once that period is past, it is conceivable that a more stable environment will result as people come to grips with globalization and either learn to embrace it, cope with it, or keep it at bay.

The literature on globalization is growing rapidly, much of it centering on defining its parameters and evaluating its impact. One important work that presents a deeper understanding of globalization is Friedman's best-seller *The Lexus and the Olive Tree* (Farrar, Straus & Giroux, 1999), which provides a positive view of the globalization movement. Two counterperspectives are William Greider's *One World Ready or Not: The Manic Logic of Global Capitalism* (Simon & Schuster, 1997) and David Korten's *When Corporations Rule the World* (Kumarian Press, 1996). Also see Jan Nederveen Pieterse, ed., *Global Futures: Shaping Globalization* (Zed Books, 2001).

The literature that will help us to understand the full scope of globalization has not yet been written. Also, the determination of globalization's positive or negative impact on the international system has yet to be decided. For certain, globalization will bring profound changes that will cause people from America to Zimbabwe to rethink assumptions and beliefs about how the world works. And equally certain is the realization that globalization will empower some people but that it will also leave others out, helping to maintain and perhaps exacerbate the divisions that already exist in global society.

ISSUE 15

Will the Digital/Computer World Lead to Greater Individual Freedom?

YES: Paul Starr, from "Cyberpower and Freedom," *The American Prospect* (July/August 1997)

NO: Jerry Kang, from "Cyberspace Privacy: A Primer and Proposal," *Human Rights* (Winter 1999)

ISSUE SUMMARY

YES: Professor of sociology Paul Starr argues that the information revolution will help to create a period of individual freedom empowered by computer technology.

NO: Professor of law Jerry Kang argues that privacy, which is fundamental to the concept of freedom, is severely threatened by the current love of cyberspace. He suggests that the U.S. Congress and, by extension, individual states should take action to ensure individual privacy in cyberspace.

Scanning the Internet for products, services, friends, lovers, and information has become an obsession for some people, a necessity for others, and a desire for still others. More than 60 percent of all U.S. households have a computer. Most of these have Internet access. Experts estimate that by the end of the decade, over 80 percent of U.S. households will have a computer and Internet access. In an increasingly wireless world, these numbers will explode in the years to come. New technologies, such as the all-in-one device (computer, television, and telephone), will solidify the presence of the computer in the Western home. In economic terms, the largest growth businesses are Web-based, with estimates running from a conservative $25 billion to an astounding $1 trillion in financial impact over the next decade.

The ease with which people can find information, purchase products, and seek out communication and companionship makes the computer and cyberspace appealing. From information storage and retrieval to communicating around the globe, the computer is fast becoming a necessary tool for many. Educators and students with computer access can search and retrieve information, explore new and heretofore inaccessible data, and make rare the age-old complaint heard by many professors that "I can't write on this topic because there is not enough information on the subject."

Yet not all people have access to Internet services, and those who do often do not know how to utilize the technology to empower themselves. In fact, most of the world does not have widespread access to this technology and are therefore left out of its potential benefits. Many who do have access become caught in a growing web of information, making their personal needs common knowledge and, indeed, public record. As people surf around the Internet, they leave trails of information about their money, buying habits, personal tastes, and thoughts. In essence, people leave volumes of personal material that can be collected, stored, and used to market, investigate, or monitor those on the information superhighway.

The digital/computer age brings with it a variety of interesting and conflicting experiences. One can scan the Internet for products and items not found elsewhere. A person can obtain valuable information, research his or her ancestry, and experience virtual tours of faraway places. A computer-active person can speak with people in far-off lands, talk to his or her leaders, and solicit information from a growing variety of sources. One can also have one's bank accounts searched and drained, personal information collected and distributed, private messages recorded and read, and rights as an individual violated without one's knowledge. In some ways, logging on and being wired in have become a form of gambling with information, privacy, and freedom.

Privacy, which is fundamental to the conduct of political, economic, and social relations, is at risk in a cyberworld with little law and numerous technological ways of peeping into others' lives. The incredible speed of the growth of this technology has outstripped laws and ethics about its use and misuse. As a result, cyberspace is like a Wild West of information, ideas, and activities, where one can make a fortune or have one's life torn asunder with a few keystrokes. People's conception of individual freedom and expression will be severely tested in this new world.

In the following selections, Paul Starr argues that the digital computer world will empower people. It will give them more choices, greater freedom of access and action, more knowledge, and greater control over their lives. He concludes that people will ultimately benefit in a host of ways as a consequence. Jerry Kang contends that the nature of cyberspace makes it a dangerous highway where millions of people are capable of peering into others' bank accounts, journals, and files; compiling information; and making people hostage to unseen forces. Kang maintains that these forces can manipulate individuals' buying habits, examine their political views, and track their whereabouts, thus inhibiting privacy and freedom.

Paul Starr  **YES**

Cyberpower and Freedom

In politics and the public imagination, computers have gone from symboliz-
ing our vulnerability to embodying our possibilities. In their early days during
the 1950s and 1960s, computers seemed destined to increase the power of government and big corporations, and the great worry was how to protect privacy
and individual freedom. Then the advent of the personal computer and other
low-cost electronics suggested that information technology might be the ultimate tool of decentralization and individual empowerment, and the rise of
global telecommunications and the Internet promised to annihilate national
borders. Now many of us sit at keyboards easily connecting to computers all
over the world, and to some people the thought suggests itself: "Why do we
need national government at all?" Things have swung around so completely
that influential analysts, especially on the right, see the information revolution
as a great historical reversal of power, ushering in a new age of individualism on
the digital frontier.

This high-tech libertarianism can be found in *Wired* magazine, the writings
of George Gilder, the publications of Newt Gingrich's Progress and Freedom
Foundation, and countless sites on the Net itself. Optimistic and forward-looking, high-tech libertarianism is more appealing than the older variety that
looked back to the devil-take-the-hindmost individualism of the nineteenth century. But the new version shares the same old misunderstanding of the bases of
freedom—that liberty will prosper the more government is diminished. To this it
adds a new illusion—that technology is now freedom's reliable shield.

Advanced communications networks offer "the ultimate shopping experience: shopping for better government," writes one exponent of these ideas, the
communications lawyer Peter Huber, in an article on "cyberpower" in *Forbes*
last December [1996]. By "better" government, he chiefly means lower taxes
and less stringent regulation. Now it's not just corporations that can shop
around and buy, say, a legal home in Delaware; "ordinary investors," Huber observes, can shop on the Net for the tax environment, monetary policy, or regulations of their choice. If you think that your government's economic policies
are wrong, you too can put your money into another currency. If the Food and

From Paul Starr, "Cyberpower and Freedom," *The American Prospect,* vol. 8, no. 33 (July/August
1997). Copyright © 1997 by *The American Prospect.* Reprinted by permission of *The American Prospect*
and the author.

Drug Administration (FDA) bars sale of a drug, you can buy it from an online pharmacy in another country and have it delivered by Federal Express. The American consumer who buys shoes made abroad will soon "shop for life insurance in London and health insurance in Geneva, and the offshore actuaries will discriminate fiercely in favor of the healthy."

To Huber, this is all to the good. Here is his happy vision: "As managers, workers and consumers," he writes, "we buy government in much the same way we buy shoes" and when one government is too costly or inconvenient, we can just switch to another. In thinking about the relations of people and governments, he has thus nicely dispensed with the old-fashioned concept of a citizen encumbered with obligations and loyalties, along with the idea of a people collectively addressing their problems through democratic institutions—all this was just a "political carnival" that is thankfully over. Governments in the new global market will be disciplined by money moving offshore, not by voters and public deliberation, and public policies will be improved because politicians will be impotent to do anything except what markets allow them to do.

Huber conceives of this system as a superior version of democracy, as if we all had equal "votes" in the global government market. But, of course, not all of us "buy government" with equal facility or effect. Managers can move plants abroad, and consumers can buy tradable foreign goods; but except for the limited class of professionals who can sell their services to foreign clients, workers need to uproot themselves to make use of foreign labor markets. As savers and investors, we also have unequal offshore opportunities since the principal asset of most families is their home. In other words, inasmuch as we are limited by our physical existence, social connections, and political loyalties, we cannot simply search and switch among global markets. The issue here is not simply income inequality, but also an imbalance of human interests. Why feel any obligation to people or places, community or nation, if you're just shopping?

Membership in a democratic state has historically implied a "bundle" of relationships. We enjoy the security and other services a state provides us and, if we fulfill our part of the bargain, we obey its rules, pay its taxes, and have our say in elections. Huber is saying technology makes it easier and cheaper to unbundle that package. The difficulty is that many people would like to enjoy the services of effective government but not pay the costs. They might like to live under one government, do business under another, and park their assets under a third. But in a world of unlimited opportunism, the government would lack the effective authority and resources to provide collective benefits that the majority of us want from it—including . . . the very rights we prize.

There is no denying that the combined force of advanced technology and global markets has undermined the regulatory capacity of government. But it is important to distinguish what technology has changed and what it hasn't. Corporations can now more easily organize production on a global basis, and investors can more readily move financial assets across national borders. Where products and services consist of bits rather than atoms, the Net may make their origins, literally, immaterial. But economic activities that require local knowledge and trust as well as physical presence are not so easily moved. And governments still have effective means of enforcing laws and regulations even when one of the parties to a transaction is offshore.

Consider two of the examples mentioned by Huber: pharmaceuticals and health insurance. His offshore alternatives confront a minor difficulty—they're illegal and for good reason. Anyone in the United States ordering drugs banned by the FDA from an offshore pharmacy would be importing them in violation of federal law. No matter where domiciled, any insurer doing business with people residing in a state needs to be licensed to do so and must comply with federal requirements under the recent Kennedy-Kassebaum legislation—technology doesn't change that. If there is any justification for pharmaceutical regulation (some minimum standard of safety) or insurance regulation (some minimum standard of solvency), the laws cannot simply be nullified by the ruse of an offshore location. As a practical matter, few people are likely to use online pharmacies located abroad because of delivery costs; and if the drugs have not received FDA approval, insurers won't reimburse them. Foreign health insurers, lacking contracts with providers, would be in a poor position to compete with domestic insurers in the age of managed care. If a market in the U.S. for offshore pharmacies and insurers existed, they could long ago have done business by snail mail and telephone, but they haven't, and the new technology doesn't affect the sources of hesitation among both sellers and buyers—the illegality of the transactions and lack of detailed knowledge about the other party and relevant market conditions.

Huber assumes that people can evaluate regulatory protections in other countries and that they will knowingly opt for more lax regulations. In some cases, that may be so. But here he underestimates the importance of trust in economic as well as social relationships and the role of government in promoting trust by assuring common ground rules and effective recourse in the event of fraud and negligence. Without a state strong enough to enforce rules and assure confidence, people are reluctant to come into the market and do business with strangers.

And so it is with global electronic commerce. The economic development of cyberspace is lagging because governments have not yet established laws and regulations that promote trust and confidence. Consumers continue to be "wary" of the Internet, a recent White House paper on the Internet and global commerce points out, "because of the lack of a predictable legal environment governing transactions. . . . This is particularly true for international commercial activity where concerns about enforcement of contracts, liability, intellectual property protection, privacy, security, taxation and other matters have caused businesses and consumers to be cautious."

In other words, what we have today on the Net is not so much individuals shopping for governments as both individuals and firms fearing to do business globally for lack of consistent laws and regulations that their governments need to agree upon. The libertarians see government as always restricting markets; they miss the positive role government plays in creating the foundations of social trust that markets (and nonmarket institutions) require.

Far from taking trust for granted in the information age, we are going to face especially serious problems sustaining it because of the growing use of the new technology for deception. I am not only referring to the difficulty of authenticating electronic exchanges—for example, determining whether e-mail is actually from the ostensible sender, whether a Web site originates from the group or person listed as responsible for it, or whether e-cash is genuine. New technology is also undermining the integrity of older forms of communication and exchange by generating improved techniques and lower costs for altering photographs, counterfeiting documents, and stealing identities (reproducing forms of identification, often for credit card fraud). Recently a company was reported to have figured out how to recombine an individual's recorded speech into new sentences that sound exactly as if the original person spoke them—a kind of audio counterfeiting that may soon, in effect, make us say anything. Information technology alone cannot provide us an absolute shield against its evil twin—disinformation technology. Our only protection is law, and that protection is available to us only if legitimate governments have the power to govern.

Another problem with high-tech libertarianism is the belief that the development of information technology and electronic markets on their own will promote maximum choice. In its current form, the Internet clearly does expand the ability of people with minimal resources to originate communication as well as to receive it. In this sense, the Web is the best rejoinder to A.J. Liebling's old complaint that "freedom of the press belongs to the man who owns one." More generally, the new era of digital communications has the potential to end the old scarcity of bandwidth in the radio spectrum that limited the number of broadcast channels. But there are also powerful forces favoring concentrated power in the new era of communications, and it would be a mistake just to rely on technology and the marketplace to curb abuses of that power.

One source of concentrated power is developing around control of the "interface"—the menu of alternatives that first pops up on the screen. Interfaces like the current browsers for navigating the Web are relatively open (they connect to nearly all sites), but more closed interfaces, like the menu of options on television sets in hotel rooms, suggest how a more closed regime might be reestablished. An interface is a choke point and potentially of enormous economic value. Why should choke points emerge? Because under the current, open regime on the Web, hardly anyone providing "content" is making money; and because of the demand for simplicity of use as the Web (or some alternative) turns into a more universal system of communication and entertainment. Companies are already creating more limited interfaces to sort out options for consumers, deliver audiences to advertisers, and assure originators of content a dependable stream of income. "Webcasting" and "push" media—which simplify the complexity of the Web by automatically downloading preselected information to the user—exemplify this shift. But to simplify is to exclude. Think of such a system not as censorship but rather as control of the means of marginalization.

The information revolution also tends to concentrate power in the firms that control dominant "architectures," such as IBM in the mainframe era and Microsoft and Intel today. Architectural dominance is endemic in industries that have rapidly evolving proprietary technical standards vital to every firm. As Charles H. Ferguson and Charles R. Morris write in their book *Computer Wars,* struggles over architectural control are a powerful stimulus to technological progress. But as Microsoft's victims can testify, the firm in control of the dominant architecture has extraordinary leverage over any potential competitor in its core or related businesses. Without government restraint in the form of antitrust, markets of these kinds easily end in monopoly.

The rapid improvement in price-performance ratios of computers, software, and other technology today seems to validate the faith in free markets. But to say that the information revolution proves the inevitable superiority of markets requires a monumental failure of short-term historical memory. After all, not just the Internet, but the computer sciences and computer industry represent a spectacular success of public investment. As late as the 1970s and early 1980s, according to Kenneth Flamm's 1987 study *Targeting the Computer,* the federal government was paying for 40 percent of all computer-related research and probably 60 to 75 percent of basic research. The motivation was national security, but the result has been the creation of comparative advantage in information technology for the United States that private firms have happily exploited and extended. When the returns were uncertain and difficult to capture, private firms were unwilling to invest, and government played the decisive role. But when the market expanded and the returns were more definite, the government receded, which is exactly the path it should have followed.

The view of government favored by the high-tech libertarians would make it impossible for us ever to repeat the success of the United States with the computers and the Net (or more generally to respond to many other examples of market failure). One of the ironies of the information revolution is that government's hand has apparently become the invisible one. There we sit at our computers, and we feel like Masters of the Universe. We press the keys and command distant computers, and we think everyone else should be able to manage their affairs as we do. But, of all spaces, cyberspace is most singularly the product of political invention and social agreement, and only law will give us the security to use it freely.

NO

Jerry Kang

Cyberspace Privacy:
A Primer and Proposal

Human ingenuity has provided us a great gift, cyberspace. This blooming network of computing-communication technologies is quickly changing the world and our behavior in it. Already, it has become cliche to catalog cyberspace's striking benefits, its endless possibilities. But great gifts often come with a great price. Congress thinks that the price will be sexual purity due to easy access to pornography. Industry thinks it will be our economy due to easy copying of Hollywood's and Silicon Valley's programs. I worry that it will be our privacy.

When I mention "privacy," lawyers naturally think of privacy as used in the historic case, *Roe v. Wade*. Others think of the privacy of their own homes and backyards, largely in territorial terms. I use "privacy" differently. Instead of emphasizing privacy in a decisional or spatial sense, I mean it in an information sense. Information privacy is an individual's claim to control the terms under which personal information is acquired, disclosed, and used.

My thesis is that cyberspace threatens information privacy in extraordinary ways, and without much thought or collective deliberation, we may be in the process of surrendering our privacy permanently as we enter the next century.

Why Care About Privacy?

Some people do not understand what the big deal is about privacy. They assume that privacy is important only for those who have something to hide. This view is misguided. Let me articulate why individuals should enjoy meaningful control over the acquisition, disclosure, and use of their personal information.

Use of personal data. Personal data are often misused. For example, personal data can be used to commit identity theft, in which an impostor creates fake financial accounts, runs up enormous bills, and disappears leaving only a wrecked credit report behind. Personal data, such as home addresses and telephone numbers, can be used to harass and stalk. Personal data, such as one's sexual orientation, can be used to deny employment because of unwarranted prejudice.

From Jerry Kang, "Cyberspace Privacy: A Primer and Proposal," *Human Rights,* vol. 26, no. 1 (Winter 1999). Copyright © 1999 by The American Bar Association. Reprinted by permission.

Disclosure of personal data. Sometimes, even if such data will not be "used" against us, its mere disclosure may lead to embarrassment. In any culture, certain conditions are embarrassing even when they are not blameworthy. Take impotency for example. In most cases, impotency will not affect whether one receives a job, a loan, or a promotion. In this sense, the data will not be misused in the allocation of rewards and opportunities. However, the mere disclosure of this medical condition would cause intense embarrassment for most men.

In addition to causing embarrassment, the inability to control the disclosure of personal data can hamper the building of intimate relationships. We construct many intimacies not only by sharing experiences but also by sharing secrets about ourselves, details not broadcast to a mass audience. If we have information privacy, we can regulate the outflow of such private information to others. By reducing this flow to a trickle (for example, to your boss), we maintain aloofness; by releasing a more telling stream (for example, to your former college roommate living afar) we invite and affirm intimacy. If anyone could find out anything about us, secrets would lose their ability to help construct intimacy.

Acquisition of personal data. Finally, consider the fact that personal information is acquired by observing who we are and what we do. When such observation is nonconsensual and extensive, we have what amounts to surveillance, which is in tension with human dignity. Human beings have dignity because they are moral persons—beings capable of self-determination, with the capacity to reflect upon and choose personal and political projects. Extensive, undesired observation interferes with this exercise of choice because we act differently when we are being watched. Simply put, surveillance leads to self-censorship. When we do not want to be surveilled, it disrespects our dignity to surveil us nonetheless, unless some important social justification exists. This insult to individual dignity has social ramifications. It chills out-of-the-mainstream behavior. It corrodes private experimentation and reflection. It threatens to leave us with a bland, unoriginal polity, characterized by excessive conformity.

The Difference That Cyberspace Makes

So now that we know why privacy matters, we must ask what difference does cyberspace make? My claim is that cyberspace makes broad societal surveillance possible and, if we do nothing, likely. To see the greater threat that cyberspace poses, imagine the following two visits to a mall—one in real space, the other in cyberspace.

In real space, you drive to a mall, walk its corridors, peer into numerous shops, and stroll through the aisles of inviting stores. You walk into a bookstore and flip through a few magazines. Finally, you stop at a clothing store and buy a friend a scarf with a credit card. In this narrative, numerous persons interact with you and collect information along the way. For instance, while walking through the mall, fellow visitors visually collect information about you, if for no other reason than to avoid bumping into you. But such information is general, e.g., it does not pinpoint the geographical location and time of the sighting, is not in a format that can be processed by a computer, is not indexed to your name or any unique identifier, and is impermanent, residing in short-term human memory. You remain a barely noticed stranger. One important exception is the credit card purchase.

By contrast, in cyberspace, the exception becomes the norm: Every interaction may soon be like the credit card purchase. The best way to grasp this point is to take seriously, if only for a moment, the metaphor that cyberspace is an actual place, a virtual reality. In this alternate universe, you are invisibly stamped with a bar code as soon as you venture outside your home. There are entities called "road providers" (your Internet Service Provider), who supply the streets and ground you walk on, who track precisely where, when, and how fast you traverse the lands, in order to charge you for your wear on the infrastructure. As soon as you enter the cyber-mall's domain, the mall tracks you through invisible scanners focused on your bar code. It automatically records which stores you visit, which windows you browse, in which order, and for how long. The specific stores collect even more detailed data when you enter their domain. For example, the cyberbookstore notes which magazines you skimmed, recording which pages you have seen and for how long, and notes the pattern, if any, of your browsing. It notes that you picked up a health magazine featuring an article on Viagra, read for seven minutes a newsweekly detailing a politician's sex scandal, and flipped ever-so-quickly through a tabloid claiming that Elvis lives. Of course, whenever any item is actually purchased, the store, as well as the credit, debit, or virtual cash company that provides payment through cyberspace, takes careful notes of what you bought—in this case, a silk scarf, red, expensive, a week before Valentine's Day.

All these data generated in cyberspace are detailed, computer-processable, indexed to the individual, and permanent. While the mall example may not concern data that appear especially sensitive, the same extensive data collection can take place as we travel through other cyberspace domains—for instance, to research health issues and politics; to communicate to friends, businesses, and the government; and to pay our bills and manage our finances. Moreover, the data collected in these various domains can be aggregated to produce telling profiles of who we are, as revealed by what we do and with whom we associate. The very technology that makes cyberspace possible also makes detailed, cumulative, invisible observation of ourselves possible. One need only sift through the click streams generated by our cyber-activity.

It turns out that few laws limit what can be done with this data collected in cyberspace. Unlike Europe, the United States has no omnibus privacy law covering the private sector's processing of personal information. Instead, U.S. law features a legal patchwork that regulates different types of personal information in different ways, depending on how it is acquired, by whom, and how it will be used. To be sure, there are numerous statutes that govern specific sectors, such as consumer credit, education, cable programming, electronic communications, videotape rentals, motor vehicle records, and the recently enacted Children's Online Privacy Protection Act. But it turns out that in toto, information collectors can largely do what they want.

The Market Solution

Let me restate the problem. All cyberactivity, even simply browsing a Webpage, involves a "transaction" between an individual and potential information collectors. These collectors not only include the counterparty to the transaction but also intermediaries (transaction facilitators) that support the electronic communications (telephone company, cable company, Internet service provider) and sometimes payment (credit card company, electronic cash company). In these transactions, personal information is inevitably generated as either necessary or incidental by-products. Privacy enthusiasts insist that the individual owns this data; information collectors vigorously disagree. What shall be done?

Perhaps the market might solve the problem. One might reasonably view personal information as a commodity, whose pricing and consumption can and should be governed by the laws of supply and demand. Through offers and counteroffers between individual and information collector, the market will move the correctly priced personal data to the party that values it most. Economists love this approach because it appears to be economically efficient. The private sector loves this approach because it staves off regulation. Regulators love this approach for the same reason in the current antiregulatory environment.

The problem is that in practice, individuals and information collectors do not negotiate express privacy contracts before engaging in each transaction. Although privacy notices have become more frequent on Webpages, it is a stretch to say that there is a "meeting of the minds" on privacy terms each time an individual browses a Webpage. What is necessary, then, is a clear articulation of the default rules governing personal data collected in a cyberspace transaction, when parties have not agreed explicitly otherwise.

There are two default rules that society might realistically adopt. First, there is the status quo's "plenary" default rule: Unless the parties agree otherwise, the information collector may process the personal data anyway it likes. Second, there is the "functionally necessary" default rule: Unless the parties agree otherwise, the information collector may process the personal data only in functionally necessary ways. This rule allows the information collector to process personal data on a need-only basis to complete the transaction in which the information was originally collected.

A one-size-fits-all default rule is efficient for some transactions but inefficient for others. Those parties for whom the default is inefficient will either contract around the rule—"flip"—or they will "stick" with the rule and accept the inefficiencies. Thus, the social cost of a default rule equals the sum of the transaction costs of contracting around the rule—the "flip cost"—plus the inefficiency cost of not contracting around the rule even when it would be more efficient to do so—the "stick cost." We seek the rule that minimizes social costs.

If we implement the plenary rule, most parties will stick because it is hard for a consumer to flip out of the default rule. First, she would face substantial research costs to determine what information is being collected and how it is used. That is because individuals today are largely clueless about how personal information is processed through cyberspace. Sometimes, they are deceived by the information collectors themselves, as the Federal Trade Commission recently charged against the Internet Service Provider, Geocities. Second, the individual would run into a collective action problem. Realistically, the information collector would not entertain one person's idiosyncratic request to purchase back personal information because the costs of administering such an individually tailored program would be prohibitive. Therefore, to make it worth the firm's while, the individual would have to band together with other like-minded individuals to renegotiate the privacy terms of the underlying transaction. These individuals would suffer the collective action costs of locating each other, then coming to a mutually acceptable proposal to deliver to the information collector—all the while discouraging free riders.

By contrast, the "functionally necessary" rule would not be sticky at all. With this default, if the firm valued personal data more than the individual, then the firm would have to buy permission to process the data in functionally unnecessary ways. Note, however, two critical differences in contracting around this default. First, unlike the individual who has to find out how information is processed, the collector need not bear such research costs since it already knows what its information practices are. Second, the collector does not confront collective action problems. It need not seek out other like-minded firms and reach consensus before coming to the individual with a request. This is because an individual would gladly entertain an individualized, even idiosyncratic, offer to purchase personal information.

Now, the task is to compare the costs of the equilibrium generated by each default rule. For the "plenary" equilibrium, the cost of the default rule is approximately the stick cost because few parties will flip. By contrast, for the "functionally necessary" equilibrium, the cost of the default rule is approximately the flip cost; almost all parties who care to flip will flip. Which cost is higher? We lack the data to be confident in our answer. However, we do know that given how seriously many individuals feel about their privacy, the stick cost of the plenary rule will not be trivial. Many individuals who care deeply about their privacy will not be able to get it. By contrast, the flip cost of the functionally necessary rule will be small because cyberspace makes communications cheap. The information collector can ask in a simple dialog box whether the individual will allow some unnecessary use of personal data, in exchange for some benefit. What is more, this inequality will increase over time. As information processing becomes more sophisticated, people will feel less in control of their personal information; accordingly they will value control more (making the cost of "sticking" greater). Simultaneously, the cost of communication will decrease as cyberspace improves (making the cost of "flipping" less).

In conclusion, I think it is more likely than not that a functionally necessary rule will be less costly to society than the plenary rule we currently have. Putting economic efficiency aside, the functionally necessary rule also better respects human dignity by respecting an individual's desire not to be surveilled.

A Modest Proposal

Congress should adopt a Cyberspace Privacy Act (the "Act"), that implements the "functionally necessary" default rule for all personal information collected in cyberspace. This rule is more efficient and more respectful of human dignity. Parties are, of course, free to contract around the default rule. The full proposed statute is available online at http://www.law.ucla.edu/faculty/kang/scholars, but here is a quick summary:

- First, a person who acquires personal data in the course of a cyberspace transaction must provide clear notice about what will be done with that information.
- Second, a person will not process personal information in a manner functionally unnecessary to the transaction without the prior consent of the individual.
- Third, an individual will have reasonable access to and rights of correction of personal data.
- Fourth, personal data that is no longer functionally necessary to the cyberspace transaction will be generally destroyed unless there is some legitimate pending request or the individual has given consent otherwise.
- Fifth, if compelled by court order or a dire emergency to the individual's own welfare, personal data may be disclosed as necessary.
- Finally, a person who violates this Act may be sued in federal court for civil damages. Moreover, the Federal Trade Commission will have administrative authority to enforce the Act.

Politically moderate, the proposed legislation should enjoy broad appeal. The private sector should not oppose the Act because it does not choke off electronic commerce in cyberspace.

Although the Act constrains certain forms of advertising based on detailed data collection, and the sharing of data with third parties, these constraints can be lifted simply by obtaining the customer's consent. Moreover, the Cyberspace Privacy Act would promote consumer confidence in—and thereby encourage—electronic commerce. Multinational corporations working in Europe might have an independent reason to accept the Act. By applying the Act to data received from the European Union, these corporations could credibly assert that they have begun to adopt adequate privacy protections necessary to maintain transborder flows under the recent European Union Data Protection Directive. Finally, the Act does not violate the First Amendment. In structure, the proposed Act does not differ materially from the privacy provisions of the Cable Act or the Video Privacy Protection Act. Neither act has been successfully challenged on First Amendment grounds.

Conclusion

A vision protective of information privacy in cyberspace will be singularly hard to maintain. Cyberspace's essence is the processing of information in ways and at speeds unimaginable just years ago. To retard this information processing juggernaut in the name of privacy seems antitechnology, even antiprogress. It cuts against the hackneyed cyber-proclamation that information wants to be free. Nevertheless, this intentional application of friction to personal information flows is warranted. If profit-seeking organizations are instituting such friction in the name of intellectual property, individuals should not be chastised for doing the same in the name of privacy.

Historically, privacy issues have been an afterthought. Technology propels us forward, and we react to the social consequences only after the fact. But the amount of privacy we retain is—to use a decidedly low-tech metaphor—a oneway ratchet. Once we ratchet privacy down, it will be extraordinarily difficult to get it back. More disturbingly, after a while, we might not mind so much. It may dawn on us too late that privacy should have been saved along the way.

POSTSCRIPT

Will the Digital/Computer World Lead to Greater Individual Freedom?

The computer, like the television, telephone, and automobile before it, provides millions with access to a larger world around them. They can access information, ideas, people, places, and cultures beyond their own physical and economic limitations. The computer is a tool that can empower and inform beyond the tools of any previous generation. The digital/computer world is the next great technological revolution, and it may turn out to be the greatest in many respects. Information storage, retrieval, and access will be greater than at any other time in human history. Theoretically, billions will have the collected human experience at their fingertips and will be able to educate themselves about almost any subject or issue. The impact for education, economics, politics, and society is phenomenal.

However, the digital/computer world is also a commercial world. The technology and infrastructure that support this world is privately owned and run for profit. Therefore, the technology must serve the interests of those who own it in order for it to be implemented and mass-produced. To understand the full impact of this technology on individual freedom, one must understand the goals of those who control it. For example, the phone allows people to call out as well as others to call in. This means that external access demands internal access. Likewise, the digital/computer world allows for the cataloging of information—commercial, personal, private, and political. People's tastes, buying habits, desires, and beliefs are cataloged and stored so that more and more products can be marketed to individuals based on their tastes and so that more and more actors can know who those individuals are. Anonymity will die, and with it a degree of freedom.

Is it convenient that The Gap can know your tastes in clothes, buying habits, and even your friends' and family members' birthdays so that they can send you e-mail ads at key times offering products for you to buy? Or is that kind of knowledge intrusive? Is it good that the government can track your buying habits, private communications, and travel patterns down to the minute? Or is that potentially threatening to your freedom?

Three works of note that will help you to explore this issue in greater depth are Benjamin Barber, "Three Scenarios for the Future of Technology and a Strong Democracy," *Political Science Quarterly* (Winter 1998–1999); Michael Mehta and Eric Darier, "Virtual Control and Disciplining on the Internet: Electronic Governmentality in the New Wired World," *Information Society* (April/June 1998); and Albert H. Teich, *Technology and the Future* (St. Martin's Press, 1997).

ISSUE 16

Is the Globalization of American Culture a Positive Development?

YES: David Rothkopf, from "In Praise of Cultural Imperialism?" *Foreign Policy* (Summer 1997)

NO: Jeffrey E. Garten, from "'Cultural Imperialism' Is No Joke," *Business Week* (November 30, 1998)

ISSUE SUMMARY

YES: David Rothkopf, an adjunct professor of international and public affairs, discusses the creation of a global culture and argues that the Americanization of the world along a U.S. value system is good and should be encouraged in the interests of the United States.

NO: Jeffrey E. Garten, dean of Yale University's School of Management, argues that people around the world see globalization as a form of American imperialism. He contends that such fears have a strong basis in reality and cannot be ignored.

In 1989 the Berlin Wall collapsed. Two years later the Soviet Union ceased to exist. With this relatively peaceful and monumental series of events, the cold war ended, and with it one of the most contentious and conflict-ridden periods in global history. It is easy to argue that in the wake of those events the United States is in ascendancy. The United States and its Western allies won the cold war, defeating communism politically and philosophically. Since 1990 democracies have emerged and largely flourished as never before across the world stage. According to a recent study, over 120 of the world's 190 nations now have a functioning form of democracy. Western companies, values, and ideas now sweep across the globe via airwaves, computer networks, and fiber-optic cables that bring symbols of U.S. culture and values (such as Michael Jordan and McDonald's) into villages and schools and cities around the world.

If American culture is embodied in the products sold by many multinational corporations (MNCs), such as McDonald's, Ford, IBM, The Gap, and others, then the American cultural values and ideas that are embedded in these products are being bought and sold in record numbers around the world. Globalization largely driven by MNCs and their control of technology brings with it values and ideas that are largely American in origin and expression. Values such as speed and ease of use, a strong emphasis on leisure time over work time, and a desire for increasing material wealth and comfort dominate the advertising practices of these companies. For citizens of the United States, this seems a natural part of the landscape. They do not question it; in fact, many Americans enjoy seeing signs of "home" on street corners abroad: a McDonald's in Tokyo, a Sylvester Stallone movie in Djakarta, or a Gap shirt on a student in Nairobi, for example.

While comforting to Westerners, this trend is disquieting to the hundreds of millions of people around the world who wish to partake of the globalizing system without abandoning their own cultural values. Many people around the globe wish to engage in economic exchange and develop politically but do not want to abandon their own cultures amidst the wave of values embedded in Western products. This tension is most pronounced in its effect on the youth around the world. Millions of impressionable young people in the cities and villages of the developing world wish to emulate the American icons that they see on soft drink cans or in movie theaters. They attempt to adopt U.S. manners, language, and modes of dress, often in opposition to their parents and local culture. These young people are becoming Americanized and, in the process, creating huge generational rifts within their own societies. Some of the seeds of these rifts and cultural schisms can be seen in the actions of the young Arab men who joined Al Qaeda and participated in the terrorist attacks of September 11, 2001.

David Rothkopf and Jeffrey E. Garten explore this process in the following selections, with each coming at it from very different positions. Rothkopf argues that the promotion of American culture across the globe is a good thing. Americanization, he says, will help to homogenize the world and to bring it closer together, and the United States will solidify itself as the political, economic, and cultural leader of the globe. Garten sees real problems with this process and contends that charges of American imperialism and cultural genocide cannot be ignored, lest America find itself facing a severe anti-U.S. backlash that takes a variety of nonviolent and violent forms.

David Rothkopf

 **YES**

In Praise of Cultural Imperialism?

The gates of the world are groaning shut. From marble balconies and over the airwaves, demagogues decry new risks to ancient cultures and traditional values. Satellites, the Internet, and jumbo jets carry the contagion. To many people, "foreign" has become a synonym for "danger."

Of course, now is not the first time in history that chants and anthems of nationalism have been heard. But the tide of nationalism sweeping the world today is unique. For it comes in reaction to a countervailing global alternative that—for the first time in history—is clearly something more than the crackpot dream of visionaries. It is also the first time in history that virtually every individual at every level of society can sense the impact of international changes. They can see and hear it in their media, taste it in their food, and sense it in the products that they buy. Even more visceral and threatening to those who fear these changes is the growth of a global labor pool that during the next decade will absorb nearly 2 billion workers from emerging markets, a pool that currently includes close to 1 billion unemployed and underemployed workers in those markets alone. These people will be working for a fraction of what their counterparts in developed nations earn and will be only marginally less productive. You are either someone who is threatened by this change or someone who will profit from it, but it is almost impossible to conceive of a significant group that will remain untouched by it.

Globalization has economic roots and political consequences, but it also has brought into focus the power of culture in this global environment—the power to bind and to divide in a time when the tensions between integration and separation tug at every issue that is relevant to international relations.

The impact of globalization on culture and the impact of culture on globalization merit discussion. The homogenizing influences of globalization that are most often condemned by the new nationalists and by cultural romanticists are actually positive; globalization promotes integration and the removal not only of cultural barriers but of many of the negative dimensions of culture. Globalization is a vital step toward both a more stable world and better lives for the people in it.

Furthermore, these issues have serious implications for American foreign policy. For the United States, a central objective of an Information Age foreign policy must be to win the battle of the world's information flows, dominating the airwaves as Great Britain once ruled the seas.

Culture and Conflict

Culture is not static; it grows out of a systematically encouraged reverence for selected customs and habits. Indeed, *Webster's Third New International Dictionary* defines culture as the "total pattern of human behavior and its products embodied in speech, action, and artifacts and dependent upon man's capacity for learning and transmitting knowledge to succeeding generations." Language, religion, political and legal systems, and social customs are the legacies of victors and marketers and reflect the judgment of the marketplace of ideas throughout popular history. They might also rightly be seen as living artifacts, bits and pieces carried forward through the years on currents of indoctrination, popular acceptance, and unthinking adherence to old ways. Culture is used by the organizers of society—politicians, theologians, academics, and families—to impose and ensure order, the rudiments of which change over time as need dictates. It is less often acknowledged as the means of justifying inhumanity and warfare. Nonetheless, even a casual examination of the history of conflict explains well why Samuel Huntington, in his *The Clash of Civilizations,* expects conflict along cultural fault lines, which is precisely where conflict so often erupts. Even worse is that cultural differences are often sanctified by their links to the mystical roots of culture, be they spiritual or historical. Consequently, a threat to one's culture becomes a threat to one's God or one's ancestors and, therefore, to one's core identity. This inflammatory formula has been used to justify many of humanity's worst acts.

Cultural conflicts can be placed into three broad categories: religious warfare, ethnic conflict, and conflict between "cultural cousins," which amounts to historical animosity between cultures that may be similar in some respects but still have significant differences that have been used to justify conflict over issues of proximity, such as resource demands or simple greed.

Religion-based conflicts occur between Christians and Muslims, Christians and Jews, Muslims and Jews, Hindus and Muslims, Sufis and Sunis, Protestants and Catholics, and so forth. Cultural conflicts that spring from ethnic (and in some cases religious) differences include those between Chinese and Vietnamese, Chinese and Japanese, Chinese and Malays, Normans and Saxons, Slavs and Turks, Armenians and Azerbaijanis, Armenians and Turks, Turks and Greeks, Russians and Chechens, Serbs and Bosnians, Hutus and Tutsis, blacks and Afrikaners, blacks and whites, and Persians and Arabs. Conflicts between "cultural cousins" over resources or territory have occurred between Britain and France, France and Germany, Libya and Egypt, and many others.

Another category that might be included in our taxonomy is quasi-cultural conflict. This conflict is primarily ideological and is not deeply enough rooted in tradition to fit within standard definitions of culture, yet it still exhibits most if not all of the characteristics of other cultural clashes. The best example here is the Cold War itself, a conflict between political cultures that was portrayed by its combatants in broader cultural terms: "godless communists" versus "corrupt capitalists." During this conflict, differences regarding the role of the individual within the state and over the distribution of income produced a "clash of civilizations" that had a relatively recent origin.

Finally, as a reminder of the toll that such conflicts take, one need only look at the 20th century's genocides. In each one, leaders used culture to fuel the passions of their armies and other minions and to justify their actions among their people. One million Armenians; tens of millions of Russians; 10 million Jews, Gypsies, and homosexuals; 3 million Cambodians; and hundreds of thousands of Bosnians, Rwandans, and Timorese all were the victims of "culture"—whether it was ethnic, religious, ideological, tribal, or nationalistic in its origins. To be sure, they fell victim to other agendas as well. But the provocative elements of culture were to these accompanying agendas as Joseph Goebbels was to Adolf Hitler—an enabler and perhaps the most insidious accomplice. Historians can, of course, find examples from across the ages of "superior" cultures eradicating "inferior" opponents—in the American West, among the native tribes of the Americas and Africa, during the Inquisition, and during the expansion of virtually every empire.

Satellites as Cultural Death Stars

Critics of globalization argue that the process will lead to a stripping away of identity and a blandly uniform, Orwellian world. On a planet of 6 billion people, this is, of course, an impossibility. More importantly, the decline of cultural distinctions may be a measure of the progress of civilization, a tangible sign of enhanced communications and understanding. Successful multicultural societies, be they nations, federations, or other conglomerations of closely interrelated states, discern those aspects of culture that do not threaten union, stability, or prosperity (such as food, holidays, rituals, and music) and allow them to flourish. But they counteract or eradicate the more subversive elements of culture (exclusionary aspects of religion, language, and political/ideological beliefs). History shows that bridging cultural gaps successfully and serving as a home to diverse peoples requires certain social structures, laws, and institutions that transcend culture. Furthermore, the history of a number of ongoing experiments in multiculturalism, such as in the European Union, India, South Africa, and the United States, suggests that workable, if not perfected, integrative models exist. Each is built on the idea that tolerance is crucial to social well-being, and each at times has been threatened by both intolerance and a heightened emphasis on cultural distinctions. The greater public good warrants eliminating those cultural characteristics that promote conflict or prevent harmony, even as less-divisive, more personally observed cultural distinctions are celebrated and preserved.

The realization of such integrative models on a global scale is impossible in the near term. It will take centuries. Nor can it be achieved purely through rational decisions geared toward implementing carefully considered policies and programs. Rather, current trends that fall under the broad definitional umbrella of "globalization" are accelerating a process that has taken place throughout history as discrete groups have become familiar with one another, allied, and commingled—ultimately becoming more alike. Inevitably, the United States has taken the lead in this transformation; it is the "indispensable nation" in the management of global affairs and the leading producer of information products and services in these, the early years of the Information Age.

The drivers of today's rapid globalization are improving methods and systems of international transportation, devising revolutionary and innovative information technologies and services, and dominating the international commerce in services and ideas. Their impact affects lifestyles, religion, language, and every other component of culture.

Much has been written about the role of information technologies and services in this process. Today, 15 major U.S. telecommunications companies, including giants like Motorola, Loral Space & Communications, and Teledesic (a joint project of Microsoft's Bill Gates and cellular pioneer Craig McCaw), offer competing plans that will encircle the globe with a constellation of satellites and will enable anyone anywhere to communicate instantly with anyone elsewhere without an established telecommunications infrastructure on the ground near either the sender or the recipient. (Loral puts the cost of such a call at around $3 per minute.)

Technology is not only transforming the world; it is creating its own metaphors as well. Satellites carrying television signals now enable people on opposite sides of the globe to be exposed regularly to a wide range of cultural stimuli. Russian viewers are hooked on Latin soap operas, and Middle Eastern leaders have cited CNN as a prime source for even local news. The Internet is an increasingly global phenomenon with active development under way on every continent.

The United States dominates this global traffic in information and ideas. American music, American movies, American television, and American software are so dominant, so sought after, and so visible that they are now available literally everywhere on the Earth. They influence the tastes, lives, and aspirations of virtually every nation. In some, they are viewed as corrupting.

France and Canada have both passed laws to prohibit the satellite dissemination of foreign—meaning American—content across their borders and into the homes of their citizens. Not surprisingly, in many other countries—fundamentalist Iran, communist China, and the closely managed society of Singapore—central governments have aggressively sought to restrict the software and programming that reach their citizens. Their explicit objective is to keep out American and other alien political views, mores, and, as it is called in some parts of the Middle East, "news pollution." In these countries, the control of new media that give previously closed or controlled societies virtually unlimited access to the outside world is a high priority. Singapore has sought to filter out certain things that are available over the Internet—essentially processing all information to eliminate pornography. China has set up a "Central Leading Group" under the State Planning Commission and the direct supervision of a vice premier to establish a similar system that will exclude more than just what might be considered obscene.

These governments are the heirs of King Canute, the infamous monarch who set his throne at the sea's edge and commanded the waves to go backward. The Soviet Union fell in part because a closed society cannot compete in the Information Age. These countries will fare no better. They need look no further than their own élites to know this. In China, while satellite dishes are technically against the law, approximately one in five citizens of Beijing has access to television programming via a dish, and almost half of the people of Guangzhou have access to satellite-delivered programming. Singapore, the leading entrepôt of Southeast Asia, is a hub in a global network of business centers in which the lives of the élites are virtually identical. Business leaders in Buenos Aires, Frankfurt, Hong Kong, Johannesburg, Istanbul, Los Angeles, Mexico City, Moscow, New Delhi, New York, Paris, Rome, Santiago, Seoul, Singapore, Tel Aviv, and Tokyo all read the same newspapers, wear the same suits, drive the same cars, eat the same food, fly the same airlines, stay in the same hotels, and listen to the same music. While the people of their countries remain divided by culture, they have realized that to compete in the global marketplace they must conform to the culture of that marketplace.

The global marketplace is being institutionalized through the creation of a series of multilateral entities that establish common rules for international commerce. If capital is to flow freely, disclosure rules must be the same, settlement procedures consistent, and redress transparent. If goods are also to move unimpeded, tariff laws must be consistent, customs standards harmonized, and product safety and labeling standards brought into line. And if people are to move easily from deal to deal, air transport agreements need to be established, immigration controls standardized, and commercial laws harmonized. In many ways, business is the primary engine driving globalization, but it would be a mistake to conclude that the implications of globalization will be limited primarily to the commercial arena.

In politics, for example, as international organizations arise to coordinate policy among many nations on global issues such as trade, the environment, health, development, and crisis management, a community of international bureaucrats is emerging. These players are as comfortable operating in the international environment as they would be at home, and the organizations that they represent in effect establish global standards and expectations—facilitating the progress of globalization.

The community of nations increasingly accepts that such supranational entities are demanded by the exigencies of the times; with that acceptance also comes a recognition that the principal symbol of national identity—namely sovereignty—must be partially ceded to those entities. The United States in particular seems to have problems with this trend. For example, the United States was involved in creating the World Trade Organization and now undermines its effectiveness by arbitrarily withdrawing from its efforts to blunt the effects of the Helms-Burton act. Still, the recognition that sometimes there are interests greater than national interests is a crucial step on the path to a more peaceful, prosperous world.

Toward a Global Culture

It is in the general interest of the United States to encourage the development of a world in which the fault lines separating nations are bridged by shared interests. And it is in the economic and political interests of the United States to ensure that if the world is moving toward a common language, it be English; that if the world is moving toward common telecommunications, safety, and quality standards, they be American; that if the world is becoming linked by television, radio, and music, the programming be American; and that if common values are being developed, they be values with which Americans are comfortable.

These are not simply idle aspirations. English is linking the world. American information technologies and services are at the cutting edge of those that are enabling globalization. Access to the largest economy in the world—America's—is the primary carrot leading other nations to open their markets.

Indeed, just as the United States is the world's sole remaining military superpower, so is it the world's only information superpower. While Japan has become quite competitive in the manufacture of components integral to information systems, it has had a negligible impact as a manufacturer of software or as a force behind the technological revolution. Europe has failed on both fronts. Consequently, the United States holds a position of advantage at the moment and for the foreseeable future.

Some find the idea that Americans would systematically seek to promote their culture to be unattractive. They are concerned that it implies a sense of superiority on Americans' part or that it makes an uncomfortable value judgment. But the realpolitik of the Information Age is that setting technological standards, defining software standards, producing the most popular information products, and leading in the related development of the global trade in services are as essential to the well-being of any would-be leader as once were the resources needed to support empire or industry.

The economic stakes are immense considering the enormous investments that will be made over the next 10 years in the world's information infrastructure. The U.S. government estimates that telecommunications investment in Latin America alone during this period will top $150 billion. China will spend a similar amount, as will the member states of the Association of South East Asian Nations. In fact, the market for telecommunications services is expected to top $1 trillion by the turn of the century.

During the decade ahead, not only will enormous sums be directed toward the establishment of the global network of networks that the Clinton administration has dubbed the "Global Information Infrastructure," but those sums will pay for the foundations of a system that will dictate decades of future choices about upgrades, systems standards, software purchases, and services. At the same time, new national and international laws will be written, and they will determine how smoothly information products and services may flow from one market to another. Will steps be taken to ensure that Internet commerce remains truly free? What decisions will be made about the encryption of data that will impact not only the security of information markets but the free flow of ideas and the rights of individuals in the Information Age? Will governments allow the democratizing promise of the Internet to enable virtually anyone with a computer to contact anyone else?

The establishment of the Global Information Infrastructure is not just an enormous commercial opportunity for the world's information leader. The development of the rules governing that infrastructure will shape the nature of global politics decisively, either enhancing or undermining freedoms, thereby either speeding or slowing the pace of integration, understanding, and tolerance worldwide. The nature of individual and national relations will be transformed. Those wires and constellations of satellites and invisible beams of electronic signals crisscrossing the globe will literally form the fabric of future civilization.

Consequently, it could not be more strategically crucial that the United States do whatever is in its power to shape the development of that infrastructure, the rules governing it, and the information traversing it. Moreover, even if much of this process of developing what we might call the "infosphere" is left to the marketplace (as it should be), governments will control crucial elements of it. Governments will award many of the biggest infrastructure development contracts offered in the next decade: Some will assist their national companies in trying to win those contracts, and state officials will meet to decide the trade rules that will govern international traffic in the world's telecommunications markets, the global regulatory environment, encryption standards, privacy standards, intellectual property protections, and basic equipment standards. Governments will determine whether these are open or closed markets and what portion of development dollars will be targeted at bringing the benefits of these technologies to the poor to help counteract information inequities. Already some government intercessions into this marketplace have failed. Notably, Japan's efforts to shape the development of high-definition television standards sent that nation down an analog path in what turned out to be a digital race. Yet there are many places where there is an important role for governments and where the United States should have a carefully considered overarching policy and an aggressive stance to match.

Exporting the American Model

Many observers contend that it is distasteful to use the opportunities created by the global information revolution to promote American culture over others, but that kind of relativism is as dangerous as it is wrong. American culture is fundamentally different from indigenous cultures in so many other locales. American culture is an amalgam of influences and approaches from around the world. It is melded—consciously in many cases—into a social medium that allows individual freedoms and cultures to thrive. Recognizing this, Americans should not shy away from doing that which is so clearly in their economic, political, and security interests—and so clearly in the interests of the world at large. The United States should not hesitate to promote its values. In an effort to be polite or politic, Americans should not deny the fact that of all the nations in the history of the world, theirs is the most just, the most tolerant, the most willing to constantly reassess and improve itself, and the best model for the future. At the same time, Americans should not fall under the spell of those like Singapore's Lee Kuan Yew and Malaysia's Mahathir bin-Mohamad, who argue that there is "an Asian way," one that non-Asians should not judge and that should be allowed to dictate the course of events for all those operating in that corner of the world. This argument amounts to self-interested political rhetoric. Good and evil, better and worse coexist in this world. There are absolutes, and there are political, economic, and moral costs associated with failing to recognize this fact.

Repression is not defensible whether the tradition from which it springs is Confucian, Judeo-Christian, or Zoroastrian. The repressed individual still suffers, as does society, and there are consequences for the global community. Real costs accrue in terms of constrained human creativity, delayed market development, the diversion of assets to enforce repression, the failure of repressive societies to adapt well to the rapidly changing global environment, and the dislocations, struggles, and instability that result from these and other factors. Americans should promote their vision for the world, because failing to do so or taking a "live and let live" stance is ceding the process to the not-always-beneficial actions of others. Using the tools of the Information Age to do so is perhaps the most peaceful and powerful means of advancing American interests.

If Americans now live in a world in which ideas can be effectively exported and media delivery systems are powerful, they must recognize that the nature of those ideas and the control of those systems are matters with which they should be deeply concerned. Is it a threat to U.S. interests, to regional peace, to American markets, and to the United States's ability to lead if foreign leaders adopt models that promote separatism and the cultural fault lines that threaten stability? It certainly is. Relativism is a veil behind which those who shun scrutiny can hide. Whether Americans accept all the arguments of Huntington or not, they must recognize that the greater the cultural value gaps in the world, the more likely it is that conflict will ensue. The critical prerequisite for gaining the optimum benefits of global integration is to understand which cultural attributes can and should be tolerated—and, indeed, promoted—and which are the fissures that will become fault lines.

It is also crucial that the United States recognize its limitations. Americans can have more influence than others, but they cannot assure every outcome. Rather, the concerted effort to shape the development of the Global Information Infrastructure and the ideas that flow within it should be seen merely as a single component of a well-rounded foreign and security policy. (And since it is not likely to be an initiative that is widely liked or admired or enhanced through explicit promotion, it is not an approach that should be part of American public diplomacy efforts.)

Of course, implementing such an approach is not going to be easy in an America that is wracked by the reaction to and the backlash against globalization. Today, the extreme left and right wings of both major political parties are united in a new isolationist alliance. This alliance has put the brakes on 60 years of expanding free trade, has focused on the threats rather than the promise posed by such critical new relationships as those with China and other key emerging markets, and has seized on every available opportunity to disengage from the world or to undermine U.S. abilities to engage or lead effectively. It will take a committed effort by the president and cooperation from leaders on Capitol Hill to overcome the political opposition of the economic nationalists and neoisolationists. It will not happen if those in leadership positions aim simply to take the path of least political resistance or to rest on the accomplishments of the recent past. In a time of partisan bickering, when the emphasis of top officials has shifted from governing to politicking, there is a risk that America will fail to rise to these challenges. While the Clinton administration has broken important ground in developing a Global Information Infrastructure initiative and in dealing with the future of the Internet, encryption issues, and intellectual property concerns, these efforts are underfunded, sometimes managed to suit political rather than strategic objectives, shortsighted (particularly the steps concerning encryption, in which rapid changes and the demands of the marketplace are being overlooked), and poorly coordinated. At the same time, some of America's most powerful tools of engagement—which come in the form of new trade initiatives—seemingly have been shelved. This problem is most clearly manifested in the fact that fast-track negotiating-authority approval has not yet been granted and in the real possibility that Congress will refuse to grant such approval before the turn of the century.

The Clinton administration and its successors must carefully consider the long-term implications of globalization, such as the impact of the rise of new markets on America's economic influence and how America can maintain its leadership role. Aspects of American culture will play a critical role in helping to ensure the continuation of that leadership. American cultural diversity gives the United States resources and potential links with virtually every market and every major power in the world. America's emphasis on the individual ensures that American innovation will continue to outstrip that of other nations. Working in its favor is the fact that the "Pax Americana" is a phenomenon of the early years of globalization and that the U.S. ascendancy to undisputed leadership came at the same time as the establishment of international institutions such as the United Nations, the World Bank, and the International Monetary Fund; thus, for all the challenges of adjustment, the United States has more leadership experience than any other nation in this new global environment. Also, though some may decry Americans' emphasis on "newness" and suggest that it is a result of their lack of an extensive history, it also represents a healthy lack of cultural "baggage": It is this emphasis on newness that puts the United States in the best position to deal with a world in which the rapidity of change is perhaps the greatest strategic challenge of all.

Identity Without Culture

The opportunity lies before us as Americans. The United States is in a position not only to lead in the 21st century as the dominant power of the Information Age but to do so by breaking down the barriers that divide nations—and groups within nations—and by building ties that create an ever greater reservoir of shared interests among an ever larger community of peoples. Those who look at the post-Cold War era and see the "clash of civilizations" see only one possibility. They overlook the great strides in integration that have united the world's billions. They discount the factors that have led to global consolidation and the reality that those factors grow in power with each new day of the global era—integration is a trend that builds upon itself. They argue that America should prepare for the conflicts that may come in this interim period without arguing that it should accelerate the arrival of a new era with every means at its disposal.

Certainly, it is naive to expect broad success in avoiding future conflicts among cultures. But we now have tools at our disposal to diminish the disparities that will fuel some of those conflicts. While we should prepare for conflict, we should also remember that it is not mere idealism that demands that we work for integration and in support of a unifying global culture ensuring individual rights and enhancing international stability: It is also the ultimate realpolitik, the ultimate act of healthy self-interest.

Allowing ourselves to be swept up in the backlash against globalization would undermine America's ability to advance its self-interests. Americans must recognize that those interests and the issues pertaining to them reach across the disciplines of economics, politics, science, and culture. An interdisciplinary approach to international policymaking is thus required. We must also fully understand the new tools at our disposal. We must understand the profound importance and nature of the emerging infosphere—and its potential as a giant organic culture processor, democratic empowerer, universal connector, and ultimate communicator. Moreover, it is not enough to create and implement the right policies using the new tools at our disposal. Policymakers must better communicate the promise of this new world and make clear America's stake in that promise and the role Americans must play to achieve success. The United States does not face a simple choice between integration or separation, engagement or withdrawal. Rather, the choice is between leading a more peaceful world or being held hostage to events in a more volatile and violent one.

NO

"Cultural Imperialism" Is No Joke

Washington's crusade for free trade is often seen abroad as a Trojan horse for companies, such as Walt Disney Co. and Cable News Network, that would dominate foreign lifestyles and values. Most Americans react to these fears with a shrug. That's a big mistake.

The entertainment industry, including movies, music, software, and broadcasting, is America's second-largest exporter after aircraft and has penetrated all global markets. Films such as *Lethal Weapon* are hits on every continent. *Reader's Digest* publishes in 19 languages. Windows computer programs and MTV can be found in remote corners of China.

The transmission of our culture goes beyond the arts or the media. When Washington exalts free enterprise, it promotes the notion that individual freedom has a higher value than government authority. When it advocates the rule of law overseas, it pushes a U.S.-style legal system.

Rebellion From the Roman to the Soviet empires, superpowers have aimed to spread their cultures, and from Lorenzo de' Medici to Michael Eisner, there has always been a link between commerce and culture. Still, while America's lifestyle and ideas can be liberating and uplifting, they are also often destabilizing abroad. Movies and music frequently glorify violence and rebellion. Darwinian capitalism requires societies to uproot traditional structures without adequate regulation, safety nets, or education. The U.S. legal system encourages confrontation, not conciliation.

Americans should not have difficulty empathizing with foreign fears of cultural invasion. Recall U.S. anxieties a decade ago when Sony Corp. bought Columbia Pictures and Mitsubishi Corp. purchased New York's Rockefeller Center. Now reaction against American "cultural imperialism" is building. Just a few years ago, France almost torpedoed the Uruguay Round of global trade negotiations because it wanted to limit the activities of U.S. entertainment companies. Last spring, a multilateral treaty on investment rules was derailed in part because of a spat between Brussels and Washington over protection of Europe's cultural industries. In August, Canada called together 19 other governments to plot ways to ensure their cultural independence from America. Mexico is considering legislation requiring that a certain percentage of its media programming remain in the hands of its citizens. U.N.-sponsored conferences on preserving national cultures are proliferating. In contrast to the American preference for financial liberalization, capital controls are becoming respectable in Asia.

The U.S. should do more than heed these warnings; it should recognize that strong cultures abroad are in America's self-interest. Amid the disorientation that comes with globalization, countries need cohesive national communities grounded in history and tradition. Only with these in place can they unite in the tough decisions necessary to building modern societies. If societies feel under assault, insecurities will be magnified, leading to policy paralysis, strident nationalism, and anti-Americanism.

With satellites and the Internet, the spread of American culture cannot be stopped—nor should it. But Corporate America and Washington could lessen U.S. dominance by encouraging cultural diversity around the globe. Companies such as Time Warner Inc. and PepsiCo Inc. could fund native entrepreneurs wishing to create local cultural industries. They could showcase regional film and theatrical productions and finance university research and teaching in the region's history, art, and literature.

The Clinton Administration could reverse current trade policy and permit temporary quotas and subsidies abroad to preserve certain local cultural industries, such as film and TV. It could encourage the World Bank to build up foreign countries' tourism infrastructure. It could expand assistance to U.N. efforts to restore national monuments that have been neglected or destroyed.

At a time when so many nations that have recently embraced Adam Smith are in deep recession, the Treasury and State Depts. could lower the volume on their rhetoric about the magic of the free marketplace. And when so much of U.S. society is fed up with inordinate litigation, officials could be more modest about the glories of America's legal system.

Protecting national cultures could soon become a defensive rallying point for societies buffeted by globalization and undergoing tumultuous change. Being more sensitive to foreign concerns would ease the prospect of backlash and even bolster America's ability to export its ideas and ideals for the long haul. The U.S. should at least try.

POSTSCRIPT

Is the Globalization of American Culture a Positive Development?

Globalization is a process of technological change and economic expansion under largely capitalist principles. The key actors driving the globalization process are multinational corporations like McDonald's, Coca-Cola, Nike, and Exxon Mobil. These companies are rooted in the American-Western cultural experience, and their premise is based on a materialistic world culture that is striving for greater and greater wealth. That value system is Western and American in origin and evolution. It is therefore logical to assume that as globalization goes, so goes American culture.

Evidence of "American" culture can be seen across the planet: kids in Djakarta or Lagos wearing Michael Jordan jerseys and Nike shoes, for example, and millions of young men and women from Cairo to Lima listening to Michael Jackson records. Symbols of American culture abound in almost every corner of the world, and most of that is associated with economics and the presence of multinational corporations.

As the youth of the world are seduced into an American cultural form and way of life, other cultures are often eclipsed. They lose traction and fade with generational change. Many would argue that this loss is unfortunate, but others would counter that it is part of the historical sweep of life. Social historians suggest that the cultures of Rome, Carthage, Phoenicia, and the Aztecs, while still influential, were eclipsed by a variety of forces that were dominant and historically rooted. While tragic, it was inevitable in the eyes of some social historians.

Regardless of whether this eclipse is positive or negative, the issue of cultural imperialism remains. Larger and more intrusive networks of communication, trade, and economic exchange bring values. In this world of value collision comes choices and change. Unfortunately, millions will find themselves drawn toward a lifestyle of materialism that carries with it a host of value choices. The losers in this clash are local cultures and traditions that, as so often is the case among the young, are easily jettisoned and discarded. It remains to be seen whether or not they will survive the onslaught.

Works on this subject include Benjamin Barber, "Democracy at Risk: American Culture in a Global Culture," *World Policy Journal* (Summer 1998); "Globalism's Discontents," *The American Prospect* (January 1, 2002); Seymour Martin Lipset, *American Exceptionalism: A Double-Edged Sword* (W. W. Norton, 1996); and Richard Barnet and John Cavanagh, *Global Dreams: Imperial Corporations and the New World Order* (Simon & Schuster, 1995).

On the Internet . . .

Chemical and Biological Weapons Resource Page

The Chemical and Biological Weapons Resource Page, created by the Center for Nonproliferation Studies, provides a list of documents concerning threat assessment, response, counterterrorism, domestic preparedness, and funding to combat terrorism.

http://cns.miis.edu/research/cbw/cbterror.htm

Nuclear Terrorism: How to Prevent It

This site of the Nuclear Control Institute discusses nuclear terrorism and how best to prevent it. Topics include terrorists' ability to build nuclear weapons, the threat of "dirty bombs," and whether or not nuclear reactors are adequately protected against attack. This site features numerous links to key nuclear terrorism documents and Web sites as well as to recent developments and related news items.

http://www.nci.org/nuketerror.htm

CDI Terrorism Project

The Center for Defense Information's (CDI) Terrorism Project is designed to provide insights, in-depth analysis, and facts on the military, security, and foreign policy challenges of terrorism. The project looks at all aspects of fighting terrorism, from near-term issues of response and defense to long-term questions about how the United States should shape its future international security strategy.

http://www.cdi.org/terrorism/

Exploring Global Conflict: An Internet Guide to the Study of Conflict

Exploring Global Conflict: An Internet Guide to the Study of Conflict is an Internet resource designed to provide understanding of global conflict. Information related to specific conflicts in areas such as Northern Ireland, the Middle East, the Great Lakes region in Africa, and the former Yugoslavia is included on this site. Current news and educational resource sites are listed as well.

http://www.uwm.edu/Dept/CIS/conflict/congeneral.html

PART 5

The New Global Security Agenda

*T*he end of the cold war freed the concept of security from its bipolar constraints. Security is now the purview of all peoples and, as such, our concept of security has broadened considerably. Issues that were once on the back burner of geopolitics are now front-page news. Stories about clashes of civilizations, landmines, chemical and biological terrorism, "narco-guerillas," and war crimes are covered daily.

This section examines some of the key issues shaping the security dilemma of the twenty-first century.

- Is Biological and Chemical Terrorism the Next Grave Threat to the World Community?

- Is a Nuclear Terrorist Attack on America Likely?

- Is There a Military Solution to Terrorism?

- Are Civil Liberties Likely to Be Compromised in the War Against Terrorism?

- Are Cultural and Ethnic Conflicts the Defining Dimensions of Twenty-First-Century War?

ISSUE 17

Is Biological and Chemical Terrorism the Next Grave Threat to the World Community?

YES: Al J. Venter, from "Keeping the Lid on Germ Warfare," *International Defense Review* (May 1, 1998)

NO: Ehud Sprinzak, from "The Great Superterrorism Scare," *Foreign Policy* (Fall 1998)

ISSUE SUMMARY

YES: Journalist Al J. Venter examines newly emerging germs and asserts that they will expand the threat of germ warfare in the coming years.

NO: Professor of political science Ehud Sprinzak argues that a biological or chemical terrorist attack is highly unlikely and that governments, in justifying expanding defense spending and other measures, are exploiting people's fears of such an occurrence.

Ever since the advent of nuclear and, later, biological and chemical weapons, the specter of their use on a widespread scale against military and civilian targets has hung over the globe. The realities of chemical weaponry have been with us since the crude gas bombs of World War I. Subsequent advances in chemical weaponry have been utilized on relatively small scales (although with tremendously tragic consequences) in places like Iraq, Afghanistan, and Vietnam. Such agents as sarin, cyanogen chloride, anthrax, ricin, and Ebola virus can render individuals dead or dying with little chance of survival—some within hours of contact. These and others now head the list of chemical and biological agents that can inflict great death in relatively small amounts.

America's worst fears became reality on September 11, 2001, when Al Qaeda launched attacks on New York City and Washington, D.C. Now, in the post–September 11 world, concerns regarding chemical and biological weapons are omnipresent in the Middle East, Russia, Europe, Asia, and the United States. The U.S. attack against Saddam Hussein's Iraq was based largely on the perceived threat of Hussein's using chemical and biological agents or giving them to terrorists.

With the demise of the superpower conflict and the apparent lessening of the threat of country-to-country nuclear war, chemical and biological terrorism has become one of the top defense concerns. There are many reasons for this. First, such weapons are relatively cheap to develop and manufacture when compared to their nuclear counterparts. They are also easy to conceal and deliver. Modest amounts of certain chemical agents injected into air or water supplies can kill thousands, perhaps even millions, in concentrated urban areas. Second, laws and treaties governing the production, sale, and use of such weapons have been spotty. Different governments (including that of the United States) are reluctant to ban such weapons because others already have them. Also, nations have yet to agree on the proper wording of treaties to ban chemical weapons. Third, these weapons are increasingly alluring to groups with political goals who feel impotent in the fight against larger, more powerful enemies. They believe that by striking a dramatic blow at civilian targets, they can realize their aims without the sure suicide of using conventional military or political means.

Many experts believe that these chemical and biological weapons will become more prevalent as modern science produces new chemical compounds and biological elements that lead to greater risk of toxicity. They point to increased governmental concerns over such agents, as evidenced by growing intelligence activity designed to thwart these very occurrences.

Yet others suggest that the fears raised by these agents are overblown and melodramatic. They argue that raising this fear in the aftermath of the cold war is dubious and probably indicates a desire to maintain threat levels in order to increase military and defense spending in various states. Some contend that national security structures and agencies around the world are looking for new justifications for their presence and, thus, have raised the specter of chemical and biological warfare without due cause or real threat.

The following selections debate this issue from opposite ends of the spectrum. Al J. Venter argues that the threat of biological and chemical terrorism is real and pronounced. He contends that keeping the lid on germ warfare is our greatest challenge and that it demands increased knowledge, awareness, and spending to prevent such threats from becoming catastrophic realities. Ehud Sprinzak maintains that warnings about germ warfare are part of a big scare tactic designed to allow for greater military spending and surveillance. He argues that the nature of biological and chemical weapons and potential adversaries is such that the threat is limited and not particularly credible.

Al J. Venter  **YES**

Keeping the Lid on Germ Warfare

The 21st century will be the age of the gene just as the 20th century was the age of the atom.

Molecular bio-technology will transform agriculture, energy production, health care, and microelectronics, however, it will also pose significant military and strategic challenges. Whereas recombinant DNA technology offers great benefits to humankind, it also has a darker side—the genetic engineering of microbial pathogens, toxins, and even natural brain chemicals, to create more deadly and persistent weapons of war.

Although Iraq's biological weapons (BW) program is under the microscope of those seeking to halt BW proliferation, the *Wall Street Journal* wrote that this "diverts Western attention away from the broader problems of chemical and biological weapons worldwide—and especially in Russia".

During the Cold War a consensus existed that the Soviet Union was using recombinant DNA technology for military purposes. It was also attempting the recombination of the venom-producing genes from cobras and scorpions (and even bees) with the DNA of normally harmless bacteria. Such an organism would infect the body and surreptitiously produce paralytic cobra toxin. If delivered as a respirable aerosol, such an engineered agent could infect tens of thousands of people.

Former Soviet scientists have described a jointly-operated gene warfare program between the military and the Soviet Academy of Sciences which enjoyed the full support of the Kremlin leadership. The United States, meanwhile, had unilaterally stopped military research into offensive biological and toxin warfare capability in 1972.

Some observers are worried that work in this field is continuing. However, Prof. Matthew Meselson, professor of biochemistry at Harvard University and former chairman of the Federation of American Scientists, has spent a great deal of time investigating what the Soviets, and now the Russians, are doing in the area of biological warfare. He maintains that Moscow's biological research programs are now minimal, pointing to the fact that at every one of the known BW installations, recent satellite imaging has not revealed any ultraviolet emissions.

From Al J. Venter, "Keeping the Lid on Germ Warfare," *International Defense Review,* vol. 31, no. 5 (May 1, 1998). Copyright © 1998 by Jane's Information Group Limited. Reprinted by permission.

However, biological issues continue to feature in the media. UK and US troops were inoculated against anthrax during their recent deployment in the Persian Gulf, and the UK government issued an all-ports alert following intelligence reports of an attempted smuggling of an anthrax toxin into the country concealed in duty-free merchandise.

The Greatest Threat

Iraq's research into anthrax, botulinum toxin, gas gangrene bacteria, and various chemical poisons such as VX gas, is regarded as the greatest threat. Mustard gas and nerve agents were used in combat against Iran during the 1980–88 Gulf War, and nerve agents were deployed against a Kurdish village in 1988 with casualty figures in the thousands. Moreover, a little-publicized CIA report from 1996—*Intelligence Related to Gulf War Illness*—said that: "There are no indications that any biological agent was destroyed by Coalition bombing."

Understanding the potential of biological warfare remains something of a gray area. Eric Croddy in his book *Chemical and Biological Warfare, An Annotated Bibliography* makes a persuasive case that it is actually quite difficult to kill huge numbers of people using chemical or biological warfare (CBW) agents. "In fact," says Croddy, "it is a considerable challenge to use microbes and biological toxins as weapons of any scale."

Most bacteria and protein toxins are fragile. They are thermolabile (unstable when heated), and are sensitive to acidic solutions and ultraviolet radiation from the sun. Not only must a biological agent 'jump' a number of environmental hurdles, it must also face the antibody/antigen response once inside human tissue.

"The Russians, as with ourselves in the West," says Croddy, "discovered a long time ago that the human body is remarkably resilient. The largest organ, the skin, is a selectively permeable shield against common bacteria, rickettsia and parasites that are ubiquitous in our atmosphere. Even if ingested in food or water, most microbes that would otherwise be remarkably virulent, usually die—their protein toxins denatured by acidic and enzymic action in the gut. And if they find their way in through a cut in the skin, likely as not almost all bacteria will be engulfed by the phagocytic guardians of our immune systems."

Overcoming these defensive mechanisms requires unique features: because the most 'hospitable' and vulnerable sites of entry are the lungs, any biological warfare microbes have to be delivered in some form of respirable aerosol to cause mass casualties. Consequently, microencapsulation is a possible future technology for delivery of some of the more fragile viruses.

Only the most hardy microbes survive the necessary processing in today's biological weapons, and one pathogen—anthrax—fits most criteria as an effective BW. In addition to its ability to form an aerosol, this bacterium on infection attacks the body's own defenses. Significantly, Russia has experimented with an anthrax strain that shows resistance to antibiotic treatment.

Found in domestic livestock, anthrax (Bacillus anthracis) is most commonly encountered among sheep. Shearing of these animals allows the bacteria to become airborne and is the causative agent in woolsorter's disease (a form of pneumonia).

As a spore-forming bacterium, anthrax can survive for decades in soil. This makes it ideal for freeze-drying into an exceptionally fine powder. Once ensconced in the lungs, the capsule surface of the spore resists the body's immunological response. One of three toxins released by the anthrax bacteria further reduces the body's ability to react protectively.

For decades, scientists believed that anthrax killed its victims by forming 'logjams' in the blood stream. Research has since determined that Anthrax toxin III (in combination with other factors) is arguably the most intrusive culprit. Multiplying in the lungs and then in the bloodstream, anthrax reproduces in ever greater numbers by geometric progression. Anecdotal reports of patients succumbing suddenly to anthrax following two or three days of symptoms are consistent with the release of lethal toxin.

Further investigations into anthrax reveal more unsettling discoveries, not least that some strains appear to be resistant to penicillin. This presents another dilemma: in the event of a terrorist anthrax attack, sufficient antibiotics for large population concentrations might not be immediately available. It takes time to manufacture antibiotics. If the threat is real, tens of millions of people would need to be immediately inoculated. Then there is the question of time and whether a therapeutic course could be given quickly enough.

The potential for genetic engineering of this bacterium also presents some horrific consequences: a bacteriological weapon, already well-suited to killing thousands of people within days, could be modified (as Russia has done) to make it resistant to antibiotics. It is, as one observer noted, the perfect weapon and there is every indication that Iraq possesses it in abundance. During the 1990–91 Gulf War, Iraq had large quantities of anthrax, with the intention of dispersing the bacteria over Coalition lines and across Israel.

An outbreak of anthrax in humans occurred in 1979 at Sverdlovsk (now Ekaterinburg) in the former Soviet Union. The original release of spores came from a biological warfare research laboratory in the southern suburbs of the city of 1.2 million people on the eastern slopes of the Urals. It was freeze-dried and the amount released into the atmosphere, according to Meselson, was anything from 4mg to 1g. (It is impossible to see 4mg with the naked eye.)

The resultant epidemic—96 people were infected and 64 died—provoked intense international debate. The Russians never revealed how much anthrax was involved, and it took years to get to the point where they were prepared even to admit that it was an accident.

Speculation continues as to whether the accidental release resulted from activities prohibited by the Biological and Toxin Weapons Convention of 1972. Subsequent research has shown that under such circumstances, the fatality rate—without aggressive medical treatment—would have been more than 95 per cent.

Anthrax particles are most effective when they are within a certain micron range; the accepted wisdom indicates anything between 1–10m. Extensive research that has been carried out at the United States Army Medical Research Institute of Infectious Diseases (USAMRIID), Fort Detrick, Maryland, has indicated that between 8,000 and 10,000 spores constitute an LD50 (lethal dose that will kill half of those infected) for humans. If anthrax is disseminated in sufficiently fine particles, it quickly gets into the bloodstream through the lungs.

Significantly, capsulating bacteria like Bacillus anthracis tend to defeat the body's immune system in two ways: first they resist being engulfed by white blood cells; second, they produce toxins that actually vitiate the human immune response. Then the bacteria start multiplying at a rapid rate, producing a most lethal toxin. Fortunately this symptom can be treated if caught early enough.

Sudden Death

However, if the strain is the latest Russian development and proves to be impervious to antibiotics, any such treatment is rendered ineffective. The progression in such a case is simple; as the bacteria multiply, the level of toxin dramatically increases. Death can be sudden, following a few days of incubation. At Sverdlovsk the first victim was dead in days, and the last case died six weeks after infection.

The consensus is that anyone exposed to at least 20 anthrax spores per cubic liter of air for about 30min will probably receive close to an LD50 dose. If enough infected air is inhaled in that time, sufficient spores will enter the victim's lungs to make for an infectious dose.

Iraq carried out research in this field in laboratories at Al Hakam and Salman Pak, both on the outskirts of Baghdad. Although Coalition bombing in 1991 destroyed much of this, intelligence sources indicate that a significant measure of this BW program has survived intact. United Nations inspectors with the UN Special Commission (UNSCOM) are now searching for these assets. There is no doubt that the weapons exist—the problem is finding them.

Those countries with advanced research BW programs—such as Iraq, Iran, Libya, North Korea, and Syria—have anthrax occurring naturally, usually in such places as stockyards, wool shearing depots and the like. Governments are consequently able to seek scientific aid regarding treatment from humanitarian organizations as part of their ongoing research, because anthrax is a viable threat to livestock.

Mail-Order 'Anthrax'

In 1986, before the breakdown of relations between Iraq and the US, the latter country supplied seed cultures of anthrax from the American Type Culture Collection, a laboratory in Rockville, Maryland. Spore samples were ordered by telephone and were sent, as a matter of course, by normal mail. Iraq and other renegade states working on BWs were then able to ask the relevant bodies for anthrax, claiming it was needed for research on antibiotic regimens.

This supply channel has since been blocked. According to the US Department of Defense (DoD), this is the most common method of creating a national

BW program; as such, anthrax will continue to be the prime experimental pathogen, which is why it is regarded as such a serious threat.

Similarly, botulinum toxin—another 'pathogen of choice' for some developing countries—presents a different type of threat. In humans the toxin interrupts conduction between our peripheral nervous system and muscle receptors. Botox, as it is commonly called, is prescribed by doctors to treat Strabismus dystonius and other neurological disorders. It is also in everyday cosmetic use (to counter the onset of facial or other bodily wrinkles); botox makes the muscles flaccid or paralyzes them and they take months to recover.

If this occurred on a larger scale in the human body—where the majority of functions are determined by the way muscles behave—it would be fatal. It takes only an extremely small dose of botulinum to kill by respiratory paralysis. One such example is that of a woman who was poisoned by botulinum from eating just half of a green bean (which became infected during the canning process); there are not many microbes in half a green bean, but it contained sufficient toxin for it to kill. In the case of botulinism, the LD50 is about 0.001g/kg of body weight.

It is difficult to establish which of the two—anthrax or botulinum—is the more toxic. Certainly, botulinum acts considerably faster than anthrax which takes some days to get into the system before a septicaemic reaction is manifested. It can be preceded by flu-like symptoms. Although scientists are aware of the end result, not enough is known about botulinum to describe the sequence of events in detail.

For all this, neither anthrax nor botulinum has been tested on any scale in modern warfare. The Japanese Aum Shinrikyo cult [the cult responsible for the 1995 nerve gas attack on a crowded Tokyo subway station] made an unsuccessful attempt to rig a fan in a Tokyo building in order to spray anthrax spores over the city, but errors were made either in production or delivery. Militarily, the consensus is that these biological and toxin agents, while devastating to those who are unprepared, will not make or break wars. These pathogens remain unproven entities, but that does not make them any less potent as killers.

Cult Chemists

Aum Shinrikyo had access to a considerable amount of money (estimated at between US$300 million and US$1 billion) with which it hired chemists and biologists. Members of the cult were also reported to have arrived in Kikwit in the former Zaire (now the Republic of Congo) at the time of an outbreak of the Ebola epidemic, to obtain samples for BW purposes.

Dr Jane Alexander, of the US Defense Advanced Research Projects Agency (DARPA), has studied the Japanese sect. She discovered that Aum had also undertaken research into the 0157:H7 variant of E.coli, which produces a toxin. Although E.coli is difficult to manipulate as an aerosol for BW use, such bugs can cause serious problems (in Osaka, Japan, 10,000 people were infected with a deviant form of E.coli 0157:H7).

What Aum was working on entailed the insertion of botulinum toxin inside E.coli bacteria in an attempt to manufacture a lethal carrier of this toxin. It is likely that Aum was investigating plasmids (small rings of DNA material that carry genetic information) to implant material into bacteria as well. In fact, pharmaceutical companies use E.coli plasmids to produce Vitamin C which is cheaper than trying to synthesize it—for these purposes E.coli is relatively innocuous and exists as a normal and healthy bacteria in human intestines.

In this way, says Croddy, the processing of E.coli as a weapon is an attempt to 'sneak' an E.coli strain into the body, which does not (or cannot) react immediately. This virulent new strain would then start to multiply. (It is much more serious if a toxin-generating gene has been inserted into a bacterial DNA.)

The ramifications of such genetic engineering are endless: in the case of the E.coli 0157:H7 strain, a toxin would be produced that can result in serious problems in children and the elderly, including renal failure. This strain of E.coli was responsible for an outbreak of food poisoning in Scotland during 1996, when more than a dozen people died and hundreds were taken ill.

Such effects can be taken one step further by inserting other genes to produce more virulent toxins, which is what the former Soviet Union was researching. Western intelligence agencies are aware that some work into biological warfare programs continued in the Commonwealth of Independent States until at least 1992 (the Sverdlovsk-17 BW plant was shut down in the same year).

The UK and US have monitored these developments carefully. They are in the process of determining the long-term implications, largely to ensure that the West has adequate defenses should the deployment of germ weapons become a reality.

One of the issues raised—particularly in the light of revelations about Iraq's BW program—is whether the US will ever again acquire a real offensive CBW capability. The consensus is 'absolutely not'. There are, however, numerous institutions—the military (USAMRIID [U.S. Army Medical Research Institute of Infectious Diseases] in particular), universities, and so on—that spend a great deal of money in establishing how and which microbial pathogens might be manipulated, in order to anticipate what potentially hostile states or subversive groups might be working on.

DARPA is making a serious effort to look closely at E.coli in terms of biological defense. Here, too, Russia has been working to transfer segments of genetic material from one bacterium to another, which is relevant in terms of antibiotic resistance. The upshot is whether or not a new germ manufactured by a hostile power can be effectively countered.

This has caused scientists to speculate how E.coli 0157:H7 originally came into existence, and how it became a toxin-generating microbe. There are some scientists who argue that the strain emerged naturally and mutated from bacteria exchanging genetic information. According to others, it could have been bio-engineered in a laboratory and spread from there. Most scientific favor rests with the first argument. In terms of research into E.coli, Croddy's view is that: "All the major European powers are tinkering about with it defensively. Personally, I would say that USAMRIID is further ahead than just about anybody else."

In the area of biological warfare defense, money—and a lot of it—is being spent. In the US, DARPA is co-ordinating most of the purse strings, channeling much of the money through the Biological Defense Program Office. Significant funds are also being diverted to educational institutions such as Johns Hopkins University and others in the academic research field.

Real-Time Detection

Research into a small mass-spectrometer that will fit into a standard-sized briefcase which could conceivably be used under battlefield conditions is also under way. This will be capable of identifying in real-time if a friendly force is deployed in a 'hot' (contaminated) zone. Although this development is better suited for the characterization of chemical agents, it is hoped that highly accurate readings of the primary eight BW threats will also be available.

"Then," says Croddy, "if you detect any sort of biological presence, you may not know immediately what it is and whether it is a potential threat. But you will have a pretty good idea. You will probably want to 'suit-up' anyway."

The DoD concluded recently that the biological warfare threat was one area in which the United States has found itself to be the most vulnerable. This was said repeatedly at a symposium on the subject held in Atlanta, Georgia, in March [1998]. More than 2,000 delegates from 70 countries were present, many of them military officers. This indicates the significant level of interest in a menace which could, if ever released from the confines of a laboratory, herald a global epidemic.

NO

Ehud Sprinzak

The Great Superterrorism Scare

Last March [1998], representatives from more than a dozen U.S. federal agencies gathered at the White House for a secret simulation to test their readiness to confront a new kind of terrorism. Details of the scenario unfolded a month later on the front page of the *New York Times*: Without warning, thousands across the American Southwest fall deathly ill. Hospitals struggle to rush trained and immunized medical personnel into crisis areas. Panic spreads as vaccines and antibiotics run short—and then run out. The killer is a hybrid of smallpox and the deadly Marburg virus, genetically engineered and let loose by terrorists to infect hundreds of thousands along the Mexican-American border.

This apocalyptic tale represents Washington's newest nightmare: the threat of a massive terrorist attack with chemical, biological, or nuclear weapons. Three recent events seem to have convinced the policymaking elite and the general public that a disaster is imminent: the 1995 nerve gas attack on a crowded Tokyo subway station by the Japanese millenarian cult Aum Shinrikyo; the disclosure of alarming new information about the former Soviet Union's massive biowarfare program; and disturbing discoveries about the extent of Iraqi president Saddam Hussein's hidden chemical and biological arsenals. Defense Secretary William Cohen summed up well the prevailing mood surrounding mass-destruction terrorism: "The question is no longer if this will happen, but when."

Such dire forecasts may make for gripping press briefings, movies, and bestsellers, but they do not necessarily make for good policy. As an unprecedented fear of mass-destruction terrorism spreads throughout the American security establishment, governments worldwide are devoting more attention to the threat. But as horrifying as this prospect may be, the relatively low risks of such an event do not justify the high costs now being contemplated to defend against it. Not only are many of the countermeasures likely to be ineffective, but the level of rhetoric and funding devoted to fighting superterrorism may actually advance a potential superterrorist's broader goals: sapping the resources of the state and creating a climate of panic and fear that can amplify the impact of any terrorist act.

Capabilities and Chaos

Since the Clinton administration issued its Presidential Decision Directive on terrorism in June 1995, U.S. federal, state, and local governments have heightened their efforts to prevent or respond to a terrorist attack involving weapons of mass destruction. A report issued in December 1997 by the National Defense Panel, a commission of experts created by congressional mandate, calls upon the army to shift its priorities and prepare to confront dire domestic threats. The National Guard and the U.S. Army Reserve must be ready, for example, to "train local authorities in chemical—and biological—weapons detection, defense, and decontamination; assist in casualty treatment and evacuation; quarantine, if necessary, affected areas and people; and assist in restoration of infrastructure and services." In May, the Department of Defense announced plans to train National Guard and reserve elements in every region of the country to carry out these directives.

In his 1998 State of the Union address, President Bill Clinton promised to address the dangers of biological weapons obtained by "outlaw states, terrorists, and organized criminals." Indeed, the president's budget for 1999, pending congressional approval, devotes hundreds of millions of dollars to superterrorism response and recovery programs, including large decontamination units, stockpiles of vaccines and antibiotics, improved means of detecting chemical and biological agents and analyzing disease outbreaks, and training for special intervention forces. The FBI, Pentagon, State Department, and U.S. Health and Human Services Department will benefit from these funds, as will a plethora of new interagency bodies established to coordinate these efforts. Local governments are also joining in the campaign. Last April [1998], New York City officials began monitoring emergency room care in search of illness patterns that might indicate a biological or chemical attack had occurred. The city also brokered deals with drug companies and hospitals to ensure an adequate supply of medicine in the event of such an attack. Atlanta, Denver, Los Angeles, San Francisco, and Washington are developing similar programs with state and local funds. If the proliferation of counterterrorism programs continues at its present pace, and if the U.S. army is indeed redeployed to the home front, as suggested by the National Defense Panel, the bill for these preparations could add up to tens of billions of dollars in the coming decades.

Why have terrorism specialists and top government officials become so obsessed with the prospect that terrorists, foreign or homegrown, will soon attempt to bring about an unprecedented disaster in the United States? A close examination of their rhetoric reveals two underlying assumptions:

- **The Capabilities Proposition.** According to this logic, anyone with access to modern biochemical technology and a college science education could produce enough chemical or biological agents in his or her basement to devastate the population of London, Tokyo, or Washington. The raw materials are readily available from medical suppliers, germ banks, university labs, chemical-fertilizer stores, and even ordinary pharmacies. Most policy today proceeds from this assumption.

- **The Chaos Proposition.** The post-Cold War world swarms with shadowy extremist groups, religious fanatics, and assorted crazies eager to launch a major attack on the civilized world—preferably on U.S. territory. Walter Laqueur, terrorism's leading historian, recently wrote that "scanning the contemporary scene, one encounters a bewildering multiplicity of terrorist and potentially terrorist groups and sects." Senator Richard Lugar agrees: "fanatics, small disaffected groups and subnational factions who hold various grievances against governments, or against society, all have increasing access to, and knowledge about the construction of, weapons of mass destruction. . . . Such individuals are not likely to be deterred . . . by the classical threat of overwhelming retaliation."

There is, however, a problem with this two-part logic. Although the capabilities proposition is largely valid—albeit for the limited number of terrorists who can overcome production and handling risks and develop an efficient means of dispersal—the chaos proposition is utterly false. Despite the lurid rhetoric, a massive terrorist attack with nuclear, chemical, or biological weapons is hardly inevitable. It is not even likely. Thirty years of field research have taught observers of terrorism a most important lesson: Terrorists wish to convince us that they are capable of striking from anywhere at anytime, but there really is no chaos. In fact, terrorism involves predictable behavior, and the vast majority of terrorist organizations can be identified well in advance.

Most terrorists possess political objectives, whether Basque independence, Kashmiri separatism, or Palestinian Marxism. Neither crazy nor stupid, they strive to gain sympathy from a large audience and wish to live after carrying out any terrorist act to benefit from it politically. As terrorism expert Brian Jenkins has remarked, terrorists want lots of people watching, not lots of people dead. Furthermore, no terrorist becomes a terrorist overnight. A lengthy trajectory of radicalization and low-level violence precedes the killing of civilians. A terrorist becomes mentally ready to use lethal weapons against civilians only over time and only after he or she has managed to dehumanize the enemy. From the Baader-Meinhoff group in Germany and the Tamil Tigers in Sri Lanka to Hamas and Hizballah in the Middle East, these features are universal.

Finally, with rare exceptions—such as the Unabomber—terrorism is a group phenomenon. Radical organizations are vulnerable to early detection through their disseminated ideologies, lesser illegal activities, and public statements of intent. Some even publish their own World Wide Web sites. Since the 1960s, the vast majority of terrorist groups have made clear their aggressive intentions long before following through with violence.

We can draw three broad conclusions from these findings. First, terrorists who threaten to kill thousands of civilians are aware that their chances for political and physical survival are exceedingly slim. Their prospects for winning public sympathy are even slimmer. Second, terrorists take time to become dangerous, particularly to harden themselves sufficiently to use weapons of mass destruction. Third, the number of potential suspects is significantly less than doomsayers would have us believe. Ample early warning signs should make effective interdiction of potential superterrorists easier than today's overheated rhetoric suggests.

The World's Most Wanted

Who, then, is most likely to attempt a superterrorist attack? Historical evidence and today's best field research suggest three potential profiles:

- Religious millenarian cults, such as Japan's Aum Shinrikyo, that possess a sense of immense persecution and messianic frenzy and hold faith in salvation via Armageddon. Most known religious cults do not belong here. Millenarian cults generally seclude themselves and wait for salvation; they do not strike out against others. Those groups that do take action more often fit the mold of California's Heaven's Gate, or France's Order of the Solar Temple, seeking salvation through group suicide rather than massive violence against outsiders.

- Brutalized groups that either burn with revenge following a genocide against their nation or face the prospect of imminent destruction without any hope for collective recovery. The combination of unrestrained anger and total powerlessness may lead such groups to believe that their only option is to exact a horrendous price for their loss. "The Avengers," a group of 50 young Jews who fought the Nazis as partisans during World War II, exemplifies the case. Organized in Poland in 1945, the small organization planned to poison the water supply of four German cities to avenge the Holocaust. Technical problems foiled their plan, but a small contingent still succeeded in poisoning the food of more than 2,000 former SS storm troopers held in prison near Nuremberg.

- Small terrorist cells or socially deranged groups whose alienated members despise society, lack realistic political goals, and may miscalculate the consequences of developing and using chemical or biological agents. Although such groups, or even individual "loners," cannot be totally dismissed, it is doubtful that they will possess the technical capabilities to produce mass destruction.

Groups such as Hamas, Hizballah, and Islamic Jihad, which so many Americans love to revile—and fear—do not make the list of potential superterrorists. These organizations and their state sponsors may loathe the Great Satan, but they also wish to survive and prosper politically. Their leaders, most of whom are smarter than the Western media implies, understand that a Hiroshima-like disaster would effectively mean the end of their movements.

Only two groups have come close to producing a superterrorism catastrophe: Aum Shinrikyo and the white supremacist and millenarian American Covenant, the Sword and the Arm of the Lord, whose chemical-weapons stockpile was seized by the FBI in 1985 as they prepared to hasten the coming of the Messiah by poisoning the water supplies of several U.S. cities. Only Aum Shinrikyo fully developed both the capabilities and the intent to take tens of thousands of lives. However, this case is significant not only because the group epitomizes the kind of organizations that may resort to superterrorism in the future, but also because Aum's fate illustrates how groups of this nature can be identified and their efforts preempted.

Although it comes as no comfort to the 12 people who died in Aum Shinrikyo's attack, the cult's act of notoriety represents first and foremost a colossal Japanese security blunder. Until Japanese police arrested its leaders in May 1995, Aum Shinrikyo had neither gone underground nor concealed its intentions. Cult leader Shoko Asahara had written since the mid-1980s of an impending cosmic cataclysm. By 1995, when Russian authorities curtailed the cult's activities in that country, Aum Shinrikyo had established a significant presence in the former Soviet Union, accessed the vibrant Russian black market to obtain various materials, and procured the formulae for chemical agents. In Japan, Asahara methodically recruited chemical engineers, physicists, and biologists who conducted extensive chemical and biological experiments in their lab and on the Japanese public. Between 1990 and 1994, the cult tried six times—unsuccessfully—to execute biological-weapons attacks, first with botulism and then with anthrax. In June 1994, still a year before the subway gas attack that brought them world recognition, two sect members released sarin gas near the judicial building in the city of Matsumoto, killing seven people and injuring 150, including three judges.

In the years preceding the Tokyo attack, at least one major news source provided indications of Aum Shinrikyo's proclivity toward violence. In October 1989, the *Sunday Mainichi* magazine began a seven-part series on the cult that showed it regularly practiced a severe form of coercion on members and recruits. Following the November 1989 disappearance of a lawyer, along with his family, who was pursuing criminal action against the cult on behalf of former members, the magazine published a follow-up article. Because of Japan's hypersensitivity to religious freedom, lack of chemical—and biological—terrorism precedents, and low-quality domestic intelligence, the authorities failed to prevent the Tokyo attack despite these ample warning signs.

Anatomy of an Obsession

If a close examination reveals that the chances of a successful superterrorist attack are minimal, why are so many people so worried? There are three major explanations:

Sloppy Thinking

Most people fail to distinguish among the four different types of terrorism: mass-casualty terrorism, state-sponsored chemical- or biological-weapons (CBW) terrorism, small-scale chemical or biological terrorist attacks, and superterrorism. Pan Am 103, Oklahoma City, and the World Trade Center are all examples of conventional terrorism designed to kill a large number of civilians. The threat that a "rogue state," a country hostile to the West, will provide terrorist groups with the funds and expertise to launch a chemical or biological attack falls into another category: state-sponsored CBW terrorism. The use of chemical or biological weapons for a small-scale terrorist attack is a third distinct category. Superterrorism—the strategic use of chemical or biological agents to bring about a major disaster with death tolls ranging in the tens or hundreds of thousands must be distinguished from all of these as a separate threat.

Today's prophets of doom blur the lines between these four distinct categories of terrorism. The world, according to their logic, is increasingly saturated with weapons of mass destruction and with terrorists seeking to use them, a volatile combination that will inevitably let the superterrorism genie out of the bottle. Never mind that the only place where these different types of terrorism are lumped together is on television talk shows and in sensationalist headlines.

In truth, the four types of terrorism are causally unrelated. Neither Saddam Hussein's hidden bombs nor Russia's massive stockpiles of pathogens necessarily bring a superterrorist attack on the West any closer. Nor do the mass-casualty crimes of Timothy McVeigh in Oklahoma City or the World Trade Center bombing. The issue is not CBW quantities or capabilities but rather group mentality and psychological motivations. In the final analysis, only a rare, extremist mindset completely devoid of political and moral considerations will consider launching such an attack.

Vested Interests

The threat of superterrorism is likely to make a few defense contractors very rich and a larger number of specialists moderately rich as well as famous. Last year, Canadian-based Dycor Industrial Research Ltd. unveiled the CB Sentry, a commercially available monitoring system designed to detect contaminants in the air, including poison gas. Dycor announced plans to market the system for environmental and antiterrorist applications. As founder and president Hank Mottl explained in a press conference, "Dycor is sitting on the threshold of a multi-billion dollar world market." In August, a *New York Times* story on the Clinton administration's plans to stockpile vaccines around the country for civilian protection noted that two members of a scientific advisory panel that endorsed the plan potentially stood to gain financially from its implementation. William Crowe, former chair of the joint chiefs of staff, is also bullish on the counterterrorism market. He is on the board of an investment firm that recently purchased Michigan Biologic Products Institute, the sole maker of an anthrax vaccine. The lab has already secured a Pentagon contract and expects buyers from around the world to follow suit. As for the expected bonanza for terrorism specialists, consultant Larry Johnson remarked last year to *U.S. News & World Report,* "It's the latest gravy train."

Within the U.S. government, National Security Council experts, newly created army and police intervention forces, an assortment of energy and public-health units and officials, and a significant number of new Department of Defense agencies specializing in unconventional terrorism will benefit from the counterterrorism obsession and megabudgets in the years ahead. According to a September 1997 report by the General Accounting Office, more than 40 federal agencies have been involved already in combating terrorism. It may yet be premature to announce the rise of a new "military-scientific-industrial complex," but some promoters of the superterrorism scare seem to present themselves as part of a coordinated effort to save civilization from the greatest threat of the twenty-first century.

SETTING THE FBI FREE

When members of the Japanese cult Aum Shinrikyo went shopping in the United States, they were not looking for cheap jeans or compact discs. They were out to secure key ingredients for a budding chemical-weapons program—and they went unnoticed. Today, more FBI agents than ever are working the counterterrorism beat: double the number that would-be superterrorists had to contend with just a few years ago. But is the FBI really better equipped now than it was then to discover and preempt such terrorist activity in its earliest stages?

FBI counterterrorism policy is predicated on guidelines issued in 1983 by then-U.S. attorney general William French Smith: The FBI can open a full investigation into a potential act of terrorism only "when facts or circumstances reasonably indicate that two or more persons are engaged in activities that involve force or violence and a violation of the criminal laws of the United States." Short of launching a full investigation, the FBI may open a preliminary inquiry if it learns from any source that a crime might be committed and determines that the allegation "requires some further scrutiny." This ambiguous phrasing allows the FBI a reasonable degree of latitude in investigating potential terrorist activity.

However, without a lead—whether an anonymous tip or a public news report—FBI agents can do little to gather intelligence on known or potential terrorists. Agents cannot even download information from World Wide Web sites or clip newspapers to track fringe elements. The FBI responds to leads; it does not ferret out potential threats. Indeed, in an interview with the Center for National Security Studies, one former FBI official griped, "You have to wait until you have blood on the street before the Bureau can act."

CIA analysts in charge of investigating foreign terrorist threats comb extensive databanks on individuals and groups hostile to the United States. American citizens are constitutionally protected against this sort of intrusion. A 1995 presidential initiative intended to increase the FBI's authority to plant wiretaps, deport illegal aliens suspected of terrorism, and expand the role of the military in certain kinds of cases was blocked by Congress. Critics have argued that the costs of such constraints on law enforcement may be dangerously high—reconsidering them would be one of the most effective (and perhaps least expensive) remedies against superterrorism.

Morbid Fascination

Suspense writers, publishers, television networks, and sensationalist journalists have already cashed in on the superterrorism craze. Clinton aides told the *New York Times* that the president was so alarmed by journalist Richard Preston's depiction of a superterrorist attack in his novel *The Cobra Event* that he passed the book to intelligence analysts and House Speaker Newt Gingrich for review. But even as media outlets spin the new frenzy out of personal and financial interests, they also respond to the deep psychological needs of a huge audience. People love to be horrified. In the end, however, the tax-paying public is likely to be the biggest loser of the present scare campaign. All terrorists—even those who would never consider a CBW attack—benefit from such heightened attention and fear.

Counterterrorism on a Shoestring

There is, in fact, a growing interest in chemical and biological weapons among terrorist and insurgent organizations worldwide for small-scale, tactical attacks. As far back as 1975, the Symbionese Liberation Army obtained instructions on the development of germ warfare agents to enhance their "guerrilla" actions. More recently, in 1995, four members of the Minnesota Patriots Council, an antitax group that rejected all forms of authority higher than the state level, were convicted of possession of a biological agent for use as a weapon. Prosecutors contended that the men conspired to murder various federal and county officials with a supply of the lethal toxin ricin they had developed with the aid of an instruction kit purchased through a right-wing publication. The flourishing mystique of chemical and biological weapons suggests that angry and alienated groups are likely to manipulate them for conventional political purposes. And indeed, the number of CBW threats investigated by the FBI is increasing steadily. But the use of such weapons merely to enhance conventional terrorism should not prove excessively costly to counter.

The debate boils down to money. If the probability of a large-scale attack is extremely small, fewer financial resources should be committed to recovering from it. Money should be allocated instead to early warning systems and preemption of tactical chemical and biological terrorism. The security package below stresses low-cost intelligence, consequence management and research, and a no-cost, prudent counterterrorism policy. Although tailored to the United States, this program could form the basis for policy in other countries as well:

- **International deterrence.** The potential use of chemical and biological weapons for enhanced conventional terrorism, and the limited risk of escalation to superterrorism, call for a reexamination of the existing U.S. deterrence doctrine—especially of the evidence required for retaliation against states that sponsor terrorism. The United States must relay a stern, yet discreet message to states that continue to support terrorist organizations or that disregard the presence of loosely affiliated terrorists within their territory: They bear direct and full responsibility for any future CBW attack on American targets by the organizations they sponsor or shelter. They must know that any use of weapons of mass destruction by their clients against the United States will constitute just cause for massive retaliation against their countries, whether or not evidence proves for certain that they ordered the attack.

- **Domestic deterrence.** There is no question that the potential use of chemical and biological weapons for low-level domestic terrorism adds a new and dangerous dimension to conventional terrorism. There is consequently an urgent need to create a culture of domestic deterrence against the nonscientific use of chemical and biological agents. The most important task must be accomplished through legislation. Congress should tighten existing legislation against domestic production and distribution of biological, chemical, and radiological agents and devices.

The Anti-Terrorism Act of 1996 enlarged the federal criminal code to include within its scope a prohibition on any attempts, threats, and conspiracies to acquire or use biological agents, chemical agents, and toxins. It also further redefined the terms "biological agent" and "toxin" to cover a number of products that may be bioengineered into threatening agents. However, the legislation still includes the onerous burden of proving that these agents were developed for use as weapons. Take the case of Larry Wayne Harris, an Ohio man arrested in January [1998] by the FBI for procuring anthrax cultures from an unknown source. Harris successfully defended his innocence by insisting that he obtained the anthrax spores merely to experiment with vaccines. He required no special permit or license to procure toxins that could be developed into deadly agents. The FBI and local law enforcement agencies should be given the requisite authority to enforce existing laws as well as to act in cases of clear and present CBW danger, even if the groups involved have not yet shown criminal intent. The regulations regarding who is allowed to purchase potentially threatening agents should also be strengthened.

A campaign of public education detailing the dangers and illegality of nonscientific experimentation in chemical and biological agents would also be productive. This effort should include, for example, clear and stringent university policies regulating the use of school laboratories and a responsible public ad campaign explaining the serious nature of this crime. A clear presentation of the new threat as another type of conventional terrorism would alert the public to groups and individuals who experiment illegitimately with chemical and biological substances and would reduce CBW terrorism hysteria.

- **Better intelligence.** As is currently the case, the intelligence community should naturally assume the most significant role in any productive campaign to stop chemical and biological terrorism. However, new early warning CBW indicators that focus on radical group behavior are urgently needed. Analysts should be able to reduce substantially the risk of a CBW attack if they monitor group radicalization as expressed in its rhetoric, extralegal operations, low-level violence, growing sense of collective paranoia, and early experimentation with chemical or biological substances. Proper CBW intelligence must be freed from the burden of proving criminal intent.

- **Smart and compact consequence management teams.** The threat of conventional CBW terrorism requires neither massive preparations nor large intervention forces. It calls for neither costly new technologies nor a growing number of interagency coordinating bodies. The decision to form and train joint-response teams in major U.S. cities, prompted by the 1995 Presidential Decision Directive on terrorism, will be productive if the teams are kept within proper proportions. The ideal team would be streamlined so as to minimize the interagency rivalry that has tended to make these teams grow in size and complexity. In addition to FBI agents, specially trained local police, detection and decontamination experts, and public-health specialists, these compact units should include psychologists and public-relations experts trained in reducing public hysteria.

- **Psychopolitical research.** The most neglected means of countering CBW terrorism is psychopolitical research. Terrorism scholars and U.S. intelligence agencies have thus far failed to discern the psychological mechanisms that may compel terrorists to contemplate seriously the use of weapons of mass destruction. Systematic group and individual profiling for predictive purposes is almost unknown. Whether in Europe, Latin America, the Middle East, or the United States, numerous former terrorists and members of radical organizations are believed to have considered and rejected the use of weapons of mass destruction. To help us understand better the considerations involved in the use or nonuse of chemical and biological weapons, well-trained psychologists and terrorism researchers should conduct a three-year, low-cost, comprehensive project of interviewing these former radicals.

- **Reducing unnecessary superterrorism rhetoric.** Although there is no way to censor the discussion of mass-destruction terrorism, President Clinton, his secretaries, elected politicians at all levels, responsible government officials, writers, and journalists must tone down the rhetoric feeding today's superterrorism frenzy.

There is neither empirical evidence nor logical support for the growing belief that a new "postmodern" age of terrorism is about to dawn, an era afflicted by a large number of anonymous mass murderers toting chemical and biological weapons. The true threat of superterrorism will not likely come in the form of a Hiroshima-like disaster but rather as a widespread panic caused by a relatively small CBW incident involving a few dozen fatalities. Terrorism, we must remember, is not about killing. It is a form of psychological warfare in which the killing of a small number of people convinces the rest of us that we are next in line. Rumors, anxiety, and hysteria created by such inevitable incidents may lead to panic-stricken evacuations of entire neighborhoods, even cities, and may produce many indirect fatalities. It may also lead to irresistible demands to fortify the entire United States against future chemical and biological attacks, however absurd the cost.

Americans should remember the calls made in the 1950s to build shelters, conduct country-wide drills, and alert the entire nation for a first-strike nuclear attack. A return to the duck-and-cover absurdities of that time is likely to be as ineffective and debilitating now as it was then. Although the threat of chemical and biological terrorism should be taken seriously, the public must know that the risk of a major catastrophe is extremely minimal. The fear of CBW terrorism is contagious: Other countries are already showing increased interest in protecting themselves against superterrorism. A restrained and measured American response to the new threat may have a sobering effect on CBW mania worldwide.

POSTSCRIPT

Is Biological and Chemical Terrorism the Next Grave Threat to the World Community?

It is easy to predict that terrorist organizations will strike at their enemies. It is their stated goal and part of their tactical approach to achieving political aims. However, it is hard to predict how and when terrorist organizations will act. Weapons choice is a highly idiosyncratic issue fraught with many variables, including available cash, training, expertise, understanding, and determination. It is reasonable to speculate that certain weapons will be tempting to groups that want to exercise maximum effect. For example, the bomb that Timothy McVeigh detonated outside an Oklahoma City federal building in April 1995 was simple and easy to develop, reflecting a lack of resources coupled with a desire to have maximum effect. The explosive used in the World Trade Center bombing in February 1993 was more sophisticated but also a function of resources and training.

It is important to note that the line between wanting to employ more sophisticated weapons and actual action is difficult to establish. Many states have pursued the development of nuclear weapons, for example, but few (Libya, North Korea, Syria, and Iraq) have achieved such capability. Still others with the capability (such as Germany, Sweden, Italy, Brazil, and Spain) have decided not to develop such weapons. The assumption that all terrorist groups would use such weapons if they were available is not necessarily true unless one subscribes to the view that all terrorists are irrational, which is not supported by the facts.

Most terrorist organizations engage in conflict for political goals: self-determination of a people, removal of a particular regime, or ideological objectives. As a recent study by the RAND Corporation shows, terrorists do have goals, and the consequences of their actions are calculated. Potential impacts, world opinion, media coverage, and retaliation are factored into the decision-making process. Therefore, it is not logical to assume that with new weapons comes the likelihood of their use. What may be more likely is that different types of organizations may push toward or away from certain weapons based on their stated political goals and also on the quality of their leadership.

Literature on this subject includes *The Next War* by Caspar Weinberger and Peter Schweizer (Regnery, 1996) and John F. Sopko, "The Changing Proliferation Threat," *Foreign Policy* (Winter 1996–1997). In both publications, the authors argue that the United States is ignoring the threat of chemical and biological warfare at its own peril. John Mueller and Karl Mueller counter this argument in "Sanctions of Mass Destruction," *Foreign Affairs* (May/June 1999). Also see Michael T. Osterholm and John Schwartz, *Living Terrors: What America Needs to Know to Survive the Coming Bioterrorist Catastrophe* (Delta, 2001).

ISSUE 18

Is a Nuclear Terrorist Attack on America Likely?

YES: Graham Allison, from "We Must Act As If He Has the Bomb,"
The Washington Post (November 18, 2001)

NO: Jessica Stern, from "A Rational Response to Dirty Bombs,"
Financial Times (June 11, 2002)

ISSUE SUMMARY

YES: Professor of government Graham Allison contends that the United States must assume that terrorist groups like Al Qaeda have dirty bombs and nuclear weapons and that it must act accordingly. He argues that to assume that such groups do not possess such weapons invites disaster.

NO: Jessica Stern, a lecturer in public policy, argues that Americans are in danger of overestimating terrorist capabilities and thus creating a graver threat than actually exists. She warns that the United States must not overreact in its policy response and that prudent security measures will greatly reduce such threats now and in the future.

Since the terrorist attacks of September 11, 2001, much has been written about the specter of nuclear terrorism and the releasing of a dirty bomb (one loaded with radioactive material) in an urban/civilian setting. The events of September 11 have all but ensured the world's preoccupation with such an event for the foreseeable future. Indeed, the arrest of a U.S. man with dirty bomb materials indicates that such plans may indeed be in the works between Al Qaeda and other terrorist cells. When this horror is combined with the availability of elements of nuclear-related material in places like the states of the former Soviet Union, Pakistan, India, Iraq, Iran, North Korea, and many other states, one can envision a variety of sobering scenarios.

Hollywood feeds these views with such films as *The Sum of All Fears* and *The Peacemaker,* in which nuclear terrorism is portrayed as all too easy to carry out and likely to occur. It is difficult in such environments to separate fact from fiction and to ascertain objectively the probabilities of such events. So many factors go into a successful initiative in this area. One needs to find a committed cadre of terrorists, sufficient financial backing, technological know-how, intense security and secrecy, the means of delivery, and many other variables, including luck. In truth, such acts may have already been advanced and thwarted by governments, security services, or terrorist mistakes and incompetence. We do not know, and we may never know.

Regional and ethnic conflicts of a particularly savage nature in places like Chechnya, Kashmir, Colombia, and Afghanistan help to fuel fears that adequately financed zealots will see in nuclear weapons a swift and catastrophic answer to their demands and angers. Osama bin Laden's contribution to worldwide terrorism has been the success of money over security and the realization that particularly destructive acts with high levels of coordination can be "successful." This will undoubtedly encourage others with similar ambitions against real or perceived enemies.

Conversely, many argue that fear of the terrorist threat has left us imagining that which is not likely. They point to a myriad of roadblocks to terrorist groups' obtaining all of the elements necessary for a nuclear or dirty bomb. They cite technological impediments, monetary issues, lack of sophistication, and inability to deliver. They also cite governments' universal desire to prevent such actions. Even critics of Iraqi leader Saddam Hussein have argued that were he to develop such weapons, he would not deliver them to terrorist groups nor would he use them except in the most dire of circumstances, such as his own regime's survival. They argue that the threat is overblown and, in some cases, merely used to justify increased security and the restriction of civil liberties.

The following selections reflect this dichotomy of views. In the wake of September 11, Graham Allison argues that America cannot afford to underestimate terrorist groups, their commitment, and their willingness to obtain the ultimate weapon. His thesis is that America must assume that Al Qaeda has a nuclear or dirty bomb and that it will use it. The implications for U.S. foreign policy include a prolonged war with Al Qaeda until it is destroyed, combined with a significant effort to secure the wealth of nuclear material in the former Soviet Union, which Allison contends is the most vulnerable to theft or black market purchase.

In the second selection, Jessica Stern argues that although a threat exists, the United States should not overplay it. She does suggest that the probability of a nuclear terrorist attack can be greatly reduced by using preventive security measures and education. Essentially, Stern contends that overreaction on America's part is an element of the terrorist arsenal; we must not overestimate the dangers.

Graham Allison

We Must Act As If He Has the Bomb

The question is suddenly urgent: Could the inconceivable happen? President [George W.] Bush has previously warned the world that Osama bin Laden is seeking to develop weapons of mass destruction. Now, bin Laden himself claims to have chemical and nuclear weapons—and "the right to use them." We cannot know for certain whether he is bluffing, but Homeland Security Director Tom Ridge has confirmed that documents detailing how to make nuclear weapons have been found in an al Qaeda safe house in Kabul. And we can certainly expect that as the noose tightens around the terrorist's neck, he and his associates will become increasingly desperate.

All of this means that, incredible as the possibility remains even in the aftermath of Sept. 11, we must now seriously contemplate that bin Laden's final act could be a nuclear attack on America.

The consequences of such an attack would far outstrip the horror we have already witnessed. Imagine that al Qaeda had struck the World Trade Center not with a van filled with explosives, as in 1993, nor with planes fully loaded with jet fuel, but with an SUV containing a nuclear device. Even a crude device could create an explosive force of 10,000 to 20,000 tons of TNT, demolishing an area of about three square miles. Not only the World Trade Center, but all of Wall Street and the financial district and the lower tip of Manhattan up to Gramercy Park would have disappeared. Hundreds of thousands of people would have died suddenly. In Washington, if such a vehicle exploded near the White House, an area reaching as far as the Jefferson Memorial would be immediately and completely destroyed, and a larger area, extending from the Pentagon to beyond the Capitol, would suffer damage equal to that caused to the Alfred P. Murrah Federal Building in Oklahoma City in 1995.

That same year, in a Post op-ed, I warned: "In the absence of a determined program of action, we have every reason to anticipate acts of nuclear terrorism against American targets before this decade is out." I was fortunately wrong about the timing, but I believe the same estimate can be made with even greater justification today. The question is whether the outrage of Sept. 11 will now motivate the United States and other governments to act urgently to minimize the risk of nuclear mega-terrorism.

Unhappily, the evidence to date is not encouraging.

As the Bush administration took office in January, a bipartisan task force, chaired by former Senate majority leader Howard Baker (now ambassador to Japan) and former White House counsel Lloyd Cutler, presented a report card on non-proliferation programs with Russia. The task force's principal finding was that "the *most urgent unmet national security threat* [my emphasis] to the United States today is the danger that weapons of mass destruction or weapons-usable material in Russia could be stolen, sold to terrorists or hostile nation states, and used against American troops abroad or citizens at home."

The danger can be summarized in three propositions. First, attempts to steal nuclear weapons or weapons-usable material are not hypothetical, but a recurring fact. The past decade has seen scores of incidents in which individuals and groups have successfully stolen weapons material from sites in Russia and sought to export it—but have been caught. Just in the past month, the chief of the Russian defense ministry directorate responsible for nuclear weapons reported two recent incidents in which terrorist groups unsuccessfully attempted to break into Russian nuclear storage sites. In the mid-1990s, more than 1,000 pounds of highly enriched uranium—enough material to allow terrorists to build more than 20 nuclear weapons—sat unprotected in Kazakhstan. Recognizing the danger, the American government purchased the material and removed it to a Department of Energy facility in Oak Ridge, Tenn.

Second, if al Qaeda or some similar group obtained 40 pounds of highly enriched uranium, or less than half that weight in plutonium, it could, with materials otherwise available off the shelf, produce a nuclear device in less than a year. Obtaining such fissionable material—an ingredient that is fortunately difficult and expensive to manufacture—is in fact the only high hurdle to creating a nuclear device. But as a director of the Lawrence Livermore Laboratories wrote a quarter of a century ago, "If the essential nuclear materials like these are in hand, it is possible to make an atomic bomb using the information that is available in the open literature." An even easier alternative is a radioactivity dispersal device, a conventional bomb wrapped in radioactive materials that disperse as fallout when the bomb explodes.

Third, terrorists would not find it difficult to sneak such a nuclear device into the United States. The nuclear material required is actually smaller than a football. Even a fully assembled device, such as a suitcase nuclear weapon, could be shipped in a container, in the hull of a ship or in a trunk carried by an aircraft. Since Sept. 11, the number of containers arriving at U.S. points of entry that are being X-rayed has increased to approximately 10 percent: 500 of the 5,000 containers currently arriving daily at the port of New York/New Jersey, for instance. But as the chief executive of CSX Lines, one of the foremost container-shipping companies, put it: "If you can smuggle heroin in containers, you may be able to smuggle in a nuclear bomb."

If bin Laden and other terrorists have not so far succeeded in acquiring nuclear weapons, or materials from which to assemble them, we should give thanks for our great good fortune. If they have acquired them—as bin Laden now claims—most people will quickly conclude that, under existing conditions, this was bound to happen.

There can be little doubt that bin Laden and his associates would carry out a nuclear assault were they capable of doing so. Last year, the CIA intercepted a message in which a member of al Qaeda boasted of plans for a "Hiroshima" against America. According to the Justice Department indictment for the 1998 bombings of American embassies in Kenya and Tanzania, "At various times from at least as early as 1993, Osama bin Laden and others, known and unknown, made efforts to obtain the components of nuclear weapons." Additional evidence supplied by a former member of al Qaeda describes the group's attempts to buy uranium of South African origin, repeated travels to three Central Asian states to try to buy a complete warhead or weapons-usable material, and discussions with Chechens in which money and drugs were offered for nuclear weapons. Bin Laden himself has declared that acquiring nuclear weapons is a "religious duty."

Preventing nuclear terrorist attacks on the American homeland will require a serious, comprehensive defense—not for months or years, but far into the future. The response must stretch from aggressive prevention and preemption to deterrence and active defenses. Strict border controls will be as important to America as ballistic-missile defenses.

To fight the immediate threat, the United States must move smartly on two fronts. First, no effort can be spared in the military, economic and diplomatic campaign to defeat and destroy al Qaeda, and in the international intelligence and law-enforcement effort to discover and disrupt al Qaeda sleeper cells and interrupt attempted shipments of weapons.

Second, the United States must seize the opportunity of a more cooperative Russia to "go to the source" of the greatest danger today: the 99 percent or more of the world's nuclear, biological and chemical weapons that are stored in Russia and the United States. The surest way to prevent nuclear assaults is to prevent terrorists from gaining control of these weapons or materials from which to make them. President Bush acknowledged this in his joint news conference with Russian President Vladimir Putin last Thursday, declaring that "Our highest priority is to keep terrorists from acquiring weapons of mass destruction."

What the two presidents failed to announce, however, are concrete actions to achieve this objective. While their success in agreeing to cut the number of operational strategic nuclear weapons cannot be gainsaid, the stark reality is that this reduction has no effect on our most urgent unmet national security threat.

Bush and Putin should have announced that the United States and Russia would lead a new joint international undertaking to minimize the risks of nuclear terrorism, as well as terrorism by means of other weapons of mass destruction. They should have pledged to ensure that their respective governments will do everything physically and technically possible to prevent terrorists or criminals from stealing weapons or weapons-usable material from their stockpiles. They should have instructed their governments to develop a joint plan of action to concentrate weapons and materials in the fewest possible sites, secure them by the most technically advanced means, and neutralize highly enriched uranium by blending it down for subsequent use in civilian nuclear power plants. Within Russia, such a program should be jointly financed by the United States, its allies in the war against terrorism and Moscow.

Despite the successes of the past week, the long-term goals of our war on terrorism remain elusive, and the future no doubt holds frustrations as well as celebrations. In that light, calling upon leaders to act to prevent attacks of a kind that have not yet occurred may seem overly demanding. But if we fail to act on this agenda now, how shall we explain ourselves on the morning after a nuclear Sept. 11?

Jessica Stern  **NO**

A Rational Response to Dirty Bombs

The capture of Abdullah al-Muhajir, the US citizen accused of plotting a "dirty bomb" attack on Washington DC, is another demonstration that the world is now vulnerable to unconventional terrorism.

We have known for some time that al-Qaeda has been seeking weapons of mass destruction. Operatives described their attempts to acquire uranium for use in nuclear weapons in a New York City federal court over a year ago. Plans and materials for designing unconventional weapons were found in Afghanistan. And Osama bin Laden has repeatedly boasted about his success in hiring scientists to assist him in producing nuclear and biological weapons.

The arrest of Mr al-Muhajir, a former Chicago gang member who changed his name from Jose Padilla after converting to Islam, reminds us that radical Islamists aim to turn rage, humiliation and perceived deprivation into a weapon of war. They do not recruit exclusively in Muslim countries; they seek out disaffected youth all around the world.

The UK has become a popular gathering place for radical Islamists from around the world, some of whom are recruited to join jihads from mosques. A former FBI counter-terrorism official estimates that between 1,000 and 2,000 young men left the US to become Mujahideen in purported holy wars during the 1990s. Two New York mosques sent 40 to 50 recruits per year overseas in the mid-1990s, he says.

The jihadis were never tracked as a group. Immigration officials do not keep records of US citizens travelling abroad, and a combination of legal controls and self-restraint stopped the FBI and CIA from monitoring their activities.

Nor should it be a surprise that al-Qaeda successfully recruited a former gang member who had served time in a US prison. Terrorists have long used prisons as a source of operatives, and young men join gangs for some of the same reasons they join terrorist groups. Hopelessness, the feeling of being left behind by rapidly changing societies and fear are the fuels of terror, whether practised by inner-city gangs or by global terrorist criminals.

Still, it is important to keep the threat in perspective. Dirty bombs are far more frightening than lethal. Numerous studies by government and non-government scientists have shown that a dirty bomb would kill only people in the immediate vicinity of the explosion. While people living downwind of the blast would be exposed to additional radiation, there would be very few additional cancer deaths, probably undetectable in the statistical noise.

The US government considered developing radiological weapons during the second world war, but abandoned the project as impracticable. Unlike chemical and biological agents, radioactive poisons act slowly. They are difficult to disseminate in concentrations sufficient to cause death, radiation sickness or cancer. In contrast, chemical agents can be stored for a long time, and are easier to transport. That makes them more attractive to terrorists than radiation devices if the main objective is to kill many people.

But radioactive weapons can be effective instruments of terror because of their psychological impact—the human fear of radiation makes them inherently terrifying. For more than a quarter of a century, psychologists and risk analysts have sought to identify the attributes of risks that are especially feared. Studies of perceived risk show that fear is disproportionately evoked by certain characteristics of radiation: it is mysterious, unfamiliar, indiscriminate, uncontrollable, inequitable and invisible. Exposure is involuntary, and the effects are delayed.

The media also tend to highlight terrorist incidents, heightening dread and panic still further. Because Belfast is still considered a terrorist city, many people consider it to be more dangerous than Washington DC, although there are far more murders per head of population in Washington than in Belfast.

We feel a gut-level fear of terrorism, and are prone to trying to eradicate the risk entirely, with little regard to the cost. In contrast, when risky activities are perceived as voluntary and familiar, danger is likely to be underestimated. On average, more than 100 US citizens a day die in car accidents. Yet people expose themselves to the risk because it is a voluntary act and drivers feel the illusion of control.

What can be done about the problem? First, we need to realise this is a new kind of war. Our enemies deliberately target civilians. But uncertainty, dread and disruption are their most important weapons. Our most important response, then, is an informed public that understands not only the risks we face, but also the role of fear.

But public education is only the first step. Many policy measures can reduce the likelihood and impact of such threats. Nuclear power plants must be secured. Evacuation and clean-up plans should be readied and hospitals should be prepared. Radiation detectors should be deployed at ports and borders. Tracking systems for radioactive isotopes must be improved. Despite the relatively low casualty rate for radiological attacks, the psychological impact will be far more devastating if governments are perceived to be unprepared.

Unconventional weapons, used in a total war, require an unconventional response. New agencies and organisations will have to be involved. Businesses will play an increasingly important role. The food industry needs to be aware that the enemy in this war will not be dressed as a soldier and may not carry a gun. Instead, he may be an insider working at a food processing plant aiming to steal radioactive sources or contaminate food products.

Terrorism is a form of psychological warfare, requiring a psychologically informed response. Our hardest challenge is not to overreact—the terrorists' fondest hope—and not to give in to fears. We will need to find the right balance between civil liberties and public safety. The news about Abdullah al-Muhajir makes us want to scour our cities in search of would-be killers. It would be better to act in a measured and deliberate way.

POSTSCRIPT

Is a Nuclear Terrorist Attack on America Likely?

There are many arguments to support the contention that nuclear and dirty bombs are hard to obtain, difficult to move and assemble, and even harder to deliver. There is also ample evidence to suggest that most, if not all, of the U.S. government's work is in one way or another designed to thwart such actions because of the enormous consequences were such acts to be carried out. These facts should make Americans rest easier and allay fears if only for the reasons of probability.

However, Allison's contention that failure to assume the worst may prevent the thwarting of such terrorist designs is persuasive. Since September 11 it is clear that the world has entered a new phase of terrorist action and a new level of funding, sophistication, and motivation. The attitude that because something is difficult it is unlikely to take place may be too dangerous to possess. The collapse of the USSR has unleashed a variety of forces, some positive and some more sinister and secretive. The enormous prices that radioactive material and nuclear devices can command on the black market make the likelihood of temptation strong and possibly irresistible.

If states are to err, perhaps they should err on the side of caution and preventive action rather than on reliance on the statistical probability that nuclear terrorism is unlikely. We may never see a nuclear terrorist act in this century, but it is statistically likely that the reason for this will not be for lack of effort on the part of motivated terrorist groups.

Some important research and commentary on nuclear terrorism can be found in Elaine Landau, *Osama bin Laden: A War Against the West* (Twenty-First Century Books, 2002); Jan Lodal, *The Price of Dominance: The New Weapons of Mass Destruction and Their Challenge to American Leadership* (Council on Foreign Relations Press, 2001); and Jessica Stern, *The Ultimate Terrorists* (Harvard University Press, 1999).

ISSUE 19

Is There a Military Solution to Terrorism?

YES: Wesley K. Clark, from "Waging the New War: What's Next for the U.S. Armed Forces," in James F. Hoge, Jr., and Gideon Rose, eds., *How Did This Happen? Terrorism and the New War* (PublicAffairs, 2001)

NO: Andrew Stephen, from "The War That Bush Cannot Win," *New Statesman* (September 24, 2001)

ISSUE SUMMARY

YES: Wesley K. Clark, a retired general of the U.S. Army, argues that with the proper flexible strategy, the U.S. military can eliminate groups like Al Qaeda and help the United States win the war on terrorism.

NO: Andrew Stephen, a columnist for *New Statesman,* maintains that the problems behind terrorism are deep and that U.S. patriotism and bravado will not win the war against it.

The terrorist attacks of September 11, 2001, ushered in a new set of issues and concerns for the United States and, indeed, much of the rest of the world. These issues center around the specific causes of the attacks, the underlying causes of terrorism in general, and the ability of the United States or any other nation to combat such acts and, if so, how. All this discussion is taking place within an emotionally charged atmosphere of shock, patriotism, vengeance, and self-righteousness, as the United States faces yet another international challenge.

Although the post–September 11 war in Afghanistan was a success for the United States insofar as the Taliban government was eliminated and Al Qaeda forces were sent running, the larger questions of military force and terrorism have yet to be settled. Central to this debate is the question of whether or not a military solution is even possible in the fight to prevent terrorism.

Terrorist cells exist in a host of states, including Somalia, Pakistan, Iraq, Sudan, Iran, and Yemen. Yet the ability of the United States or any other state to strike such targets and eliminate cells therein is different than was the case with

Afghanistan. Although military force was successfully used against terrorism in the case of Afghanistan, this example is unique in several key respects:

- The Taliban regime was held in low regard by most states, including the majority of Islamic states, which feared the Taliban's radical brand of fundamentalism.
- Al Qaeda cells and camps were easily identifiable, and Osama bin Laden and the presence of thousands of non-Afghan fighters made their operations easier targets.
- Al Qaeda was boastful of its role in the September 11 attacks and thus engendered worldwide condemnation.

These realities made striking Afghanistan politically and militarily possible. In the other cases (with the possible exception of Somalia), these conditions do not exist. Therefore, military operations in these and other locations become trickier, both politically and militarily. President George W. Bush's labeling of Iran, Iraq, and North Korea an "axis of evil" in his January 2002 State of the Union Address illustrates the multiplicity of opinions regarding the military option against terrorism. The contentious debate within the United Nations on the issue of war with Iraq heightens these differences even more.

Proponents of a military solution argue that once the United States adopts the "correct" strategic mindset and uses special forces, air bombardment, local allies, and deception, the military can liquidate terrorist cells and terrorism itself. This view is predicated on the notion that it is a matter of finding the right tools and methods with which to eliminate terrorism. It might be a compelling argument in cases like Afghanistan and similar locations, but it might not be as convincing when dealing with such situations as those in the Israeli-occupied West Bank and Gaza.

The inherent problems that lead to terrorism as a form of conflict behavior are deep, difficult to gauge, and borne of long-standing societal issues of alienation, governmental breakdown, and, in the case of some groups, intense hatred of external, societal, cultural forces that are perceived as invaders and, by definition, illegitimate. Amidst the growing poverty and lack of opportunity in the urban areas of the developing world, there is ample opportunity for terrorist groups to recruit young people who have little hope and growing anger and resentment. The discussion of potential solutions to terrorism is vital at this time because how the world responds to September 11 in the short and long term will determine the nature of terrorist conflict for years to come.

In the following selections, Wesley K. Clark and Andrew Stephen challenge us to examine our analyses of and responses to terrorism and, hopefully, to calculate the consequences of our choices. Clark supports utilizing a military strategy to eliminate terrorist groups and, ultimately, terrorism itself. Stephen contends that the military strategy does not take into account the deeper causes of terrorism and therefore cannot eliminate it.

Wesley K. Clark

Waging the New War: What's Next for the U.S. Armed Forces

Since the end of the Cold War commentators have called for the transformation of the U.S. military, but the purpose, scope, and essential nature of this transformation have never been clear. The events of September 11 have given a new impetus to calls for a transformation, and they have also helped set some parameters for the changes required. Although no military leaders believe that military action alone can resolve the problem of terrorism, the military clearly has a role to play in winning the fight.

At the conclusion of the Persian Gulf War in 1991, the U.S. military consisted of just over 2 million active personnel and some 1.2 million reserves. These troops were organized along the traditional lines of U.S. forces, built on the experience of World War II and modified after the Korean and Vietnam wars. With its more than 700,000 men and women in uniform, the Army was the largest force, designed to fight and win a sustained land battle with support from the Air Force. The Army's combat capabilities were vested in 14 divisions, each comprising between 10,000 and 17,000 soldiers, with additional support tucked into higher-echelon commands that commanded and supported several divisions simultaneously.

The Air Force was organized into squadrons and wings, each distinguished by the type of aircraft flown and by function. Squadrons usually consisted of 18 aircraft with the necessary crews and support, and wings were composed of two or more squadrons, with some dedicated to tactical fighters and others to bombers or reconnaissance Wings, in turn, were grouped into "numbered" air forces, roughly the equivalent of an Army division or corps. At the end of the Gulf War, the Air Force comprised some 24 tactical fighter wings, 231 heavy bombers, several intercontinental ballistic missile wings, hundreds of tanker aircraft, and other support organizations, with a total of 510,000 men and women in uniform.

The Navy was organized by ships, squadrons, and fleets, with its actual operations grouped into task forces—collections of ships of various types formed for a specific mission. At the end of the Gulf War, the Navy encompassed 527

From Wesley K. Clark, "Waging the New War: What's Next for the U.S. Armed Forces," in James F. Hoge, Jr., and Gideon Rose, eds., *How Did This Happen? Terrorism and the New War* (PublicAffairs, 2001). Copyright © 2001 by The Council on Foreign Relations, Inc. Reprinted by permission of PublicAffairs, a member of Perseus Books, LLC.

ships of various types (including more than 100 attack submarines), 15 aircraft carriers, and 570,000 sailors.

The Marine Corps, with 194,000 personnel, was the smallest of the services, composed of three divisions. It usually operated in marine expeditionary units, built around a reinforced marine battalion augmented with aircraft and support and launched on a handful of ships capable of supporting amphibious landings or helicopter operations.

And by 1991 there was a new command, created by the Defense Authorization Act of 1987, that had its own, independent congressional funding: the Special Operations Command, which comprised a collection of Army, Navy, and Air Force units and was located at MacDill Air Force Base in Florida. Commands from within the other military services, such as the Army Special Operations Command, managed the unique special operations forces. In addition, a special hostage-rescue and counterterrorist strike force known as the Joint Special Operations Command was created at Fort Bragg, North Carolina. There were also five special forces groups, small organizations of under 1,000 people who are regionally oriented in culture and languages, which provided training and assistance to foreign military forces. And finally, there was the 75th Ranger Regiment, consisting of three 600-person battalions of specially skilled airborne troops who conducted the most demanding combat tasks.

The fundamental military problem the United States faced in the early 1990s was to prevent the exhilaration of having won the Cold War and the resulting cry for a "peace dividend" from causing the collapse of the American military. Spending cuts were mandatory; the only question was how large they would be.

Transforming the Military

Commentators and critics argued that in addition to spending cuts, some kind of fundamental transformation was required, sometimes citing the lessons of the Gulf War, and at other times arguing for some new kind of approach. The idea of transformation seemed to be connected as much with saving resources as with new missions and requirements, although some continued to use the Soviet term, "revolution in military affairs" (RMA), to highlight the opportunities for transformation. Proponents of an RMA could argue that high technology, particularly the so-called precision strike weapons, had made large ground forces obsolete and enabled sizable reductions in force structures. Some spoke of the need to skip a generation of weaponry to pursue even more advanced (though never fully defined) weapons technologies, a step that would also, it was argued, reduce the enormous personnel costs of maintaining the Cold War–era force.

Struggling to cope with these continuing pressures to reduce the defense budget, General Colin Powell, then Chairman of the Joint Chiefs of Staff and the nation's seniormost military leader, expounded the idea of the "Base Force." This was to be an irreducible minimum size for the military, which was justified in only the most general ways as having been derived from the worldwide interests and security responsibilities of the United States. Under Powell's leadership the military significantly reduced its assets overseas, cutting the European-based force, for example, from around 300,000 troops to a little more than 100,000 by 1996. The congressionally mandated Base Realignment and Closure Commission was created in 1988 to solve the political puzzle of how to distribute impending military reductions around the United States, and four rounds of cuts were undertaken, resulting in the closure of some 97 installations.

In the early years of the Clinton administration, it was recognized that the defense budget simply could not be sustained at existing levels unless a more specific case was made for why the military required the forces and assets it had. Moving from "illustrative planning scenarios" through a series of internal studies known as the "Bottom-Up Review," the administration eventually decided on requirements sufficient to fight and win two nearly simultaneous "major regional contingencies," or MRCs, such as new wars in Iraq or Korea.

Although this two-MRC standard provided a benchmark that generally halted the shrinkage of the size and structure of the armed forces, it failed to answer fully the fundamental questions about military requirements: What exactly do we need and why? To guide thinking in this area, in 1996 the Pentagon published the first of a series of new documents titled "Joint Vision 2010." This vision was heavily influenced by the technological successes of precision weaponry and modern communications in the Gulf War, but it also tried to project requirements for forces and capabilities across the "full spectrum of conflict" that could be imagined, ranging from peacekeeping to major war.

Despite all this overarching policy thinking, by and large the military services were going their own ways. The Army clung to its division-based force, making gradual and largely marginal changes in the size and composition of its divisions to accommodate personnel reductions down to an active strength of 491,000 men and women. The Army also found itself trying to emulate the Air Force's tighter integration of active and reserve forces. The Navy, for its part, moved dramatically to cut its fleet to 359 vessels by 1996 and explored new, more efficient designs for surface ships and submarines. The Air Force reorganized itself twice during the 1990s, first forming "composite wings" composed of a tailored mix of aircraft aligned more closely with the needs of Army units, and then creating the air expeditionary forces, which were packaged for rapid overseas deployments. In the process the Air Force shrank by some 20 fighter squadrons, by more than 100 heavy bombers, and by 121,000 uniformed troops. Finally the Marine Corps—protected by legislation that ensured that its minimum strength would not fall below 172,500 personnel—worked to enlarge the scope of its activities and the range of its capabilities. No reduction was made in the strength or capabilities of the Special Operations Command.

In all of the services, the powerful lure of publicity and high technology drove intense interest in "precision" weaponry, which uses lasers or video technology to help it pinpoint targets. The Army improved the accuracy and range of its large, tactical ballistic-missile system; the Navy modernized its stable of existing aircraft to enable precision strikes in a ground attack; and the Air Force, already the leader in this field, developed enhanced munitions with longer ranges and, using the global positioning system, the capability to strike in any weather conditions. The Air Force also fielded its new "stealth" bomber, the B-2, to augment the squadrons of F-117 stealth fighters that had been used in the Gulf War. Other technologies began to appear as well, including very high-powered, long-range electro-optical systems mounted on aircraft, as well as increasingly sophisticated pilotless "drones" to be used for reconnaissance. Even some Army attack helicopters, tanks, and armored fighting vehicles were enhanced with new optics, computer communications, and, in some cases, radar systems to pinpoint targets. This entire process came to be known as "platform modernization" (as opposed to platform replacement), meaning that older airframes, vehicles, and ships could be enhanced with new technology and capabilities for a fraction of the cost of procuring new high-tech airframes, tanks, or other complete systems.

Nevertheless, the services also scrambled to acquire new platforms. The chief contenders were three different tactical fighters: the supersonic-cruise, stealthy, and very expensive F-22 for the Air Force; the robust, lower-cost, joint strike fighter, which could be used in multiple roles; and the Navy's enhanced F-18E/F plane. Additional demands were made for the V-22 aircraft, capable of taking off and landing vertically, which would potentially extend the Marines' reach much farther inland, as well as provide enhanced capabilities for the Special Operations Command's forces. Other claimants on the defense procurement budget were Army helicopter modernization, a new self-propelled Army artillery piece known as the Crusader, a new amphibious assault vehicle for the Marines, a new submarine, a new class of destroyers, and even additional C-17 transport aircraft. And after 1999, embarrassed by its painful failure to meet public expectations in its slow deployment to Kosovo, the Army began a broad effort to transform its heavy forces into a lighter, more readily deployable organization that could reach far-flung theaters beyond the reach of forward-positioned heavy forces. Adding to the budgetary pressures were the plans for development of theater and national missile defenses, which in turn involved the development of new radars, missile systems, and communications systems; as well as the increasing demands for defenses against chemical and biological warfare.

There simply wasn't enough money in the defense budgets to buy all that was sought, even with the enlargements of the defense budget that began during the second Clinton administration. Small wonder that the Joint Chiefs of Staff were seeking an additional $50 billion per year or so from the incoming Bush administration, even before the full scope of an enhanced missile-defense program was laid out. Despite all the efforts that had been made to rationalize military requirements, it appeared once more that some sort of transformation was going to be driven as much by budgetary constraints as by changes of mission.

Fighting a New Kind of War

The events of September 11, 2001, have dramatically altered the resources available to the Department of Defense, while at the same time clarifying how the U.S. military itself must be transformed. For the most part, the military operations of this war must be fought using the tools we already have: the Army divisions, the B-2 bomber, the ships, and the special operations forces that have been carefully built up over the past fifteen years. The challenge will be to adapt the existing organizations, procedures, and capabilities to meet the somewhat different requirements of this new kind of war.

The requirements, of course, will be driven by the mission, our adversaries, and the environment within which any military operations must be conducted. Some missions may appear familiar, evoking memories of the air campaigns over Iraq or Yugoslavia, and intended to gain freedom of the skies or, as it is usually termed by the military, "air superiority." Missions may even require amphibious assaults or some conventional land battles. But much of what will occur will require that forces and capabilities be tailored to a specific need, a kind of "mix and match," to fit the objectives and constraints of the various missions at hand. Some of these arrangements may appear innovative; others may simply seem bizarre.

The war against terrorism is not a war that can be won from the skies; it will require troops on the ground in many different roles. The first steps in this war have, of course, involved air-power. In taking the war first to the Taliban regime in Afghanistan, for example, the United States relied on its formidable air forces to destroy the limited air-defense capabilities in that country. In a now-familiar orchestration, cruise missiles and stealth bombers swept in at night to attack early-warning radars, antiaircraft missile sites, and airfields. The attack force was tailored, of course, with Air Force B-1 and B-52 bombers taking off from Diego Garcia in the Indian Ocean, B-2 bombers beginning their missions from the United States itself, and Navy aircraft based on carriers at sea.

But from this point onward, the forces will have to be adapted more specifically for the kind of war they are fighting. As the Air Force learned in its operations over Kosovo, it is difficult to spot dispersed and hidden enemy forces from the air. By augmenting manned air patrols with unmanned drones, which are able to loiter and revisit suspected areas for several hours while transmitting to their controllers pictures of the landscape below for detailed examination by specialists in rear areas, the Air Force can facilitate the detection of enemy forces on the ground. During the Kosovo campaign, both the Predator and the Hunter drones were used in this way with some success.

Once these measures have been taken from the sky, the next step may be to deploy forward reconnaissance elements, small teams made up of special forces or long-range surveillance units able to hide among enemy positions and then call in strike aircraft to attack enemy forces. These small groups can infiltrate on foot or by vehicle, or they can be inserted by helicopters or parachute drops far beyond friendly positions. These techniques, though little discussed publicly, were pioneered during the Cold War. With satellite radios and laser designators, these special groups can provide critical directions for air attack on a continuing basis. Another technique will be to use attack helicopters, either the small special operations type with highly experienced crews, or the larger, more heavily armed Apache helicopters.

These techniques were quietly explored, though never fully implemented, in Kosovo. In that conflict in the spring of 1999, highly experienced special forces teams maintained liaison with the Albanian forces looking into Serbia, and in some cases positioned themselves so that they could directly observe enemy forces. Intelligence was the critical ingredient for their success, and every effort was made to exploit the knowledge of the host country and of those fighting on the ground. Counterbattery radars, capable of detecting artillery and mortar fire and calculating firing positions, were also deployed and put into action in Kosovo. And, of course, the Apache helicopters were organized, trained, and rehearsed for the mission of flying over the mountains into Kosovo to detect and strike Serb forces. The conflict in Afghanistan is likely to make use of the kind of tailored forces created during the bombing campaign against Yugoslavia.

Another strategy that is being used in the conflict in Afghanistan, and one whose use was under active consideration when Yugoslav president Slobodan Milosevic surrendered in the summer of 1999, would involve dispatching into the theater special teams of commando forces or even reinforced Ranger companies to conduct raids and to strike at enemy forces using heavy firepower from the air.

And, as was the case in Kosovo, standard ground forces will also be required in Afghanistan to secure bases and provide reaction forces in the case of trouble. It appears that these bases will be located initially in the states surrounding Afghanistan, such as Uzbekistan, but eventually it will be possible to move some small forward-operating bases into the areas of Afghanistan that are under the control of the Northern Alliance. These bases will use I specially organized forces consisting of reinforced infantry units augmented with mortars, radars, and perhaps a few armored vehicles.

Organizing such forces is challenging because typically the situation calls for conflicting strategies: keeping the force small helps to minimize the logistical burden, and yet there is also a requirement for robust and versatile capabilities. These trade-offs will demand deploying the most modern technologies in support of the force, including more modern communications systems, wheel-mounted rocket artillery, and perhaps even new infantry weapons capable of firing exploding projectiles. Forces must be capable of delivering direct fire at extended ranges using sniper rifles and automatic grenade-launchers to accommodate the rough terrain and long-range fields of fire that are characteristic of arid Afghanistan. And because of the need for a small "forward footprint," much of the parent division-sized force stationed in the United States is apt to be left behind.

With these small and variable forces, the combat in Afghanistan is likely to be of unpredictable duration. A collapse of the Taliban forces is a distinct possibility, but the terrain and ethnic composition of the country also encourage persistent factional fighting that could, under the right circumstances, continue for some time. In the meantime, a pattern of raids, strikes, and offensive maneuvers is likely to provide episodic glimpses of a remote theater of war. Still, if the United States succeeds in isolating the Taliban from all significant outside assistance while retaining the support of the majority of the Afghan people, builds on its air dominance, and moves ahead patiently with the insertion of airborne and ground-based reconnaissance and small numbers of other special ground forces, the ultimate outcome cannot be in doubt.

Less visible will be some of the activities in friendly states seeking assistance in dealing with terrorists inside their own borders. In these states—in Southeast Asia or perhaps North Africa—U.S. forces will function as trainers and will take direct action only in exceptional circumstances. But again, the forces for such direct action would chiefly be those already organized and trained: elite special operations forces. Operating from the sea or from friendly forward bases, such forces can launch strikes hundreds of miles inland against terrorist-training camps and facilities, or, with the consent of the host government, can call in air strikes that can simply wipe out terrorist-training camps. In these locations the most important requirement for success will be surprise. Not only will the plans for such activities be extremely closely held—so will be the results.

In Afghanistan and in those states seeking assistance, the real key to effective operations will be information about the terrorists: details about their identities, locations, habits, logistics, and aims. The most valuable information is predictive in nature; it enables security forces to determine where the terrorists will be, not just where they have been. But information of this nature is also the most difficult to acquire. In theory, it may be gained through electronic means, pilfered plans, or personal contact. In practice, it is best gained by well-positioned observers, including informants and infiltrators of terrorist cells—so-called "human intelligence." Often working from the inside, human intelligence is the best way to track an enemy's intentions and to provide critical assistance in targeting individuals and small groups. Human intelligence may come from national sources or from allied states, and it will often be last-minute, quickly perishable information. This means that the special forces involved in operations must have exquisite linkages to intelligence agencies, enabling them to receive pertinent information with virtually no delay. It also implies a rapid-response capability that argues for greater decentralization in decision-making during counterterrorist operations.

These operations will also run counter to the conventional conservatism of the post-Vietnam military, which has proved time and again that it is risk- and casualty-averse. If the targeting of forces in Afghanistan is done from the air by unmanned vehicles or helicopters, there will always be the risks of engagement by antiaircraft fire and missiles. Special forces teams will also face risks; the battle of October 3, 1993, in Mogadishu (when 18 U.S. soldiers were killed in the streets of that city) remains a stark reminder that even superb troops with overwhelming firepower can still get themselves in trouble. More risks will be imposed by the need to identify, detain, and interrogate terrorists, than by operations to simply strike and kill them Accepting such risks will prove challenging for U.S. high-level commanders on three counts: first, because any tactical failures will give the enemy a moral advantage that can be exploited by an adroit information campaign; second, because highly trained, experienced troops are difficult to replace; and finally, because the strong bonds between military commanders and their soldiers favor the exploration of lower-risk alternatives. Still, there should be no doubt as to the readiness of the military at all levels ultimately to accept the potential sacrifices required in these operations.

Confronting Rogue States

The force and operations described above will most likely be used against terrorist organizations and within failed states like Afghanistan that lack the developed infrastructure and military organizations of functioning states. But in some instances the United States may confront functioning states that themselves sponsor terrorism. Particularly difficult cases will be those states believed to possess nuclear, chemical, or biological weapons. No doubt all alternative means will be considered to remove the possibility of terrorism before considering the use of force, which would in these cases be a last resort. The influence of an aroused, determined America should not be underestimated in these circumstances.

But should such states persist in harboring terrorists, U.S. counterterrorist operations will look much more like conventional military operations, and they will draw heavily on all the existing force structures, doctrines, and procedures developed during the Cold War and modified slightly over the past decade. Experience with Iraq has indicated that limited strikes from the air are of little value against a regime determined to acquire and maintain weapons of mass destruction. Hence, more complete combat operations, starting with a campaign to gain air superiority and followed by overland, amphibious, and air-delivered ground incursions, could be expected. In the worst case, another Desert Storm–like operation might be mounted but aimed this time at regime change. Lesser operations could be aimed at defeating the enemy's ground forces, eliminating terrorist infrastructure and networks, and destroying facilities for the production and storage of weapons of mass destruction.

In any operations in such states, as well as in Afghanistan, there must be an endgame, a strategy involving some sort of international administration or support, substantial demilitarization, and long-term arrangements for security likely to be assured by an international force. Such missions will be familiar to all North Atlantic Treaty Organization member armies, including that of the United States, that have participated in missions of a similar nature in Bosnia or Kosovo. These missions require land forces in significant numbers, equipped with substantial heavy weaponry, helicopters, military police, engineers, logistics, and robust communications. Such missions also pose special requirements for interacting with civilian populations and supporting deployed international law-enforcement organizations, and the work after entry could continue for years. Surely, the most important lesson to have emerged from the fiasco that is Afghanistan is that theaters of past military operations must not be quickly abandoned, as the United States did in that country after the pull-out of the Soviet troops in 1989.

Throughout the counterterrorist campaigns, which may well stretch on for months and even years, the U.S. armed forces will gradually transform themselves into lighter, more deployable structures. They will implement new technologies to complement precision strike weaponry: theater missile-defense capabilities from high-speed interceptor missiles and airborne lasers; more capable drones and ground vehicles for reconnaissance, fighting, and engineering tasks; more capable, more lethal, and lighter-weight infantry weaponry with longer ranges; more sophisticated automated scanning, identification, and targeting systems that rely on multispectral, seismic, and perhaps even biometric technologies; new logistics featuring electric propulsion, automated prefailure diagnostics, and significantly reduced forward logistics footprints.

Thus the events of September 11 mark a crucial turning point in the debate about defense transformation. There is a practical answer now to the cry for the RMA. The "readiness debate" is over: the point at which forces must be ready is now, or within a few months, not at some indefinite time in the future. We will fight largely with what we have, inserting as much new technology as we can, tailoring organizations to emphasize more teamwork among the armed services, and reducing the numbers of troops deployed overseas. The armed forces are likely to leave some equipment and elements behind as they "skinny-down" for actual operations. And at the margins they will seek to create additional capabilities like those of the special forces. Ultimately, however, the United States will be grateful for the relatively robust structures left over from the Army's stubborn reluctance to depart from its traditional divisional structure. The two-MRC strategy has served its purpose: it led U.S. forces through an ambiguous post–Cold War period with substantial capabilities and largely ready to fight.

Looking ahead, defense analysts will find no easy answer in the next, postwar round of defense cutbacks. Fundamental questions about the ultimate structure of the international system and potential competitors will likely remain; new technologies will be required, either to take advantage of new opportunities or to offset emerging vulnerabilities; and sizable numbers of ground forces are likely to be engaged in numerous postwar activities abroad. As a global power, the United States will need to retain the capacities to respond simultaneously to more than one crisis or in more than one region. And defense analysts will have to create the strategic formulations to justify this need to the public.

NO

Andrew Stephen

The War That Bush Cannot Win

There is something ineffably poignant living here in the eye of the storm. Perhaps because it is the calm before an even worse storm? Intelligence reports genuinely suggest that anthrax, smallpox and perhaps even a suitcase nuclear bomb could yet be unleashed on Washington from, say, some anonymous flat across the river in Arlington; certainly many Washingtonians, hitherto wrapped in complacent cocoons of self-importance, now fear themselves to be in personal danger.

Friends who are also senior members of the Bush administration are working 20 hours a day at the White House or the National Security Agency, failing even to find time to obey President Bush's injunctions to go to church. Everyone has their own horror stories, ranging from the defence secretary Donald Rumsfeld's account of how he was flying to Washington and could see palls of smoke rising from his own Pentagon from 50 miles away; to how a fellow soccer parent, Michele Heidenberger, was the flight attendant on the American Airlines plane that flew into the Pentagon and was invested with the normally routine task of holding the cockpit-door key.

In the grounds of Washington's National Cathedral, electronic bells ring out on Sunday lunchtime as a schoolgirls' soccer match gets under way; very small children gather for a "walk for the homeless", and each is given an American flag to wave as they walk. A military helicopter circles incongruously overhead. Parents, fresh from church and most wearing red, white and blue lapel ribbons, voice their feelings: "Afghanistan," one father announces with grim satisfaction, "will be a parking lot in a few days." Outside, beefy young men drive around honking their horns in jeeps festooned with the Stars and Stripes. Signs saying "Don't mess with the US!" are stapled to trees. One retired US military officer is saying on National Public Radio: "We've got to do whatever it takes to make the rest of the world think like we do." A parent chimes in: "The sermon in the cathedral was about turning the other cheek." He is shaking his head in disbelief.

What is so poignant is that these are all decent, sane, educated Americans—some of them touched personally by the fearsome events of 11 September. But . . .

they have yet to assimilate the realisation that America lost its innocence on that terrible day—that it became as vulnerable to disagreeable outside forces as the rest of us. But they still believe wholeheartedly that American might must always be right. In the war that so many now anticipate with relish, they believe the US will once again confirm its omnipotence. They just have to await confirmation of this, when they will finally settle down in front of their televisions to see satisfying fireworks lighting up Kabul or Jalalabad or Baghdad, live from CNN; their faith in the Bush administration to deliver on its oratorical undertakings of past days is both touching and alarming for those of us who feel more detached from such nationalistic fervour.

There is genuine bewilderment—which quickly fans into anger—over why anybody would want to do something so horrible to so beautiful and free a country as the US. Americans have had it drilled into them by their schools and churches since infancy that theirs is the land of the free and home of the brave. And doesn't that make America both uniquely different and superior to all other countries? It is something those of us who regard ourselves as friends of America have to face: that the vast majority of Americans see themselves and their country as superior to the remaining 95 per cent of the world. The logic is thus inescapable: to Americans, any terrorists attacking their country must be evil crackpots consumed by envy and jealousy of US lifestyles. And these crackpots can and must be eradicated (or "taken out" in the again fashionable machospeak) so that America can triumphantly carry on being the land of the free and home of the brave. The innate goodness of America is such that any outbreak of anti-American violence must just be a weird aberration.

This is the alarmingly simplistic position that the US has got itself into since 11 September. Because American newspapers and television (with a handful of exceptions) have such woeful news coverage of the rest of the world, Americans tend to be ignorant of foreign affairs; they have no conception whatsoever of how much their country and what it stands for are despised by scores of millions of people, especially in the Arabic and billion-strong Muslim countries.

In some cases, that may indeed be the result of envy and jealousy. The world's deprived see a window into an apparently more seductive and affluent lifestyle via televisions, yet then find that mighty government departments such as the US Immigration and Naturalisation Service are devoted to keeping them out of a land that is supposed to be so free to the wretched, yearning masses.

But in other cases—even, we must presume, in that of the wretched multi-millionaire Osama Bin Laden—the despising is based on a genuine, if fanatical, repulsion towards America, its ideals and practices, and its self-satisfaction.

The moment this fault line in the American approach to its terrorist attackers is exposed—the belief that they are confined to a relatively easily identifiable network of evildoers, shunned instinctively by the rest of the world and whose whereabouts can therefore ultimately be firmly established—both the approach of the administration and the jingoistic bellicosity of the American public become positively dangerous.

History, after all, is not on America's side: even if we take the example of Northern Ireland as a microcosm, terrorism that emanated from a tiny geographical area and an equally small population could not be contained over a generation by a modern NATO army. The Provisional IRA's bombers and shooters comprised a small number of usually young men who lived within a population that would, if put to the test, shelter and protect those same young men because gut sympathies did not lie with Britain.

In a war with any Arab and Muslim communities or countries, this same phenomenon will multiply exponentially. The US (and its "allies", including Britain) would, in effect, be fighting the Vietcong again—but this time the enemy would be scattered throughout the world, rather than being confined to small pockets of Northern Ireland or Vietnam. What is now being proposed here, put simply, is a war America cannot win.

There is another dangerous phenomenon that is now being increasingly fostered by the media here. Precisely because routine coverage of world affairs is so dire, Americans tend only to be able to comprehend the world outside as comprising goodies and baddies, those who are with America or against. This approach is then further simplified into identifying symbolic individual monsters: a couple of decades ago, the man every American loved to hate was Colonel Gaddafi; ten years later, it was Saddam Hussein; now Osama Bin Laden is the devil incarnate about whom every American has become an instant expert, courtesy of CNN.

Yet the problem is not one evil mastermind in the way Hollywood would have it, but the major fissures and tensions within the world that have developed over centuries—long before America's existence, even. (It still does not occur to the vast majority of Americans, I suspect, that there is any connection between Arab and Muslim opposition to the US and Washington's staunchly pro-Israel policies.) Would American rage be assuaged, I wonder, if the Taliban presented Bin Laden on a plate to the US so that he could be strapped down on the same executioners' gurney as Timothy McVeigh? Certainly not: much more blood has to be spilt to avenge this act against a country whose collective philosophy is based much more on Old Testament vengeance than the Christian forgiveness it nominally supports in such huge numbers.

I talked to an American mother at the kids' walk for the homeless at the National Cathedral who had lived in Britain for ten years, and thus had a perspective denied most Americans. "We've got to stop being the braggarts on the international playground," she said. "We're too full of ourselves. If we go in, we must darned well get it right. Two wrongs don't make a right."

This was one of the few sane, restrained conversations I have had with anybody here since 11 September. But in an unreported comment, Colin Powell —despite having been a hawkish general, he is now the politically most moderate voice at the top of the Bush administration—said that he realised that many of the Taliban want to be made martyrs, and that in Afghanistan "life is a vale of tears where death can be a joyful deliverance".

But . . . that moderate voice of Powell is increasingly being marginalised by the cold-war mentality and ferocity of the troika of Rumsfeld, Cheney and

Bush himself ("We're gonna smoke 'em out"). There have now been literally thousands of attacks on Muslim institutions and on people looking vaguely Arab. A petrol bomb was thrown at a mosque in Texas, while dozens of others have been attacked or daubed by graffiti ("sand niggers"). A large proportion of taxi drivers in DC are Iraqi or Iranian; most now stay at home. A man with a pistol drove his car on to the lawn of the Saudi embassy here, close to me. Even Sikhs and Sikh businesses, totally removed from the conflict, have been attacked: after all, they wear turbans, so they must surely be wogs up to no good?

To his credit, Bush visited the Islamic Center in DC last Monday. But the same day, he could not resist that bragging which seems perfectly natural to Americans, but which so irritates nearly all the rest of the world: "We're the greatest entrepreneurial society in the world. We've got the best farmers and ranchers in the world." And so on. In the days following the attacks, Bush gradually grew into his tough-guy oratorical role: instead of the hijackers being "folks", they metamorphosed into "evildoers . . . barbaric people".

Predictably, Bush's popularity polls have now soared beyond even those of his father in the midst of the Gulf war—but that underlying doubt about his capacity to handle the crisis, and his evocation of "Wanted—Dead or Alive" posters in Westerns (vis-a-vis Bin Laden), remain a topic of sotto voce conversations here; there is a tacit assumption among senior Republicans that Dick Cheney is the one in charge, and was on the day of the hijackings and Bush's peculiar absence. But an official who had three hours of "face time" with Bush on the day after the hijackings insisted to me that—notwithstanding his nervy appearances in public—Bush was calm. We will see. In the meantime, Rudy Giuliani, the mayor of New York, has won the most public admiration for remaining on Ground Zero (as the TV stations, irritatingly, now refer to Lower Manhattan) and being a figure of visible, even heroic, authority when that was most needed.

But Bush's attempt to put together the same kind of symbolic international coalition that fought the Gulf war—his father's main achievement as president—was faltering within a week of the outrages. Belgium, Norway, Italy and Germany all made noises distancing themselves from Lord Robertson's almost immediate invocation of NATO solidarity; Americans preferred to look away when Pakistanis said that their country would become the graveyard of the US army if it invaded.

The very first international visitor who happened to be in Washington at the time was John Howard, the Australian prime minister; Bush was momentarily buoyed by Australian pronouncements of solidarity, especially when Howard took the opportunity to point out that Australia is the only country in the world to have fought alongside the US in every war it has ever fought.

The next important international visitor was President Megawati Sukarnoputri of Indonesia, the largest Muslim nation in the world. On Wednesday came Jacques Chirac—and then, finally, Tony Blair was allowed his would-be Churchillian photo-op meeting with Bush on the understanding that Britain would offer unconditional support, a blank cheque of solidarity for whatever plans the troika is cooking up. Despite the assumption in London that Britain is always considered to be the US's most important ally, that is not the universal view here. Even in the Gulf war, in which the British government and media would have us believe Britain played a militarily crucial role, only 5.59 per cent of troops were actually British. We can be confident that any British involvement in Bush Jr.'s self-proclaimed "war against terrorism" will be equally symbolic and minimal, though that is not how it will be presented to the baying British public.

And so there it is: this country, and in particular the capital, remain locked in a curious mix of sombreness, anger and a certainty that the outrages will be satisfactorily avenged. I have two friends who watched people jump out of the World Trade Center, and one of them working in an adjoining financial sky-scraper saw a giant airliner pass literally feet from her 54th-floor window before tearing into one of the 110-storey towers. Both are tough and each happens to have been a veteran of Belfast: but I suspect the impact of such traumas will live with them for ever, as it will for thousands of others. Heaven knows what effect their non-stop work will have on the rescue workers in New York and the Penta-gon, too. The television channels, meanwhile, continue with their relentlessly hyped-up, often hysterical and frequently inaccurate coverage ("America United" is the mantra of Rupert Murdoch's Fox news channel). I'm sure it would help the nation calm down if only such channels voluntarily shut down for a week and let us all mourn in peace.

The flags, the flags, those ever-proliferating flags. By now, they have be-come a symbol not so much of grief over what has happened, but of bellicosity over what their wielders hope will soon happen—revenge and war. They have thus become symbols of aggression rather than of half-mast, tranquil sadness.

It remains all so poignant here in the eye of the storm, because the rage can only grow and yet cannot adequately ever be assuaged. America has em-barked on a course it cannot win and which will bring it only yet more grief, at a time in its history when it can least bear it. It is a combustible combination of circumstances, and it will give all of us uncertain futures for years to come.

POSTSCRIPT

Is There a Military Solution to Terrorism?

There is a combined sense of frustration and anger around the world regarding terrorism and its impact, along with the realization that this form of conflict behavior is growing in intensity. September 11 and the subsequent preoccupation with security in the United States has brought home for most Americans the fact that terrorism is difficult to defend against. This may be why many U.S. citizens support attacking areas where terrorists are known to train and thrive. Yet opinion polls and analyses suggest that there is also a common belief that terrorism will not be eradicated by military means alone. Even the leaders of the United States, Great Britain, and Israel, who are on the front line against terrorism, are sober in their assessments of the strategy, how long it would take, and whether or not it could ever be entirely successful.

Stephen points out that terrorism has sprung up from a series of complex sociocultural, political, and economic factors. Many of these are present in large urban areas of the developing world, where the growing demographic reality that millions of young people are unemployed and uneducated breeds disenchantment and despair. The question of who is to blame for that despair might make for an interesting debate, but it does little to resolve the issue. Westerners are no more likely to believe that they are responsible for creating the conditions of September 11 than disenfranchised youth are likely to accept total responsibility for their own plight, particularly as the gap between the rich and the poor continues to grow.

This leaves us with the fundamental question of whether or not the military solution will eradicate terrorism, and the answer remains uncertain. A coordinated global strategy similar to the one carried out by the United States may significantly impede terrorists' ability to strike. Over time, we could see the gradual elimination of such behavior as its costs become too great. Equally possible is the notion that this same strategy might breed more terrorism as a backlash. The brothers and sisters of captured terrorists might take up the cause and actually increase such attacks. What is clear is that the war against terrorism seems to be part of a new form of battle that is not unlike "the long twilight struggle" of which President John F. Kennedy spoke in his inaugural address over 40 years ago. It may be years or even decades before a clear answer to the question of how to fight terrorism is determined.

ISSUE 20

Are Civil Liberties Likely to Be Compromised in the War Against Terrorism?

YES: Morton H. Halperin, from "Less Secure, Less Free," *The American Prospect* (November 19, 2001)

NO: Kim R. Holmes and Edwin Meese III, from "The Administration's Anti-Terrorism Package: Balancing Security and Liberty," *The Heritage Foundation Backgrounder* (October 3, 2001)

ISSUE SUMMARY

YES: Morton H. Halperin, a senior fellow of the Council on Foreign Relations, cites the USA Patriot Act, as well as the strategy employed by the George W. Bush administration to enact it, in sounding an alarm about the abuses of civil liberties that he feels will likely result from this new legislation.

NO: Kim R. Holmes, vice president and director of the Kathryn and Shelby Cullom Davis Institute for International Studies at the Heritage Foundation, and Edwin Meese III, Ronald Reagan Distinguished Fellow in Public Policy at the foundation, argue that the Bush administration's approach to fighting terrorism, including its legislative package, strikes the proper balance between liberty and security.

T wo thousand years ago, the Roman statesman Marcus Tullius Cicero said, "Inter arma silent leges," which translates into "In times of war the law is silent." Cicero's point was that governments must act differently during wartime and that one of its likely casualties is the rule of law. Given Western democracies' commitment to the latter principle, particularly to the precept that individuals possess a long list of rights because they are members of the human race, Cicero's words represent a chilling reminder of the inherent tension between security and liberty.

It is not surprising that an overwhelming number of Americans were quick to criticize the George W. Bush administration's approach to winning the war against terrorism—the USA Patriot Act, which gives the executive branch greater power and more tools with which to fight terrorism. Specifically, citizens are concerned about the potential for the curtailment of civil rights.

Efforts to address potential civil rights violations brought about by the war on terrorism have been varied. The MacArthur Foundation awarded over $1 million to 10 groups to investigate the presumed assault on civil liberties by new antiterrorism laws. Three hundred law professors signed a letter of protest against the creation of military tribunals. A Web site has been established for the specific purpose of demanding that the president maintain civil liberties; six months after the terrorist attack on the World Trade Center, over 200,000 e-mails with signatures had been forwarded to the White House. And editorial after editorial echoes the fear that the war on terrorism's first victim would be freedom.

The average American citizen seems far less concerned about the impact of the USA Patriot Act on individual liberties. For example, the NPR/Kaiser/ Kennedy School National Survey on Civil Liberties, released in November 2001, concluded that despite holding strong opinions on civil liberties in principle, a "vast majority of Americans are willing to forgo some civil liberties to fight terrorism." A *Newsweek* poll taken in November 2001 found that 72 percent of those polled thought that the level of restrictions on civil liberties imposed by the White House is "about right." And an article in the December 20, 2001, issue of *The Christian Science Monitor* reports that most Americans are happy to have Big Brother watching.

At the same time, scholars have reexamined history to ascertain how the tension between security and liberty was resolved in earlier times of trouble. Not surprisingly, critics on opposite sides of the issue reach different conclusions about what history teaches us. Those who oppose the Bush administration's policies point to the Alien and Sedition Acts (which, among other things, made it a crime to criticize the government), a variety of questionable steps taken by President Abraham Lincoln during the Civil War, legislation passed during World War I curtailing criticism of America's war efforts, McCarthyism of the 1950s, and President Richard Nixon's vigorous pursuit of anti–Vietnam War protesters as prime examples of how the government has used external threats to ignore the norms of civil liberties. Others suggest that America has underestimated past threats and has typically resisted tilting the balance away from liberty and toward higher security until almost too late.

In the following selections, Morton H. Halperin criticizes the Bush administration for the strategies it employed in having antiterrorism legislation passed and the content of that legislation, including the lack of distinction between foreign-intelligence investigations and criminal investigations. Kim R. Holmes and Edwin Meese III, while agreeing that civil liberties in American political life are of paramount importance, argue that critics of the administration have ignored two important points: that one must distinguish between constitutional liberties on the one hand and mere privileges and conveniences on the other, and that liberty depends on security. They conclude that the administration's antiterrorism measures draw these distinctions properly.

Morton H. Halperin    **YES**

Less Secure, Less Free

For civil libertarians, there was one extra nightmare when we finally got to sleep on that awful day of September 11, 2001. We knew that the Washington bureaucracy's wish list of additional powers to conduct surveillance of Americans would not be based on a careful analysis of what went wrong. We feared that in the new climate, Congress would rush through the Bush administration's request without reading the text. The result would be less liberty but no greater capability to prevent terrorist acts. Within a few days, civil-liberties advocates formed a broad coalition—Organizations in Defense of Freedom—and produced a statement emphasizing our support for necessary changes in how the government conducts surveillance and our confidence that, with hard work and goodwill, compromises can be found that protect both our security and our liberty (see www.indefenseoffreedom.org).

On the eve of the press conference announcing our coalition and its statement, an antiterrorism bill, described as what the attorney general wanted, immaculately appeared on Capitol Hill. Two things about the legislation are deeply troubling: its substance and the manner in which it was rammed through Congress. The bill was not formally submitted by the administration, since that would have required a review process coordinated by the Office of Management and Budget—a procedure that would have provided an opportunity for all concerned agencies to provide comments.

As we had feared, the attorney general demanded that Congress pass the bill within a week and without change. He set an example for others by not reading the bill himself. (At the only hearing for outside witnesses held in either house, Republican Senator Arlen Specter of Pennsylvania noted that although Mr. Ashcroft kept asserting that the draft legislation permitted him to detain an alien indefinitely only in deportation proceedings, the text itself had no such limitation.) It looked at first as if the attorney general would get his way: The bill would be taken to both floors and passed without anyone knowing what was in it.

Then things seemed to change. With the help of his staff, Democrat Patrick Leahy of Vermont, the chairman of the Senate Judiciary Committee, started an intense negotiation with the Justice Department and won agreement on some changes. In the House, Congressman Jim Sensenbrenner of Wisconsin, the Republican who chairs the Judiciary Committee, and Michigan's John Conyers, the ranking Democrat, began negotiations that led to substantial improvements in the bill. The committee held an actual markup, at which further changes were made, and promised more revisions before the bill went to the floor.

But all of this turned out to be a sham. In meetings with the House and Senate leadership, Attorney General Ashcroft warned that additional terrorist acts were imminent and that Congress would be to blame if the bill were not passed immediately. This was nonsense, but Congress could not withstand such pressure.

In the Senate, an all-night negotiation between leaders of the Senate and committee leaders and their staffs led to a bipartisan bill that took back most of the concessions previously made to Senator Leahy and ignored the House compromise version. The majority leader, Senator Thomas Daschle of South Dakota, then took this draft to the floor and sought unanimous consent to bring up the proposal and pass it without debate or amendment. Only Senator Russell Feingold, a Wisconsin Democrat, objected, and he was allowed to offer three amendments. These were quickly tabled (rejected) on the motion of Senator Leahy and the bill passed with only Senator Feingold voting no.

Then it was the turn of the House. After another all-night drafting session, a text was produced that had only minor changes from the Senate-passed bill. It was rushed to the floor and passed with only three Republican and 75 Democratic votes in opposition. Thus by Friday, October 12, both houses had passed nearly identical antiterrorism bills. Despite the assertion that the powers granted by the law were urgently needed, the Congress headed off for the weekend without leaving conferees behind to reconcile the two versions, and later the House stopped working for a week in the face of concerns about anthrax without completing the final language.

⋅⦿⋅

All of us in this country—civil libertarians included—understand that we face a ruthless and diabolical opponent who flies civilian airplanes into buildings and is dedicated to killing Americans at home and abroad. Law-enforcement officials and intelligence agencies do need new authority, and we may well have to permit greater intrusions into our privacy in order to prevent horrendous acts of terror.

But the administration's bill was not developed in response to the events of September 11 or by an analysis of why there was such a monumental intelligence failure: Its measures were grabbed off the shelf and in many cases had nothing to do with what happened on that unforgettable Tuesday. Moreover, the administration resisted every reasonable effort to find an accommodation between the requirements of security and those of liberty. During the course of the negotiations, participants concerned about protecting civil liberty as well as security put forward a number of important suggestions, none of which were taken seriously.

For example, the administration was asked to identify its urgent needs so that Congress could provide emergency interim authority pending examination of more-permanent solutions. Officials responded by hinting that witnesses before the grand jury investigating the September 11 terrorist acts were providing information that should urgently be passed to the Central Intelligence Agency or to foreign governments to prevent imminent terrorist acts but could not be because existing law bans such sharing of information. It is true that a literal reading of the grand-jury secrecy provisions seems to allow disclosure to outsiders only for purposes of "law enforcement." But surely a judge would find that the courts have inherent power to order disclosure to save lives. Even so, legislative leaders offered to immediately pass a provision allowing intelligence officials dealing with international terrorism to receive pertinent information disclosed by any grand-jury witnesses concerning September 11 or other terrorist events. The administration expressed no interest.

Moreover, most of the administration's provisions were not limited to international terrorists seeking to kill or harm Americans; they covered "terrorists" loosely defined. Critics of the bill were willing to accept most, if not all, of the proposed expansions of surveillance powers—for both intelligence and criminal investigations, including the plans for sharing information across that line—as long as the new powers were limited to terrorism and foreign-intelligence information as carefully defined in the Foreign Intelligence Surveillance Act of 1978 (FISA). The administration bitterly resisted any effort to restrict the bill as a whole or any specific provisions (including relaxing the rules of grand-jury secrecy) to this situation.

Critics repeatedly sought meetings with executive-branch officials to explain concerns and seek common ground. In almost every case, such meetings were refused. In the past, they have led to agreements between civil-liberties groups and administrations of both political parties—for example, on legislation creating the FISA court and exempting certain CIA files from the Freedom of Information Act.

Why does this matter? While this is not the place to rehash all the past abuses of the FBI and the intelligence agencies, it is worth remembering that powers granted with one purpose in mind are often used for others. Thus, antiwar protestors have been investigated as agents of a foreign power and abortion-clinic protestors have been monitored under RICO, the Racketeer Influence and Corrupt Organizations Act. The definitions in this new bill will permit groups such as Greenpeace to be investigated as terrorist organizations and supporters of the political activities of the African National Congress to be spied on or incarcerated as "terrorists." Anyone protesting the attacks on Afghanistan, especially noncitizens, could be investigated under a counterintelligence rubric.

⋅◈⋅

All in all, what is so troubling about the bill? Basically, it breaks down the distinction between foreign-intelligence investigations and criminal investigations by permitting—indeed, encouraging—the sharing of information between intelligence investigations and criminal ones. It also vastly expands the power of the government to gather information in an intelligence investigation and then proceed under the veil of intelligence even if the primary purpose is to gather information for a criminal investigation.

In order to understand why this is a problem, one must understand the origins of FISA, legislation that authorizes the government to conduct investigations that invade areas protected by the Fourth Amendment (which prohibits unreasonable searches and seizures) without the normal probable-cause requirements. FISA originated in a request from the Ford administration for authority to conduct electronic surveillance for national-security purposes. The government explained that it needed to gather foreign-intelligence information even when no crime was suspected, and was unwilling to provide after-the-fact notice to a subject that it had conducted a surveillance.

Congress debated long and hard about FISA. It passed legislation that was substantially different from the original draft, which had been submitted with the usual demand that it be enacted immediately, without changes. In the end, Congress and the administration struck a deal that had the support of some civil libertarians, including me (at the time, I spoke for the ACLU on these issues).

The basic compromise was this: Congress gave the executive branch the authority to conduct electronic surveillance for national-security purposes under a lesser standard than the probable cause that it would need to gather evidence of a crime. Equally important, the government was given permission to keep the surveillance secret. In return law enforcers agreed to judicial supervision and provisions to minimize the interception of non-germane information. Most important, it was agreed that the government would not use the FISA procedures if its investigators were conducting a criminal inquiry but would switch to the usual criminal procedures. The agreement also included a set of definitions that prevented the government from conducting intelligence investigations of Americans unless there is a relatively tight nexus to criminal activity, and it provided a high barrier to the dissemination of information about these subjects. Subsequently, in 1994 Congress broadened FISA to include physical searches even against the homes of Americans without a warrant, without advance knock or notice, and without ever informing the individual that the government surreptitiously acquired information from his or her home.

It is from this perspective that the proposed amendments to FISA must be examined. The most disturbing provision in the administration bill is the one permitting the government to initiate a FISA surveillance even when the investigation's primary purpose is to gather evidence for a criminal prosecution. As noted, FISA granted special surveillance authority for times when the government was gathering foreign intelligence rather than seeking to indict individuals for crimes. To now permit these procedures to be used in a criminal investigation would almost certainly be unconstitutional and would certainly be dangerous, because it would allow the government to avoid all of the safeguards that the Fourth Amendment and existing criminal law provide. The executive branch will be able to use FISA to conduct surveillance whenever it alleges that the targets were agents of a foreign power, thus circumventing the notice and probable-cause requirements of the Fourth Amendment. In situations not covered by the Fourth Amendment, the government could use FISA procedures to compel disclosure of business records, telephone logs, and other sensitive information, including details about people that it does not suspect of being agents of a foreign power.

Equally troubling are the provisions that permit the government to share information from grand-jury proceedings and from criminal wiretaps with intelligence agencies. These sections use a very broad definition of foreign intelligence that is designed explicitly to permit the sharing of information about the First Amendment activities of American citizens. As with the other sharing provisions, this new and sweeping authority is not limited to true terrorism investigations but covers a much broader range of activity.

The immigration provisions of the bill, although improved through negotiation, still sweep within the definition of terrorism individuals who provide support to a group designated as a terrorist organization or viewed as such by the attorney general—even if the organization does not operate in the United States or target Americans, and even if the support is for humanitarian purposes. It also covers people who speak out in favor of "terrorism" in any circumstance; this could include insurrections or paramilitary operations that the United States government supports. Noncitizens who fall within these categories could be detained indefinitely.

Most alarming to supporters of democratic principles was the way the bill was enacted: the absence of public hearings, of any markup in the Senate (coupled with the sham markup conducted in the House), of meaningful floor debate, of committee reports that explain the bill, and of a real conference between the two houses. One can only hope that Congress will conduct rigorous oversight of the new powers it is granting to the president—and that it will refuse to follow the same procedures or to be intimated when the next anti-terrorism bill is sent to Capitol Hill.

NO

Kim R. Holmes and Edwin Meese III

The Administration's Anti-Terrorism Package: Balancing Security and Liberty

The Bush Administration's anti-terrorist proposal has been the only issue since the terrorist attacks on September 11 to spark serious disagreement in Congress. Some fear that adopting the Administration's proposed new law enforcement measures to increase security would endanger civil liberties.

This is a necessary debate; however, it is also one that could easily be misunderstood. Before the debate endangers the unity needed to fight the war on terrorism, it is appropriate for policymakers to stop and think more carefully about the relationship between liberty and security. Above all, care must be taken to avoid artificially polarizing the discussion into two hostile camps—one favoring security and the other favoring civil liberties.

This can be accomplished in two ways. First, policymakers must distinguish between constitutional liberties on the one hand, and mere privileges and conveniences on the other. Second, they must understand that liberty depends on security and that freedom in the long run depends on eliminating the threat of terrorism as soon as possible.

Indeed, policymakers must do everything in their power to preserve the basic liberties protected by the U.S. Constitution, such as the due process of law (including the need to show probable cause and judicial review for issuing warrants and the right to a hearing); the right to be free of unreasonable search and seizures; the right of free speech and religion; and the right to assembly. While it may be permissible to suspend some rights temporarily in a state of emergency—as in a formal declaration of war by Congress—so far this has not been done.

However committed Americans must be to civil liberties, they do *not* have a constitutional right to complete privacy if it endangers the lives of others. Investigators should not be denied access to potentially critical information gained overseas by foreign intelligence sources that could save lives, merely because the methods by which it was obtained do not conform to the U.S. Constitution. Nor should sensitive intelligence information on terrorists be compromised by disclosure in open court proceedings. There must be a reasonable balance between privacy and security.

From Kim R. Holmes and Edwin Meese III, "The Administration's Anti-Terrorism Package: Balancing Security and Liberty," *The Heritage Foundation Backgrounder,* no. 1484 (October 3, 2001). Copyright © 2001 by The Heritage Foundation. Reprinted by permission.

The Administration's proposed anti-terrorism package for the most part draws these distinctions properly. Specifically, it would:

- *Update* wiretapping laws to keep up with changing technologies concerning cell phones, voice mail, and e-mail surveillance;
- *Permit* the sharing of information between law enforcement agencies and intelligence services;
- *Give* courts the authority to review terrorism cases without compromising classified information; and
- *Enable* the government to detain alien individuals who are found to pose a threat to national security until they are actually removed or until the Attorney General determines the person no longer poses a threat. These people would be detained while charges (such as violation of immigration laws) are pending against them. Once those charges were resolved in favor of the individual, he or she would go free.

Reasonable people may disagree with some of these measures and seek improvements. For example, it seems that some time limitations could be established on the detention of foreigners suspected of terrorist ties. Moreover, some provisions could be modified to ensure that information collected on U.S. citizens is not leaked or shared with someone in or out of government who has no need or right to see it. Finally, every effort should be made to ensure that changes made in criminal law are focused as much as possible on the threat of terrorism and could not be abused or used to broaden the investigative powers of the government for non-terrorist–related cases. The new legal tools, developed as emergency measures to fight terrorism, should not necessarily apply to routine criminal investigations. Information collected on people in terrorist investigations should not be used in criminal cases unrelated to terrorism—in tax evasion cases, for example.

Stop Terrorism to Protect Civil Liberties

The attacks of September 11 inevitably will effect changes in some aspects of criminal law. Foreign groups are attacking Americans within the United States, which cannot help but blur the previously bright lines of distinction between national security and criminal law. Compromises will have to be made. People flying on airplanes will have to be subjected to more intrusive questioning, and while any information an airline gathers should not be used for purposes other than security, people must realize that flying is a voluntary act. Some amount of privacy will have to be sacrificed for the sake of the common safety of the crew and passengers.

If the Administration's anti-terrorism proposal is adopted by Congress and enacted into law, there will be cases in which notice of a warrant on suspected terrorists may have to be delayed in order to avoid tipping off a suspect. And there will be times when aliens suspected of terrorism will be detained by rules and legal procedures that are more restrictive. All of these new provisions are reasonable and necessary. None of them represents an infringement on the constitutional rights of American citizens.

The nation must realize that the nature of the terrorist threat has changed and that some laws must change to deal with this new threat. The distinction between domestic and international terrorism—indeed, even between domestic and foreign threats—is now artificial. The foreign threat is here at home, and the laws that normally protect citizens from criminal activity are not sufficient to deal with the complexities of a threat that is both foreign and domestic. Facing a threat as serious as this, civil liberties are in greater danger from a "business as usual" attitude than they are from the minor changes proposed in the Administration's anti-terrorist package.

The Need to Act

However moderate the Administration's proposal may be, the future could be very different if terrorism is not stopped. The longer Americans remain insecure and vulnerable to terrorist attacks, the greater the likelihood that their constitutional liberties will be eroded in the long run. Surely, the nation needs a balanced approach to fighting terrorism at home, but the government must be able to take whatever measures are necessary and consistent with constitutional liberties to quickly re-establish an environment of security.

Until the federal government takes measures that re-establish Americans' confidence in air travel, the economy will be hampered. Until the government makes Americans feel safe and secure, Muslim Americans may be wrongly attacked. Until the government demonstrates that terrorist cells have been eradicated inside the United States, many Americans will be demanding the kind of excessive law enforcement measures that civil libertarians decry. And until the government takes decisive action against terrorist networks and states abroad, Americans will not feel confident that the threat has been removed from their shores.

Imagine what would happen if the war against terrorism fails. Repeated attacks would create panic, and a terrible backlash against civil liberties would ensue. As the casualty toll grew, the calls for draconian measures would make the rather modest provisions in the Administration's anti-terrorist package pale by comparison. A long twilight struggle against terrorism that proves ineffective would chip away at the Constitution in ways Americans can scarcely imagine. Over time, fear and loathing—particularly if America were victim to an attack from weapons of mass destruction that killed millions—would create a tremendous demand from the American people to restrict liberties in the name of security.

That is why the United States must act quickly and decisively now to destroy terrorism at home and abroad. While the President is correct in saying that the anti-terrorist campaign may take many years, the war on terrorism should not become a permanent feature of American life. The goal should be to destroy terrorism as quickly as possible. The United States needs to be patient, to be sure, but not too patient, especially as far as foreign operations against terrorism are concerned. It must be forceful and determined in the short run in order to protect Americans' constitutional liberties in the long run.

This means that the foreign campaign must be as broad as necessary to ensure that the threat of terrorism is not merely contained, but defeated. It means going after the regimes that harbor terrorists. If the international effort merely strikes at one man, one group, or even one network of terrorists without changing or fundamentally altering the pro-terrorist policies of the regimes that protect them, another terrorist leader, group, or network will arise in their place. The more forceful and effective the foreign anti-terrorist campaign, the less pressure there will be against civil liberties at home.

Americans should not forget that there is no more basic civil liberty than the right not to be blown to bits. Civil governments are formed not only to protect liberties, but to protect the lives of their citizens—from each other and from foreign attacks. The Preamble to the U.S. Constitution states that among the purposes of the federal government are to "insure domestic Tranquility, provide for the common defense, [and] promote the general Welfare" of the American people. There would be no "Blessings of Liberty"—another constitutional goal—if it were not for the order and security provided by the federal government.

Conclusion

The changes in law contained in the Administration's anti-terrorist proposal would be a small price to pay to enhance the nation's capabilities to apprehend terrorists. Whatever limited sacrifice in privacy and privileges there may be in these proposed measures is small in comparison to the long-term risks posed to civil liberties by terrorism.

John Adams said in 1765 that

> Liberty must at all hazards be supported. We have a right to it, derived from our Maker. But if we had not, our fathers have earned and bought it for us, at the expense of their ease, their estates, their pleasure, and their blood.

While American troops may be asked to pay for liberty in blood, most Americans will be asked merely to give up a few privileges and conveniences. Surely, this is a sacrifice they can afford to make.

Americans will never be free so long as terrorists are threatening their homeland. It would be ironic indeed if an inordinate fear of losing some rights were sufficient to deny the nation the tools it needs to stop the very thing that would doom the Constitution—the scourge of terrorism. Americans cannot be free unless they are secure any more than they, in the long run, can be secure unless they are free. The United States must stop terrorism in America if it is to preserve freedom.

POSTSCRIPT

Are Civil Liberties Likely to Be Compromised in the War Against Terrorism?

Halperin's article is representative of many that appeared in the months following September 11. In his opinion, the Bush administration was too eager to have its antiterrorism agenda enacted into law. Whereas the administration would respond that the sense of urgency required shortcuts in the legislative process, its detractors saw only an executive branch that believed itself suddenly freed from the system of checks and balances. Moreover, in arguing that because of past transgressions the government cannot be trusted with any additional power, Halperin ignores the dramatically different nature of the current situation.

Holmes and Meese assert that critics confuse liberties on the one hand and privileges and conveniences on the other. For Holmes and Meese, the bottom line is that the administration made the proper distinction. Their second point—that liberty depends on security, which in turn depends on the elimination of terrorism as soon as possible—is less defensible in the eyes of neutral observers. Nowhere in history can a definitive case be made that the abdication of civil liberties was a prerequisite for security and victory in wartime.

While much of the literature takes the Bush administration to task for presumably failing to emphasize the protection of civil liberties in its quest, there are a number of sources that argue either that freedoms are not in jeopardy at all or that history shows that any transgressions tend to be temporary and less evasive than previously assumed. A good place to begin additional reading is the Senate Judiciary Committee hearing of November 6, 2001 (http://judiciary.senate.gov/hearing.cfm?id=121), in which Attorney General John Ashcroft makes a strong case in support of the government's intention to uphold civil liberties in its pursuit of terrorists. A particularly thoughtful essay is James V. DeLong's "Liberty and Order: A Delicate but Clear Balance," National Review Online, http://www.nationalreview.com/comment/comment-delong100201.shtml (October 2, 2001). The article "Liberty v Security," *The Economist* (September 29, 2001) addresses the question of where the balancing point between security and liberty must be set. An excellent discussion of the negative implications for civil liberties can be found in the December 4, 2001, testimony of Nadine Strossen, president of the American Civil Liberties Union, before the Senate Committee on the Judiciary (http://judiciary.senate.gov/hearing.cfm?id=128). Finally, a global report addressing the liberty/security issue can be found in the Amnesty International report *Rights at Risk* (January 18, 2002).

ISSUE 21

Are Cultural and Ethnic Conflicts the Defining Dimensions of Twenty-First-Century War?

YES: Samuel P. Huntington, from "The Clash of Civilizations?" *Foreign Affairs* (Summer 1993)

NO: John R. Bowen, from "The Myth of Global Ethnic Conflict," *Journal of Democracy* (October 1996)

ISSUE SUMMARY

YES: Political scientist Samuel P. Huntington argues that the emerging conflicts of the twenty-first century will be cultural and not ideological. He identifies the key fault lines of conflict and discusses how these conflicts will reshape global policy.

NO: John R. Bowen, a professor of sociocultural anthropology, rejects the idea that ethnic and cultural conflicts are decisive. He argues that political choices made by governments and nations, not cultural divides and intercultural rivalry, dictate much of global international affairs.

Ethnic conflicts seem to be flaring up around the world with greater and greater frequency. The last few years have witnessed ethnic fighting in places such as Northern Ireland, the Middle East, Southeast Asia, and southern Africa. Ethnic clashes have also broken out in places like Bosnia, Kosovo, Rwanda, East Timor, and Chechnya. Certainly, such ethnic conflicts have flared up throughout the centuries in various places. Yet is it possible that ethnic conflict and clashes between cultures are on the rise and will dominate understanding of conflict in the twenty-first century?

For most of the twentieth century, ideological battles between nations took center stage. From the growth of communism and fascism in the 1920s and 1930s came ideological battles centered on notions of governance, economics, race, and the role of people in relation to the state. In that major battle communism and capitalist democracy won out. In the subsequent years the ideological bipolar conflict dominated, as the United States, the USSR, and their surrogates fought on battlefields from Angola to El Salvador and from Vietnam to Afghanistan. At least in part, the battle centered over which system and method of governance would achieve preeminence. In the end the United States won, and capitalist democracy triumphed over communism.

Today scholars and policymakers grapple with the new dynamics of the global system and search for unifying elements that will help to explain why and how groups engage in conflict. Since ideology has lost its zest, and since no apparent philosophies stand ready to directly challenge capitalism, other rallying cries are being uttered and adopted. This period has witnessed the rise or reinvigoration of such philosophies as Islamic fundamentalism, environmentalism, national self-determination (omnipresent in global politics), and ethnic identity as movements and rallying cries for groups to challenge perceived or real oppressors.

Combined with this development is a technological revolution that has brought people in various parts of the world closer than ever before. Ethnic groups traditionally divided by closed borders or reduced contact are now increasingly thrust together by political, economic, and social factors. While this development is benign enough on the surface, many who feel that their cultures are threatened, their identities challenged, and their rights usurped have reacted with disdain and distrust for "the other." As a result, ethnic conflicts have flared in the Balkans, the Middle East, the former Soviet Union, the Great Lakes region of Africa, South America, and the Indian subcontinent.

With increased tensions, greater amounts of weaponry, and less restraint offered by a bipolar world, these conflicts have raged with devastating human consequences. Issues of ethnic cleansing, genocide, land mines, nuclear proliferation, and narcoterrorism have all sprung from or been fueled by these conflicts.

In the following selection, Samuel P. Huntington argues that cultural rivalries are the wave of the coming age. He contends that fault lines between civilizations (where dominant cultures meet) will be "the flash points for crisis and bloodshed" in the coming decades. He contends that cultures will predominate in the battle for hearts and minds and that groups will engage in conflict to defend against challenges to their cultures as they perceive them.

In the second selection, John R. Bowen argues that cultural conflict is borne of a myth about people and diversity in that, by definition, they cannot live with one another. He contends that political, economic, and social issues predominate in issues of conflict and that to focus on ethnicity as the central variable obscures that reality.

Samuel P. Huntington  **YES**

The Clash of Civilizations?

The Next Pattern of Conflict

World politics is entering a new phase, and intellectuals have not hesitated to
proliferate visions of what it will be—the end of history; the return of traditional
rivalries between nation states, and the decline of the nation state from the conflicting pulls of tribalism and globalism, among others. Each of these visions
catches aspects of the emerging reality. Yet they all miss a crucial, indeed a central, aspect of what global politics is likely to be in the coming years.

It is my hypothesis that the fundamental source of conflict in this new
world will not be primarily ideological or primarily economic. The great divisions among humankind and the dominating source of conflict will be cultural.
Nation states will remain the most powerful actors in world affairs, but the principal conflicts of global politics will occur between nations and groups of different civilizations. The clash of civilizations will dominate global politics. The
fault lines between civilizations will be the battle lines of the future.

Conflict between civilizations will be the latest phase in the evolution of
conflict in the modern world. For a century and a half after the emergence of
the modern international system with the Peace of Westphalia [1648], the conflicts of the Western world were largely among princes—emperors, absolute
monarchs and constitutional monarchs attempting to expand their bureaucra-
cies, their armies, their mercantilist economic strength and, most important,
the territory they ruled. In the process they created nation states, and beginning
with the French Revolution the principal lines of conflict were between nations
rather than princes. In 1973, as R. R. Palmer put it, "The wars of kings were over;
the wars of peoples had begun." This nineteenth-century pattern lasted until
the end of World War I. Then, as a result of the Russian Revolution and the reaction against it, the conflict of nations yielded to the conflict of ideologies, first
among communism, fascism-Nazism and liberal democracy, and then between
communism and liberal democracy. During the Cold War, this latter conflict
became embodied in the struggle between the two superpowers, neither of
which was a nation state in the classical European sense and each of which
defined its identity in terms of its ideology.

These conflicts between princes, nation states and ideologies were primarily conflicts within Western civilization, "Western civil wars," as William Lind has labeled them. This was as true of the Cold War as it was of the world wars and the earlier wars of the seventeenth, eighteenth and nineteenth centuries. With the end of the Cold War, international politics moves out of its Western phase, and its centerpiece becomes the interaction between the West and non-Western civilizations and among non-Western civilizations. In the politics of civilizations, the peoples and governments of non-Western civilizations no longer remain the objects of history as targets of Western colonialism but join the West as movers and shapers of history.

The Nature of Civilizations

During the Cold War the world was divided into the First, Second and Third Worlds. Those divisions are no longer relevant. It it far more meaningful now to group countries not in terms of their political or economic systems or in terms of their level of economic development but rather in terms of their culture and civilization.

What do we mean when we talk of a civilization? A civilization is a cultural entity. Villages, regions, ethnic groups, nationalities, religious groups, all have distinct cultures at different levels of cultural heterogeneity. The culture of a village in southern Italy may be different from that of a village in northern Italy, but both will share in a common Italian culture that distinguishes them from German villages. European communities, in turn, will share cultural features that distinguish them from Arab or Chinese communities. Arabs, Chinese and Westerners, however, are not part of any broader cultural entity. They constitute civilizations. A civilization is thus the highest cultural grouping of people and the broadest level of cultural identity people have short of that which distinguishes humans from other species. It is defined both by common objective elements, such as language, history, religion, customs, institutions, and by the subjective self-identification of people. People have levels of identity: a resident of Rome may define himself with varying degrees of intensity as a Roman, an Italian, a Catholic, a Christian, a European, a Westerner. The civilization to which he belongs is the broadest level of identification with which he intensely identifies. People can and do redefine their identities and, as a result, the composition and boundaries of civilizations change.

Civilizations may involve a large number of people, as with China, ("a civilization pretending to be a state," as Lucian Pye put it), or a very small number of people, such as the Anglophone Caribbean. A civilization may include several nation states, as is the case with Western, Latin American and Arab civilizations, or only one, as is the case with Japanese civilization. Civilizations obviously blend and overlap, and may include subcivilizations. Western civilization has two major variants, European and North American, and Islam has its Arab, Turkic and Malay subdivisions. Civilizations are nonetheless meaningful entities, and while the lines between them are seldom sharp, they are real. Civilizations are dynamic; they rise and fall; they divide and merge. And, as any student of history knows, civilizations disappear and are buried in the sands of time.

Westerners tend to think of nation states as the principal actors in global affairs. They have been that, however, for only a few centuries. The broader reaches of human history have been the history of civilizations. In *A Study of History,* Arnold Toynbee identified 21 major civilizations; only six of them exist in the contemporary world.

Why Civilizations Will Clash

Civilization identity will be increasingly important in the future, and the world will be shaped in large measure by the interactions among seven or eight major civilizations. These include Western, Confucian, Japanese, Islamic, Hindu, Slavic-Orthodox, Latin American and possibly African civilization. The most important conflicts of the future will occur along the cultural fault lines separating these civilizations from one another.

Why will this be the case?

First, differences among civilizations are not only real; they are basic. Civilizations are differentiated from each other by history, language, culture, tradition and, most important, religion. The people of different civilizations have different views on the relations between God and man, the individual and the group, the citizen and the state, parents and children, husband and wife, as well as differing views of the relative importance of rights and responsibilities, liberty and authority, equality and hierarchy. These differences are the product of centuries. They will not soon disappear. They are far more fundamental than differences among political ideologies and political regimes. Differences do not necessarily mean conflict, and conflict does not necessarily mean violence. Over the centuries, however, differences among civilizations have generated the most prolonged and the most violent conflicts.

Second, the world is becoming a smaller place. The interactions between peoples of different civilizations are increasing; these increasing interactions intensify civilization consciousness and awareness of differences between civilizations and commonalities within civilizations. North African immigration to France generates hostility among Frenchmen and at the same time increased receptivity to immigration by "good" European Catholic Poles. Americans react far more negatively to Japanese investment than to larger investments from Canada and European countries. Similarly, as Donald Horowitz has pointed out, "An Ibo may be . . . an Owerri Ibo or an Onitsha Ibo in what was the Eastern region of Nigeria. In Lagos, he is simply an Ibo. In London, he is a Nigerian. In New York, he is an African." The interactions among peoples of different civilizations enhance the civilization-consciousness of people that, in turn, invigorates differences and animosities stretching or thought to stretch back deep into history.

Third, the processes of economic modernization and social change throughout the world are separating people from longstanding local identities. They also weaken the nation state as a source of identity. In much of the world religion has moved in to fill this gap, often in the form of movements that are labeled "fundamentalist." Such movements are found in Western Christianity, Judaism, Buddhism and Hinduism, as well as in Islam. In most countries and most religions the people active in fundamentalist movements are young, college-educated, middle-class technicians, professionals and business persons. The "unsecularization of the world," George Weigel has remarked, is one of the dominant social facts of life in the late twentieth century." The revival of religion, "la revanche de Dieu," as Gilles Kepel labeled it, provides a basis for identity and commitment that transcends national boundaries and unites civilizations.

Fourth, the growth of civilization-consciousness is enhanced by the dual role of the West. On the one hand, the West is at a peak of power. At the same time, however, and perhaps as a result, a return to the roots phenomenon is occurring among non-Western civilizations. Increasingly one hears references to trends toward a turning inward and "Asianization" in Japan, the end of the Nehru legacy and the "Hinduization" of India, the failure of Western ideas of socialism and nationalism and hence "re-Islamization" of the Middle East, and now a debate over Westernization versus Russianization in Boris Yeltsin's country. A West at the peak of its power confronts non-Wests that increasingly have the desire, the will and the resources to shape the world in non-Western ways.

In the past, the elites of non-Western societies were usually the people who were most involved with the West, had been educated at Oxford, the Sorbonne or Sandhurst, and had absorbed Western attitudes and values. At the same time, the populace in non-Western countries often remained deeply imbued with the indigenous culture. Now, however, these relationships are being reversed. A de-Westernization and indigenization of elites is occurring in many non-Western countries at the same time that Western, usually American, cultures, styles and habits become more popular among the mass of the people.

Fifth, cultural characteristics and differences are less mutable and hence less easily compromised and resolved than political and economic ones. In the former Soviet Union, communists can become democrats, the rich can become poor and the poor rich, but Russians cannot become Estonians and Azeris cannot become Armenians. In class and ideological conflicts, the key question was "Which side are you on?" and people could and did choose sides and change sides. In conflicts between civilizations, the question is "What are you?" That is a given that cannot be changed. And as we know, from Bosnia to the Caucasus to the Sudan, the wrong answer to that question can mean a bullet in the head. Even more than ethnicity, religion discriminates sharply and exclusively among people. A person can be half-French and half-Arab and simultaneously even a citizen of two countries. It is more difficult to be half-Catholic and half-Muslim.

Finally, economic regionalism is increasing. The proportions of total trade that were intraregional rose between 1980 and 1989 from 51 percent to 59 percent in Europe, 33 percent to 37 percent in East Asia, and 32 percent to 36 percent in North America. The importance of regional economic blocs is likely to continue to increase in the future. On the one hand, successful economic regionalism will reinforce civilization-consciousness. On the other hand, economic regionalism may succeed only when it is rooted in a common civilization. The European Community rests on the shared foundation of European culture and Western Christianity. The success of the North American Free Trade Area depends on the convergence now underway of Mexican, Canadian and American cultures. Japan, in contrast, faces difficulties in creating a comparable economic entity in East Asia because Japan is a society and civilization unique to itself. However strong the trade and investment links Japan may develop with other East Asian countries, its cultural differences with those countries inhibit and perhaps preclude its promoting regional economic integration like that in Europe and North America.

Common culture, in contrast, is clearly facilitating the rapid expansion of the economic relations between the People's Republic of China and Hong Kong, Taiwan, Singapore and the overseas Chinese communities in other Asian countries. With the Cold War over, cultural commanalities increasingly overcome ideological differences, and mainland China and Taiwan move closer together. If cultural commonality is a prerequisite for economic integration, the principal East Asian economic bloc of the future is likely to be centered on China. This bloc is, in fact, already coming into existence. As Murray Weidenbaum has observed,

> Despite the current Japanese dominance of the region, the Chinese-based economy of Asia is rapidly emerging as a new epicenter for industry, commerce and finance. This strategic area contains substantial amounts of technology and manufacturing capability (Taiwan), outstanding entrepreneurial, marketing and services acumen (Hong Kong), a fine communications network (Singapore), a tremendous pool of financial capital (all three), and very large endowments of land, resources and labor (mainland China). . . . From Guangzhou to Singapore, from Kuala Lumpur to Manila, this influential network—often based on extensions of the tranditional clans—has been described as the beckbone of the East Asian economy.[1]

Culture and religion also form the basis of the Economic Cooperation Organization, which brings together ten non-Arab Muslim countries: Iran, Pakistan, Turkey, Azerbaijan, Kazakhstan, Kyrgyzstan, Turkmenistan, Tadjikistan, Uzbekistan and Afghanistan. One impetus to the revival and expansion of this organization, founded originally in the 1960s by Turkey, Pakistan and Iran, is the realization by the leaders of several of these countries that they had no chance of admission to the European Community. Similarly, Caricom, the Central American Common Market and Mercosur rest on common cultural foundations. Efforts to build a broader Caribbean-Central American economic entity bridging the Anglo-Latin divide, however, have to date failed.

As people define their identity in ethnic and religious terms, they are likely to see an "us" versus "them" relation existing between themselves and people of different ethnicity or religion. The end of ideologically defined states in Eastern Europe and the former Soviet Union permits traditional ethnic identities and animosities to come to the fore. Differences in culture and religion create differences over policy issues, ranging from human rights to immigration to trade and commerce to the environment. Geographical propinquity gives rise to conflicting territorial claims from Bosnia to Mindanao. Most important, the efforts of the West to promote its values of democracy and liberalism as universal values, to maintain its military predominance and to advance its economic interests engender countering responses from other civilizations. Decreasingly able to mobilize support and form coalitions on the basis of ideology, governments and groups will increasingly attempt to mobilize support by appealing to common religion and civilization identity.

The clash of civilizations thus occurs at two levels. At the micro-level, adjacent groups along the fault lines between civilizations struggle, often violently, over the control of territory and each other. At the macro-level, states from different civilizations compete for relative military and economic power, struggle over the control of international institutions and third parties, and competitively promote their particular political and religious values.

The Fault Lines Between Civilizations

The fault lines between civilizations are replacing the political and ideological boundaries of the Cold War as the flash points for crisis and bloodshed. The Cold War began when the Iron Curtain divided Europe politically and ideologically. The Cold War ended with the end of the Iron Curtain. As the ideological division of Europe has disappeared, the cultural division of Europe between Western Christianity, on the one hand, and Orthodox Christianity and Islam, on the other, has reemerged. The most significant dividing line in Europe, as William Wallace has suggested, may well be the eastern boundary of Western Christianity in the year 1500. This line runs along what are now the boundaries between Finland and Russia and between the Baltic states and Russia, cuts through Belarus and Ukraine separating the more Catholic western Ukraine from Orthodox eastern Ukraine, swings westward separating Transylvania from the rest of Romania, and then goes through Yugoslavia almost exactly along the line now separating Croatia and Slovenia from the rest of Yugoslavia. In the Balkans this line, of course, coincides with the historic boundary between the Hapsburg and Ottoman empires. The peoples to the north and west of this line are Protestant or Catholic; they shared the common experiences of European history—feudalism, the Renaissance, the Reformation, the Enlightenment, the French Revolution, the Industrial Revolution; they are generally economically better off than the peoples to the east; and they may now look forward to increasing involvement in a common European economy and to the consolidation of democratic political systems. The peoples to the east and south of this line are Orthodox or Muslim; they historically belonged to the Ottoman or Tsarist empires and were only lightly touched by the shaping events in the rest of Europe; they are generally less advanced economically; they seem much less likely to develop stable democratic political systems. The Velvet Curtain of culture has replaced the Iron Curtain of ideology as the most significant dividing line in Europe. As the events in Yugoslavia show, it is not only a line of difference; it is also at times a line of bloody conflict.

Conflict along the fault line between Western and Islamic civilizations has been going on for 1,300 years. After the founding of Islam, the Arab and Moorish surge west and north only ended at Tours in 732. From the eleventh to the thirteenth century the Crusaders attempted with temporary success to bring Christianity and Christian rule to the Holy Land. From the fourteenth to the seventeenth century, the Ottoman Turks reversed the balance, extended their sway over the Middle East and the Balkans, captured Constantinople, and twice laid siege to Vienna. In the nineteenth and early twentieth centuries as Ottoman power declined Britain, France, and Italy established Western control over most of North Africa and the Middle East.

After World War II, the West, in turn, began to retreat; the colonial empires disappeared; first Arab nationalism and then Islamic fundamentalism manifested themselves; the West became heavily dependent on the Persian Gulf countries for its energy; the oil-rich Muslim countries became money-rich and, when they wished to, weapons-rich. Several wars occurred between Arabs and Israel (created by the West). France fought a bloody and ruthless war in Algeria for most of the 1950s; British and French forces invaded Egypt in 1956; American forces went into Lebanon in 1958; subsequently American forces returned to Lebanon, attacked Libya, and engaged in various military encounters with Iran; Arab and Islamic terrorists, supported by at least three Middle Eastern governments, employed the weapon of the weak and bombed Western planes and installations and seized Western hostages. This warfare between Arabs and the West culminated in 1990, when the United States sent a massive army to the Persian Gulf to defend some Arab countries against aggression by another. In its aftermath NATO planning is increasingly directed to potential threats and instability along its "southern tier."

This centuries-old military interaction between the West and Islam is unlikely to decline. It could become more virulent. The Gulf War left some Arabs feeling proud that Saddam Hussein had attacked Israel and stood up to the West. It also left many feeling humiliated and resentful of the West's military presence in the Persian Gulf, the West's overwhelming military dominance, and their apparent inability to shape their own destiny. Many Arab countries, in addition to the oil exporters, are reaching levels of economic and social development where autocratic forms of government become inappropriate and efforts to introduce democracy become stronger. Some openings in Arab political systems have already occurred. The principal beneficiaries of these openings have been Islamist movements. In the Arab world, in short, Western democracy strengthens anti-Western political forces. This may be a passing phenomenon, but it surely complicates relations between Islamic countries and the West.

Those relations are also complicated by demography. The spectacular population growth in Arab countries, particularly in North Africa, has led to increased migration to Western Europe. The movement within Western Europe toward minimizing internal boundaries has sharpened political sensitivities with respect to this development. In Italy, France and Germany, racism is increasingly open, and political reactions and violence against Arab and Turkish migrants have become more intense and more widespread since 1990.

On both sides the interaction between Islam and the West is seen as a clash of civilizations. The West's "next confrontation," observes M. J. Akbar, an Indian Muslim author, "is definitely going to come from the Muslim world. It is in the sweep of the Islamic nations from the Maghreb to Pakistan that the struggle for a new world order will begin." Bernard Lewis comes to a similar conclusion:

> We are facing a mood and a movement far transcending the level of issues and policies and the governments that pursue them. This is no less than a clash of civilizations—the perhaps irrational but surely historic reaction of an ancient rival against our Judeo-Christian heritage, our secular present, and the worldwide expansion of both.[2]

Historically, the other great antagonistic interaction of Arab Islamic civilization has been with the pagan, animist, and now increasingly Christian black peoples to the south. In the past, this antagonism was epitomized in the image of Arab slave dealers and black slaves. It has been reflected in the on-going civil war in the Sudan between Arabs and blacks, the fighting in Chad between Libyan-supported insurgents and the government, the tensions between Orthodox Christians and Muslims in the Horn of Africa, and the political conflicts, recurring riots and communal violence between Muslims and Christians in Nigeria. The modernization of Africa and the spread of Christianity are likely to enhance the probability of violence along this fault line. Symptomatic of the intensification of this conflict was the Pope John Paul II's speech in Khartoum in February 1993 attacking the actions of the Sudan's Islamist government against the Christian minority there.

On the northern border of Islam, conflict has increasingly erupted between Orthodox and Muslim peoples, including the carnage of Bosnia and Sarajevo, the simmering violence between Serb and Albanian, the tenuous relations between Bulgarians and their Turkish minority, the violence between Ossetians and Ingush, the unremitting slaughter of each other by Armenians and Azeris, the tense relations between Russians and Muslims in Central Asia, and the deployment of Russian troops to protect Russian interests in the Caucasus and Central Asia. Religion reinforces the revival of ethnic identities and restimulates Russian fears about the security of their southern borders. This concern is well captured by Archie Roosevelt:

> Much of Russian history concerns the struggle between the Slavs and the Turkic peoples on their borders, which dates back to the foundation of the Russian state more than a thousand years ago. In the Slavs' millennium-long confrontation with their eastern neighbors lies the key to an understanding not only of Russian history, but Russian character. To understand Russian realities today one has to have a concept of the great Turkic ethnic group that has preoccupied Russians through the centuries.[3]

The conflict of civilizations is deeply rooted elsewhere in Asia. The historic clash between Muslim and Hindu in the subcontinent manifests itself now not only in the rivalry between Pakistan and India but also in intensifying religious strife within India between increasingly militant Hindu groups and India's substantial Muslim minority. The destruction of the Ayodhya mosque in December 1992 brought to the fore the issue of whether India will remain a secular democratic state or become a Hindu one. In East Asia, China has outstanding territorial disputes with most of its neighbors. It has pursued a ruthless policy toward the Buddhist people of Tibet, and it is pursuing an increasingly ruthless policy toward its Turkic-Muslim minority. With the Cold War over, the underlying differences between China and the United States have reasserted themselves in areas such as human rights, trade and weapons proliferation. These differences are unlikely to moderate. A "new cold war," Deng Xiaoping reportedly asserted in 1991, is under way between China and America. . . .

Civilization Rallying: The Kin-Country Syndrome

Groups or states belonging to one civilization that become involved in war with people from a different civilization naturally try to rally support from other members of their own civilization. As the post–Cold War world evolves, civilization commonality, what H. D. S. Greenway has termed the "kin-country" syndrome, is replacing political ideology and traditional balance of power considerations as the principal basis for cooperation and coalitions. It can be seen gradually emerging in the post–Cold War conflicts in the Persian Gulf, the Caucasus and Bosnia. None of these was a full-scale war between civilizations, but each involved some elements of civilizational rallying, which seemed to become more important as the conflict continued and which may provide a foretaste of the future.

First, in the Gulf War one Arab state invaded another and then fought a coalition of Arab, Western and other states. While only a few Muslim governments overtly supported Saddam Hussein, many Arab elites privately cheered him on, and he was highly popular among large sections of the Arab publics. Islamic fundamentalist movements universally supported Iraq rather than the Western-backed governments of Kuwait and Saudi Arabia. Forswearing Arab nationalism, Saddam Hussein explicitly invoked an Islamic appeal. He and his supporters attempted to define the war as a war between civilizations. "It is not the world against Iraq," as Safar Al-Hawali, dean of Islamic Studies at the Umm Al-Qura University in Mecca, put it in a widely circulated tape. "It is the West against Islam." Ignoring the rivalry between Iran and Iraq, the chief Iranian religious leader, Ayatollah Ali Khamenei, called for a holy war against the West: "The struggle against American aggression, greed, plans and policies will be counted as a jihad, and anybody who is killed on that path is a martyr." "This is a war," King Hussein of Jordan argued, "against all Arabs and all Muslims and not against Iraq alone."

The rallying of substantial sections of Arab elites and publics behind Saddam Hussein caused those Arab governments in the anti-Iraq coalition to moderate their activities and temper their public statements. Arab governments opposed or distanced themselves from subsequent Western efforts to apply pressure on Iraq, including enforcement of a no-fly zone in the summer of 1992 and the bombing of Iraq in January 1993. The Western-Soviet-Turkish-Arab anti-Iraq coalition of 1990 had by 1993 become a coalition of almost only the West and Kuwait against Iraq.

Muslims contrasted Western actions against Iraq with the West's failure to protect Bosnians against Serbs and to impose sanctions on Israel for violating U.N. resolutions. The West, they alleged, was using a double standard. A world of clashing civilizations, however, is inevitably a world of double standards: people apply one standard to their kin-countries and a different standard to others.

Second, the kin-country syndrome also appeared in conflicts in the former Soviet Union. Armenian military successes in 1992 and 1993 stimulated Turkey to become increasingly supportive of its religious, ethnic and linguistic brethren in Azerbaijan. "We have a Turkish nation feeling the same sentiments as the Azerbaijanis," said one Turkish official in 1992. "We are under pressure. Our newspapers are full of the photos of atrocities and are asking us if we are still serious about pursuing our neutral policy. Maybe we should show Armenia that there's a big Turkey in the region." President Turgut Özal agreed, remarking that Turkey should at least "scare the Armenians a little bit." Turkey, Özal threatened again in 1993, would "show its fangs." Turkish Air Force jets flew reconnaissance flights along the Armenian border; Turkey suspended food shipments and air flights to Armenia; and Turkey and Iran announced they would not accept dismemberment of Azerbaijan. In the last years of its existence, the Soviet government supported Azerbaijan because its government was dominated by former communists. With the end of the Soviet Union, however, political considerations gave way to religious ones. Russian troops fought on the side of the Armenians, and Azerbaijan accused the "Russian government of turning 180 degrees" toward support for Christian Armenia.

Third, with respect to the fighting in the former Yugoslavia, Western publics manifested sympathy and support for the Bosnian Muslims and the horrors they suffered at the hands of the Serbs. Relatively little concern was expressed, however, over Croatian attacks on Muslims and participation of the dismemberment of Bosnia-Herzegovina. In the early stages of the Yugoslav breakup, Germany, in an unusual display of diplomatic initiative and muscle, induced the other 11 members of the European Community to follow its lead in recognizing Slovenia and Crotia. As a result of the pope's determination to provide strong backing to the two Catholic countries, the Vatican extended recognition even before the Community did. The United States followed the European lead. Thus the leading actors in Western civilization rallied behind their coreligionists. Subsequently Croatia was reported to be receiving substantial quantities of arms from Central European and other Western countries. Boris Yeltsin's government, on the other hand, attempted to pursue a middle course that would be sympathetic to the Orthodox Serbs but not alienate Russia from the West. Russian conservative and nationalist groups, however, including many legislators, attacked the government for not being more forthcoming in its support for the Serbs. By early 1993 several hundred Russians apparently were serving with the Serbian forces, and reports circulated of Russian arms being supplied to Serbia.

Islamic governments and groups, on the other hand, castigated the West for not coming to the defense of the Bosnians. Iranian leaders urged Muslims from all countries to provide help to Bosnia; in violation of the U.N. arms embargo, Iran supplied weapons and men for the Bosnians; Iranian-supported Lebanese groups sent guerrillas to train and organize the Bosnian forces. In 1993 up to 4,000 Muslims from over two dozen Islamic countries were reported to be fighting in Bosnia. The governments of Saudi Arabia and other countries felt under increasing pressure from fundamentalist groups in their own societies to provide more vigorous support for the Bosnians. By the end of 1992, Saudi Arabia had reportedly supplied substantial funding for weapons and supplies for the Bosnians, which significantly increased their military capabilities vis-à-vis the Serbs.

In the 1930s the Spanish Civil War provoked intervention from countries that politically were fascist, communist and democratic. In the 1990s the Yugoslav conflict is provoking intervention from countries that are Muslim, Orthodox and Western Christian. The parallel has not gone unnoticed. "The war in Bosnia-Herzegovina has become the emotional equivalent of the fight against fascism in the Spanish Civil War," one Saudi editor observed. "Those who died there are regarded as martyrs who tried to save their fellow Muslims."

Conflicts and violence will also occur between states and groups within the same civilization. Such conflicts, however, are likely to be less intense and less likely to expand than conflicts between civilizations. Common membership in a civilization reduces the probability of violence in situations where it might otherwise occur. In 1991 and 1992 many people were alarmed by the possibility of violent conflict between Russia and Ukraine over territory, particularly Crimea, the Black Sea fleet, nuclear weapons and economic issues. If civilization is what counts, however, the likelihood of violence between Ukrainians and Russians should be low. They are two Slavic, primarily Orthodox peoples who have had close relationships with each other for centuries. As of early 1993, despite all the reasons for conflict, the leaders of the two countries were effectively negotiating and defusing the issues between the two countries. While there has been serious fighting between Muslims and Christians elsewhere in the former Soviet Union and much tension and some fighting between Western and Orthodox Christians in the Baltic states, there has been virtually no violence between Russians and Ukrainians.

Civilization rallying to date has been limited, but it has been growing, and it clearly has the potential to spread much further. As the conflicts in the Persian Gulf, the Caucasus and Bosnia continued, the positions of nations and the cleavages between them increasingly were along civilizational lines. Populist politicians, religious leaders and the media have found it a potent means of arousing mass support and of pressuring hesitant governments. In the coming years, the local conflicts most likely to escalate into major wars will be those, as in Bosnia and the Caucasus, along the fault lines between civilizations. The next world war, if there is one, will be a war between civilizations.

The West Versus the Rest

The West is now at an extraordinary peak of power in relation to other civilizations. Its superpower opponent has disappeared from the map. Military conflict among Western states is unthinkable, and Western military power is unrivaled. Apart from Japan, the West faces no economic challenge. It dominates international political and security institutions and with Japan international economic institutions. Global political and security issues are effectively settled by a directorate of the United States, Britain and France, world economic issues by a directorate of the United States, Germany and Japan, all of which maintain extraordinarily close relations with each other to the exclusion of lesser and largely non-Western countries. Decisions made at the U.N. Security Council or in the International Monetary Fund [IMF] that reflect the interests of the West are presented to the world as reflecting the desires of the world community. The very phrase "the world community" has become the euphemistic collective noun (replacing "the Free World") to give global legitimacy to actions reflecting the interests of the United States and other Western powers.[4] Through the IMF and other international economic institutions, the West promotes its economic interests and imposes on other nations the economic policies it thinks appropriate. In any poll of non-Western peoples, the IMF undoubtedly would win the support of finance ministers and a few others, but get an overwhelmingly unfavorable rating from just about everyone else who would agree with Georgy Arbatov's characterization of IMF officials as "neo-Bolsheviks who love expropriating other people's money, imposing undemocratic and alien rules of economic and political conduct and stifling economic freedom."

Western domination of the U.N. Security Council and its decisions, tempered only by occasional abstention by China, produced U.N. legitimation of the West's use of force to drive Iraq out of Kuwait and its elimination of Iraq's sophisticated weapons and capacity to produce such weapons. It also produced the quite unprecedented action by the United States, Britain and France in getting the Security Council to demand that Libya hand over the Pan Am 103 bombing suspects and then to impose sanctions when Libya refused. After defeating the largest Arab army, the West did not hesitate to throw its weight around in the Arab world. The West in effect is using international institutions, military power and economic resources to run the world in ways that will maintain Western predominance, protect Western interests and promote Western political and economic values.

That at least is the way in which non-Westerners see the new world, and there is a significant element of truth in their view. Differences in power and struggles for military, economic and institutional power are thus one source of conflict between the West and other civilizations. Differences in culture, that is basic values and beliefs, are a second source of conflict. V. S. Naipaul has argued that Western civilization is the "universal civilization" that "fits all men." At a superficial level much of Western culture has indeed permeated the rest of the world. At a more basic level, however, Western concepts differ fundamentally from those prevalent in other civilizations. Western ideas of individualism, liberalism, constitutionalism, human rights, equality, liberty, the rule of law, democracy, free markets, the separation of church and state, often have little resonance in Islamic, Confucian, Japanese, Hindu, Buddhist or Orthodox cultures. Western efforts to propagate such ideas produce instead a reaction against "human rights imperialism" and a reaffirmation of indigenous values, as can be seen in the support for religious fundamentalism by the younger generation in non-Western cultures. The very notion that there could be a "universal civilization" is a Western idea, directly at odds with the particularism of most Asian societies and their emphasis on what distinguishes one people from another. Indeed, the author of a review of 100 comparative studies of values in different societies concluded that "the values that are most important in the West are least important worldwide."[5] In the political realm, of course, these differences are most manifest in the efforts of the United States and other Western powers to induce other peoples to adopt Western ideas concerning democracy and human rights. Modern democratic government originated in the West. When it has developed in non-Western societies it has usually been the product of Western colonialism or imposition.

The central axis of world politics in the future is likely to be, in Kishore Mahbubani's phrase, the conflict between "the West and the Rest" and the responses of non-Western civilizations to Western power and values.[6] Those responses generally take one or a combination of three forms. At one extreme, non-Western states can, like Burma and North Korea, attempt to pursue a course of isolation, to insulate their societies from penetration or "corruption" by the West, and, in effect, to opt out of participation in the Western-dominated global community. The costs of this course, however, are high, and few states have pursued it exclusively. A second alternative, the equivalent of "band-wagoning" in international relations theory, is to attempt to join the West and accept its values and institutions. The third alternative is to attempt to "balance" the West by developing economic and military power and cooperating with other non-Western societies against the West, while preserving indigenous values and institutions; in short, to modernize but not to Westernize. . . .

The Confucian-Islamic Connection

The obstacles to non-Western countries joining the West vary considerably. They are least for Latin American and East European countries. They are greater for the Orthodox countries of the former Soviet Union. They are still greater for Muslim, Confucian, Hindu and Buddhist societies. Japan has established a unique position for itself as an associate member of the West: it is in the West in some respects but clearly not of the West in important dimensions. Those countries that for reason of culture and power do not wish to, or cannot, join the West compete with the West by developing their own economic, military and political power. They do this by promoting their internal development and by cooperating with other non-Western countries. The most prominent form of this cooperation is the Confucian-Islamic connection that has emerged to challenge Western interests, values and power.

Almost without exception, Western countries are reducing their military power; under Yeltsin's leadership so also is Russia. China, North Korea and several Middle Eastern states, however, are significantly expanding their military capabilities. They are doing this by the import of arms from Western and non-Western sources and by the development of indigenous arms industries. One result is the emergence of what Charles Krauthammer has called "Weapon States," and the Weapon States are not Western states. Another result is the redefinition of arms control, which is a Western concept and a Western goal. During the Cold War the primary purpose of arms control was to establish a stable military balance between the United States and its allies and the Soviet Union and its allies. In the post–Cold War world the primary objective of arms control is to prevent the development by non-Western societies of military capabilities that could threaten Western interests. The West attempts to do this through international agreements, economic pressure and controls on the transfer of arms and weapons technologies.

The conflict between the West and the Confucian-Islamic states focuses largely, although not exclusively, on nuclear, chemical and biological weapons, ballistic missiles and other sophisticated means for delivering them, and the guidance, intelligence and other electronic capabilities for achieving that goal. The West promotes non-proliferation as a universal norm and nonproliferation treaties and inspections as means of realizing that norm. It also threatens a variety of sanctions against those who promote the spread of sophisticated weapons and proposes some benefits for those who do not. The attention of the West focuses, naturally, on nations that are actually or potentially hostile to the West.

The non-Western nations, on the other hand, assert their right to acquire and to deploy whatever weapons they think necessary for their security. They also have absorbed, to the full, the truth of the response of the Indian defense minister when asked what lesson he learned from the Gulf War: "Don't fight the United States unless you have nuclear weapons." Nuclear weapons, chemical weapons and missiles are viewed, probably erroneously, as the potential equalizer of superior Western conventional power. China, of course, already has nuclear weapons; Pakistan and India have the capability to deploy them. North Korea, Iran, Iraq, Libya and Algeria appear to be attempting to acquire them. A top Iranian official has declared that all Muslim states should acquire nuclear weapons, and in 1988 the president of Iran reportedly issued a directive calling for development of "offensive and defensive chemical, biological and radiological weapons."

Centrally important to the development of counter-West military capabilities is the sustained expansion of China's military power and its means to create military power. Buoyed by spectacular economic development, China is rapidly increasing its military spending and vigorously moving forward with the modernization of its armed forces. It is purchasing weapons from the former Soviet states; it is developing long-range missiles; in 1992 it tested a one-megaton nuclear device. It is developing power-projection capabilities, acquiring aerial refueling technology, and trying to purchase an aircraft carrier. Its military buildup and assertion of sovereignty over the South China Sea are provoking a multilateral regional arms race in East Asia. China is also a major exporter of arms and weapons technology. It has exported materials to Libya and Iraq that could be used to manufacture nuclear weapons and nerve gas. It has helped Algeria build a reactor suitable for nuclear weapons research and production. China has sold to Iran nuclear technology that American officials believe could only be used to create weapons and apparently has shipped components of 300-mile-range missiles to Pakistan. North Korea has had a nuclear weapons program under way for some while and has sold advanced missiles and missile technology to Syria and Iran. The flow of weapons and weapons technology is generally from East Asia to the Middle East. There is, however, some movement in the reverse direction; China has received Stinger missiles from Pakistan.

A Confucian-Islamic military connection has thus come into being, designed to promote acquisition by its members of the weapons and weapons technologies needed to counter the military power of the West. It may or may not last. At present, however, it is, as Dave McCurdy has said, "a renegades' mutual support pact, run by the proliferators and their backers." A new form of arms competition is thus occurring between Islamic-Confucian states and the West. In an old-fashioned arms race, each side developed its own arms to balance or to achieve superiority against the other side. In this new form of arms competition, one side is developing its arms and the other side is attempting not to balance but to limit and prevent that arms build-up while at the same time reducing its own military capabilities.

Implications for the West

This article does not argue that civilization identities will replace all other identities, that nation states will disappear, that each civilization will become a single coherent political entity, that groups within a civilization will not conflict with and even fight each other. This paper does set forth the hypotheses that differences between civilizations are real and important; civilization-consciousness is increasing; conflict between civilizations will supplant ideological and other forms of conflict as the dominant global form of conflict; international relations, historically a game played out within Western civilization, will increasingly be de-Westernized and become a game in which non-Western civilizations are actors and not simply objects; successful political, security and economic international institutions are more likely to develop within civilizations than across civilizations; conflicts between groups in different civilizations will be more frequent, more sustained and more violent than conflicts between groups in the same civilization; violent conflicts between groups in different civilizations are the most likely and most dangerous source of escalation that could lead to global wars; the paramount axis of world politics will be the relations between "the West and the Rest"; the elites in some torn non-Western countries will try to make their countries part of the West, but in most cases face major obstacles to accomplishing this; a central focus of conflict for the immediate future will be between the West and several Islamic-Confucian states.

This is not to advocate the desirability of conflicts between civilizations. It is to set forth descriptive hypotheses as to what the future may be like. If these are plausible hypotheses, however, it is necessary to consider their implications for Western policy. These implications should be divided between short-term advantage and long-term accommodation. In the short term it is clearly in the interest of the West to promote greater cooperation and unity within its own civilization, particularly between its European and North American components; to incorporate into the West societies in Eastern Europe and Latin America whose cultures are close to those of the West; to promote and maintain cooperative relations with Russia and Japan; to prevent escalation of local inter-civilization conflicts into major inter-civilization wars; to limit the expansion of the military strength of Confucian and Islamic states; to moderate the reduction of Western military capabilities and maintain military superiority in East and Southwest Asia; to exploit differences and conflicts among Confucian and Islamic states; to support in other civilizations groups sympathetic to Western values and interests; to strengthen international institutions that reflect and legitimate Western interests and values and to promote the involvement of non-Western states in those institutions.

In the longer term other measures would be called for. Western civilization is both Western and modern. Non-Western civilizations have attempted to become modern without becoming Western. To date only Japan has fully succeeded in this quest. Non-Western civilizations will continue to attempt to acquire the wealth, technology, skills, machines and weapons that are part of being modern. They will also attempt to reconcile this modernity with their traditional culture and values. Their economic and military strength relative to the West will increase. Hence the West will increasingly have to accommodate these non-Western modern civilizations whose power approaches that of the West but whose values and interests differ significantly from those of the West. This will require the West to maintain the economic and military power necessary to protect its interests in relation to these civilizations. It will also, however, require the West to develop a more profound understanding of the basic religious and philosophical assumptions underlying other civilizations and the ways in which people in those civilizations see their interests. It will require an effort to identify elements of commonality between Western and other civilizations. For the relevant future, there will be no universal civilization, but instead a world of different civilizations, each of which will have to learn to coexist with the others.

Notes

1. Murray Weidenbaum, *Greater China: The Next Economic Superpower?*, St. Louis: Washington University Center for the Study of American Business, Contemporary Issues, Series 57, February 1993, pp. 2–3.

2. Bernard Lewis, "The Roots of Muslim Rage," *The Atlantic Monthly,* vol. 266, September 1990, p. 60; *Time,* June 15, 1992, pp. 24–28.

3. Archie Roosevelt, *For Lust of Knowing,* Boston; Little, Brown, 1988, pp. 332–333.

4. Almost invariably Western leaders claim they are acting on behalf of "the world community." One minor lapse occurred during the run-up to the Gulf War. In an interview on "Good Morning America," Dec. 21, 1990, British Prime Minister John Major referred to the actions "the West" was taking against Saddam Hussein. He quickly corrected himself and subsequently referred to "the world community." He was, however, right when he erred.

5. Harry C. Triandis, *The New York Times,* Dec. 25, 1990, p. 41, and "Cross-Cultural Studies of Individualism and Collectivism," Nebraska Symposium on Motivation, vol. 37, 1989, pp. 41–133.

6. Kishore Mahbubani, "The West and the Rest," *The National Interest,* Summer 1992, pp. 3–13.

NO

John R. Bowen

The Myth of Global Ethnic Conflict

Much recent discussion of international affairs has been based on the misleading assumption that the world is fraught with primordial ethnic conflict. According to this notion, ethnic groups lie in wait for one another, nourishing age-old hatreds and restrained only by powerful states. Remove the lid, and the cauldron boils over. Analysts who advance this idea differ in their predictions for the future: some see the fragmentation of the world into small tribal groups; others, a face-off among several vast civilizational coalitions. They all share, however, the idea that the world's current conflicts are fueled by age-old ethnic loyalties and cultural differences.[1]

This notion misrepresents the genesis of conflict and ignores the ability of diverse people to coexist. The very phrase "ethnic conflict" misguides us. It has become a shorthand way to speak about any and all violent confrontations between groups of people living in the same country. Some of these conflicts involve ethnic or cultural identity, but most are about getting more power, land, or other resources. They do not result from ethnic diversity; thinking that they do sends us off in pursuit of the wrong policies, tolerating rulers who incite riots and suppress ethnic differences.

In speaking about local group conflicts we tend to make three assumptions: first, that ethnic identities are ancient and unchanging; second, that these identities motivate people to persecute and kill; and third, that ethnic diversity itself inevitably leads to violence. All three are mistaken.

Contrary to the first assumption, ethnicity is a product of modern politics. Although people have had identities—deriving from religion, birthplace, language, and so on—for as long as humans have had culture, they have begun to see themselves as members of vast ethnic groups, opposed to other such groups, only during the modern period of colonization and state-building.

The view that ethnicity is ancient and unchanging emerges these days in the potent images of the cauldron and the tribe. Out of the violence in Eastern Europe came images of the region as a bubbling cauldron of ethnonationalist sentiments that were sure to boil over unless suppressed by strong states. The cauldron image contrasts with the American "melting pot," suggesting that

From John R. Bowen, "The Myth of Global Ethnic Conflict," *Journal of Democracy*, vol. 7, no. 4 (October 1996). Copyright © 1996 by The National Endowment for Democracy and Johns Hopkins University Press. Reprinted by permission of Johns Hopkins University Press.

Western ethnicities may melt, but Eastern ones must be suppressed by the region's unlikable, but perhaps necessary, Titos and Stalins.

Nowhere does this notion seem more apt than in the former Yugoslavia. Surely the Serbs, Croats, and Bosnians are distinct ethnic groups destined to clash throughout history, are they not? Yet it is often forgotten how small the differences are among the currently warring factions in the Balkans. Serbs, Croats, and Bosnians all speak the same language (Italy has greater linguistic diversity) and have lived side by side, most often in peace, for centuries. Although it is common to say that they are separated by religion—Croats being Roman Catholic, Serbs Orthodox Christian, and Bosnians Muslim—in fact each population includes sizeable numbers of the other two religions. The three religions have indeed become symbols of group differences, but religious differences have not, by themselves, caused intergroup conflict. Rising rates of intermarriage (as high as 30 percent in Bosnia) would have led to the gradual blurring of contrasts across these lines.

As knowledgeable long-term observers such as Misha Glenny have pointed out, the roots of the current Balkan violence lie not in primordial ethnic and religious differences but rather in modern attempts to rally people around nationalist ideas. "Ethnicity" becomes "nationalism" when it includes aspirations to gain a monopoly of land, resources, and power. But nationalism, too, is a learned and frequently manipulated set of ideas, and not a primordial sentiment. In the nineteenth century, Serb and Croat intellectuals joined other Europeans in championing the rights of peoples to rule themselves in "nation-states": states to be composed of one nationality. For their part, Serbs drew on memories of short-lived Serb national states to claim their right to expand outward to encompass other peoples, just as other countries in Europe (most notably France) had done earlier. That Balkan peoples spoke the same language made these expansionist claims all the more plausible to many Serbs.[2]

At the same time, Croats were developing their own nationalist ideology, with a twist: rather than claiming the right to overrun non-Croats, it promised to exclude them. Nationalism among the Croats naturally was directed against their strong Serb neighbors. When Serbs dominated the state of Yugoslavia that was created after the First World War, Croat resentment of Serbs grew. The most militant of Croat nationalists formed an underground organization called Ustashe ("Uprising"), and it was this group, to which the Nazis gave control of Croatia, that carried out the forced conversions, expulsions, and massacres of Serbs during the Second World War. The later calls to war of the Serb leader Slobodan Milosević worked upon the still fresh memories of these tragedies.

But the events of the Second World War did not automatically lead to the slaughters of the 1990s; wartime memories could have been overcome had Yugoslavia's new leaders set out to create the social basis for a multiethnic society. But Marshal Tito chose to preserve his rule by forbidding Yugoslavs from forming independent civic groups and developing a sense of shared political values. Political opposition, whether in Croatia, Serbia, or Slovenia, coalesced instead around the only available symbols, the nationalisms of each region. Tito further fanned nationalist flames by giving Serbs and Croats privileges in each other's territories—Serbs held positions of power in Croatia, and Croats in Belgrade. In the countryside these minority presences added to nationalist resentments. Tito's short-term political cleverness—nostalgically remembered by some in the West—in fact set the stage for later slaughter. Resentments and fears generated by modern state warfare and the absence of a civil society—not ethnic differences—made possible the success of the nationalist politicians Milosević and Franjo Tudjman.

The Legacy of Colonialism

But what about Africa? Surely there we find raw ethnic conflict, do we not? Our understandings of African violence have been clouded by visions, not of boiling cauldrons, but of ancient tribal warfare. I recall a National Public Radio reporter interviewing an African UN official about Rwanda. Throughout the discussion the reporter pressed the official to discuss the "ancient tribal hatreds" that were fueling the slaughter. The official ever so politely demurred, repeatedly reminding the reporter that mass conflict began when Belgian colonial rulers gave Tutsis a monopoly of state power. But, as happens so often, the image of ancient tribalism was too deeply ingrained in the reporter's mind for him to hear the UN official's message.

What the African official had to say was right: ethnic thinking in political life is a product of modern conflicts over power and resources, and not an ancient impediment to political modernity. True, before the modern era some Africans did consider themselves Hutu or Tutsi, Nuer or Zande, but these labels were not the main sources of everyday identity. A woman living in central Africa drew her identity from where she was born, from her lineage and in-laws, and from her wealth. Tribal or ethnic identity was rarely important in everyday life and could change as people moved over vast areas in pursuit of trade or new lands. Conflicts were more often within tribal categories than between them, as people fought over sources of water, farmland, or grazing rights.

It was the colonial powers, and the independent states succeeding them, which declared that each and every person had an "ethnic identity" that determined his or her place within the colony or the postcolonial system. Even such a seemingly small event as the taking of a census created the idea of a colony-wide ethnic category to which one belonged and had loyalties. (And this was not the case just in Africa: some historians of India attribute the birth of Hindu nationalism to the first British census, when people began to think of themselves as members of Hindu, Muslim, or Sikh populations.) The colonial powers—Belgians, Germans, French, British, and Dutch—also realized that, given their small numbers in their dominions, they could effectively govern and exploit only by seeking out "partners" from among local people, sometimes from minority or Christianized groups. But then the state had to separate its partners from all others, thereby creating firmly bounded "ethnic groups."

In Rwanda and Burundi, German and Belgian colonizers admired the taller people called Tutsis, who formed a small minority in both colonies. The Belgians gave the Tutsis privileged access to education and jobs, and even instituted a minimum height requirement for entrance to college. So that colonial officials could tell who was Tutsi, they required everyone to carry identity cards with tribal labels.

But people cannot be forced into the neat compartments that this requirement suggests. Many Tutsis are tall and many Hutus short, but Hutus and Tutsis had intermarried to a such an extent that they were not easily distinguished physically (nor are they today). They spoke the same language and carried out the same religious practices. In most regions of the colonies the categories became economic labels: poor Tutsis became Hutus, and economically successful Hutus became Tutsis. Where the labels "Hutu" and "Tutsi" had not been much used, lineages with lots of cattle were simply labeled Tutsi; poorer lineages, Hutu. Colonial discrimination against Hutus created what had not existed before: a sense of collective Hutu identity, a Hutu cause. In the late 1950s Hutus began to rebel against Tutsi rule (encouraged by Europeans on their way out) and then created an independent and Hutu-dominated state in Rwanda; this state then gave rise to Tutsi resentments and to the creation of a Tutsi rebel army, the Rwandan Patriotic Front. . . .

In these cases and many others—Sikhs in India, Maronites in Lebanon, Copts in Egypt, Moluccans in the Dutch East Indies, Karens in Burma—colonial and postcolonial states created new social groups and identified them by ethnic, religious, or regional categories. Only in living memory have the people who were sorted into these categories begun to act in concert, as political groups with common interests. Moreover, their shared interests have been those of political autonomy, access to education and jobs, and control of local resources. Far from reflecting ancient ethnic or tribal loyalties, their cohesion and action are products of the modern state's demand that people make themselves heard as powerful groups, or else risk suffering severe disadvantages.

Fear From the Top

A reader might say at this point: Fine, ethnic identities are modern and created, but today people surely do target members of other ethnic groups for violence, do they not? The answer is: Less than we usually think, and when they do, it is only after a long period of being prepared, pushed, and threatened by leaders who control the army and the airwaves. It is fear and hate generated from the top, and not ethnic differences, that finally push people to commit acts of violence. People may come to fear or resent another group for a variety of reasons, especially when social and economic change seems to favor the other group. And yet such competition and resentment "at the ground level" usually does not lead to intergroup violence without an intervening push from the top.

Let us return to those two most unsettling cases, Rwanda and the Balkans. In Rwanda the continuing slaughter of the past few years stemmed from efforts by the dictator-president Juvenal Habyarimana to wipe out his political opposition, Hutu as well as Tutsi. In 1990–91 Habyarimana began to assemble armed gangs into a militia called Interahamwe. This militia carried out its first massacre of a village in March 1992, and in 1993 began systematically to kill Hutu moderates and Tutsis. Throughout 1993 the country's three major radio stations were broadcasting messages of hate against Tutsis, against the opposition parties, and against specific politicians, setting the stage for what followed. Immediately after the still unexplained plane crash that killed President Habyarimana in April 1994, the presidential guard began killing Hutu opposition leaders, human rights activists, journalists, and others critical of the state, most of them Hutus. Only then, after the first wave of killings, were the militia and soldiers sent out to organize mass killings in the countryside, focusing on Tutsis.

Why did people obey the orders to kill? Incessant radio broadcasts over the previous year had surely prepared them for it; the broadcasts portrayed the Tutsi-led Rwandan Patriotic Front as bloodthirsty killers. During the massacres, radio broadcasts promised the land of the dead to the killers. Town mayors, the militia, the regular army, and the police organized Hutus into killing squads, and killed those Hutus who would not join in. The acting president toured the country to thank those villagers who had taken part in the massacres. Some people settled personal scores under cover of the massacre, and many were carried away with what observers have described as a "killing frenzy." The killings of 1994 were not random mob violence, although they were influenced by mob psychology.[3]

In reading accounts of the Rwanda killings, I was struck by how closely they matched, point by point, the ways Indonesians have described to me their participation in the mass slaughters of 1965–66. In Indonesia the supposed target was "communists," but there, too, it was a desire to settle personal scores, greed, willingness to follow the army's orders, and fear of retaliation that drove people to do things they can even now barely admit to themselves, even though many of them, like many Hutus, were convinced that the killings stopped the takeover of the country by an evil power. In both countries, people were told to kill the children and not to spare pregnant women, lest children grow up to take revenge on their killers. Americans continue to refer to those massacres in Indonesia as an instance of "ethnic violence" and to assume that Chinese residents were major targets, but they were not: the killings by and large pitted Javanese against Javanese, Acehnese against Acehnese, and so forth.

The two massacres have their differences: Rwanda in 1993–94 was a one-party state that had carried out mass indoctrination through absolute control of the mass media; Indonesia in 1965–66 was a politically fragmented state in which certain factions of the armed forces only gradually took control. But in both cases leaders were able to carry out a plan, conceived at the top, to wipe out an opposition group. They succeeded because they persuaded people that they could survive only by killing those who were, or could become, their killers.

The same task of persuasion faced Serb and Croat nationalist politicians, in particular Croatia's Franjo Tudjman and Serbia's Slobodan Milošević, who warned their ethnic brethren elsewhere—Serbs in Croatia, Croats in Bosnia—that their rights were about to be trampled unless they rebelled. Milošević played on the modern Serb nationalist rhetoric of expansion, claiming the right of Serbs everywhere to be united. Tudjman, for his part, used modern Croat rhetoric of exclusionary nationalism to build his following. Once in power in Croatia, he moved quickly to define Serbs as second-class citizens, fired Serbs from the police and military, and placed the red-and-white "checkerboard" of the Nazi-era Ustashe flag in the new Croatian banner.

Both leaders used historical memories for their own purposes, but they also had to erase recent memories of new Yugoslav identities, tentatively forged by men and women who married across ethnic boundaries or who lived in the cosmopolitan cities. The new constitutions recognized only ethnic identity, not civil identity, and people were forced, sometimes at gunpoint, to choose who they "really" were.[5]

Contrary to the "explanations" of the war frequently offered by Western journalists, ordinary Serbs do not live in the fourteenth century, fuming over the Battle of Kosovo; nor is the current fighting merely a playing out of some kind of inevitable logic of the past, as some have written. It took hard work by unscrupulous politicians to convince ordinary people that the other side consisted not of the friends and neighbors they had known for years but of genocidal people who would kill them if they were not killed first. For Milosević this meant persuading Serbs that Croats were all crypto-Nazi Ustashe; for Tudjman it meant convincing Croats that Serbs were all Chetnik assassins. Both, but particularly Milosević, declared Bosnian Muslims to be the front wave of a new Islamic threat. Each government indirectly helped the other: Milosević's expansionist talk confirmed Croat fears that Serbs intended to control the Balkans; Tudjman's politics revived Serbs' still remembered fears of the Ustashe. Serb media played up these fears, giving extensive coverage in 1990–91 to the exhumation of mass graves from the Second World War and to stories of Ustashe terror. This "nationalism from the top down," as Warren Zimmerman, the last U.S. ambassador to Yugoslavia, has characterized it, was also a battle of nationalisms, with each side's actions confirming the other's fears.

If Rwanda and the Balkans do not conform to the images of bubbling cauldrons and ancient tribal hatreds, even less do other ongoing local-level conflicts. Most are drives for political autonomy, most spectacularly in the former Soviet Union, where the collapse of Soviet power allowed long-suppressed peoples to reassert their claims to practice their own languages and religions, and to control their own territory and resources—a rejection of foreign rule much like anti-imperial rebellions in the Americas, Europe, Asia, and Africa. Elsewhere various rebellions, each with its own history and motivations, have typically—and erroneously—been lumped together as "ethnic conflict." Resistance in East Timor to Indonesian control is a 20-year struggle against invasion by a foreign power, not an expression of ethnic or cultural identity. People fighting in the southern Philippines under the banner of a "Moro nation" by and large joined up to regain control of their homelands from Manila-appointed politicians. Zapatista rebels in Chiapas demand jobs, political reform, and, above all, land. They do not mention issues of ethnic or cultural identity in their statements—indeed, their leader is from northern Mexico and until recently spoke no Mayan. Other current conflicts are raw struggles for power among rival factions, particularly in several African countries (Liberia, Somalia, Angola) where rival forces often recruit heavily from one region or clan (giving rise to the notion that these are "ethnic conflicts") in order to make use of local leaders and loyalties to control their followers.[5]

Ethnic Diversity and Social Conflict

This brings us to the third mistaken assumption: that ethnic diversity brings with it political instability and the likelihood of violence. To the contrary, greater ethnic diversity is not associated with greater interethnic conflict. Some of the world's most ethnically diverse states, such as Indonesia, Malaysia, and Pakistan, though not without internal conflict and political repression, have suffered little interethnic violence, while countries with very slight differences in language or culture (the former Yugoslavia, Somalia, Rwanda) have had the bloodiest such conflicts. It is the number of ethnic groups and their relationships to power, not diversity per se, that strongly affect political stability. As shown in recent studies by political scientist Ted Gurr, and contrary to popular thinking, local conflicts have not sharply increased in frequency or severity during the last ten years. The greatest increase in local conflicts occurred during the Cold War, and resulted from the superpowers' efforts to arm their client states. (The sense that everything exploded after 1989, Gurr argues, comes from the reassertions of national identity in Eastern Europe and the former Soviet Union.)[6]

By and large, the news media focus on countries racked by violence and ignore the many more cases of peaceful relations among different peoples. Take Indonesia, where I have carried out fieldwork since the late 1970s. If people know of Indonesia, it is probably because of its occupation of East Timor and its suppression of political freedoms. But these are not matters of ethnic conflict, of which there is remarkably little in a country composed of more than three hundred peoples, each with its own distinct language and culture. Although throughout the 1950s and 1960s there were rebellions against Jakarta in many parts of the country, these concerned control over local resources, schooling, and religion. An on-again, off-again rebellion where I work, on the northern tip of Sumatra, has been about control over the region's vast oil and gas resources (although the Western press continues to stereotype it as "ethnic conflict").

Cultural diversity does, of course, present challenges to national integration and social peace. Why do some countries succeed at meeting those challenges while others fail? Two sets of reasons seem most important, and they swamp the mere fact of ethnic and cultural diversity.

First there are the "raw materials" for social peace that countries possess at the time of independence. Countries in which one group has been exploiting all others (such as Rwanda and Burundi) start off with scores to settle, while countries with no such clearly dominating group (such as Indonesia) have an initial advantage in building political consensus. So-called centralized polities, with two or three large groups that continually polarize national politics, are less stable than "dispersed" systems, in which each of many smaller groups is forced to seek out allies to achieve its goals. And if the major ethnic groups share a language or religion, or if they have worked together in a revolutionary struggle, they have a bridge already in place that they can use to build political cooperation.[7]

Take, again, the case of Indonesia. In colonial Indonesia (the Dutch East Indies) the Javanese were, as they are today, the most numerous people. But they were concentrated on Java and held positions of power only there. Peoples of Java, Sumatra, and the eastern islands, along with Malays and many in the southern Philippines, had used Malay as a *lingua franca* for centuries, and Malay became the basis for the language of independent Indonesia. Islam also cut across regions and ethnicities, uniting people on Sumatra, Java, and Sulawesi. Dominance was "dispersed," in that prominent figures in literature, religion, and the nationalist movement tended as often as not to be from someplace other than Java, notably Sumatra. Moreover, people from throughout the country had spent five years fighting Dutch efforts to regain control after the Second World War, and could draw on the shared experience of that common struggle.[8]

One can see the difference each of these features makes by looking next door at culturally similar Malaysia. Malays and Chinese, the largest ethnic groups, shared neither language nor religion, and had no shared memory of struggle to draw on. Malays had held all political power during British rule. On the eve of independence there was a clear fault line running between the Malay and Chinese communities.

The Importance of Political Choices

But these initial conditions do not tell the whole story, and here enters the second set of reasons for social peace or social conflict. States do make choices, particularly about political processes, that ease or exacerbate intergroup tensions. As political scientist Donald Horowitz has pointed out, if we consider only their starting conditions, Malaysia ought to have experienced considerable interethnic violence (for the reasons given above), whereas Sri Lanka, where Tamils and Sinhalese had mingled in the British-trained elite, should have been spared such violence. And yet Malaysia has largely managed to avoid it while Sri Lanka has not. The crucial difference, writes Horowitz, was in the emerging political systems in the two countries. Malaysian politicians constructed a multiethnic political coalition, which fostered ties between Chinese and Malay leaders and forced political candidates to seek the large middle electoral ground. In Sri Lanka, as we saw earlier, Sinhalese-speakers formed a chauvinist nationalist movement, and after early cooperation Tamils and Sinhalese split apart to form ethnically based political parties. Extreme factions appeared on the wings of each party, forcing party leaders to drift in their directions.

But political systems can be changed. Nigeria is a good example. Prior to 1967 it consisted of three regions—North, South, and East—each with its own party supported by ethnic allegiances. The intensity of this three-way division drove the southeast region of Biafra to attempt to break away from Nigeria in 1967, and the trauma of the civil war that followed led politicians to try a new system. They carved the country into 19 states, the boundaries of which cut through the territories of the three largest ethnic groups (Hausa, Yoruba, and

Igbo), encouraging a new federalist politics based on multiethnic coalitions. The new system, for all its other problems, prevented another Biafra. Subsequent leaders, however, continued to add to the number of states for their own political reasons. The current leader, General Sani Abacha, is now adding to an already expanded list of 30 states; this excessive fragmentation has broken up the multiethnic coalitions and encouraged ethnic politics anew. A similar direction has been pursued by Kenya's Daniel arap Moi, who has created an ethnic electoral base that excludes most Kikuyus, increasing the relevance of ethnicity in politics and therefore the level of intergroup tensions.

What the myth of ethnic conflict would say are ever-present tensions are in fact the products of political choices. Negative stereotyping, fear of another group, killing lest one be killed—these are the doings of so-called leaders, and can be undone by them as well. Believing otherwise, and assuming that such conflicts are the natural consequences of human depravity in some quarters of the world, leads to perverse thinking and perverse policy. It makes violence seem characteristic of a people or region, rather than the consequence of specific political acts. Thinking this way excuses inaction, as when U.S. president Bill Clinton, seeking to retreat from the hard-line Balkan policy of candidate Clinton, began to claim that Bosnians and Serbs were killing each other because of their ethnic and religious differences. Because it paints all sides as less rational and less modern (more tribal, more ethnic) than "we" are, it makes it easier to tolerate their suffering. Because it assumes that "those people" would naturally follow their leaders' call to kill, it distracts us from the central and difficult question of just how and why people are sometimes led to commit such horrifying deeds.

Notes

1. Two of the most widely read proponents of the view I am contesting are Robert Kaplan, in his dispatches for *The Atlantic* and in his *Balkan Ghosts: A Journey Through History* (New York: St. Martin's, 1993), and (writing mainly on large-scale conflict) Samuel P. Huntington, in "The Clash of Civilizations?" *Foreign Affairs* 72 (Summer 1993): 22–49. My concern is less with the particular difficulties of these writers' arguments, about which others have written, than with the general notion, which, as with all myths, survives the death of any one of its versions.

2. See Misha Glenny, *The Fall of Yugoslavia* (New York: Penguin, 1992), for a balanced and ethnographically rich account of the Balkan wars. On recent tendencies in European nationalisms, see especially Rogers Brubaker, *Nationalism Reframed: Nationhood and the National Question in the New Europe* (Cambridge: Cambridge University Press, 1996). Brubaker makes the important point that "nationalism" should be treated as a category of social and political ideology, and not a pre-ideological "thing."

3. Among recent overviews of massacres in Rwanda and Burundi, see Philip Gourevitch, "The Poisoned Country," *New York Review of Books,* 6 June 1996, 58–64, and René Lemarchand, *Burundi: Ethnic Conflict and Genocide* (Cambridge: Cambridge University Press, 1995).

4. That there were memories, fears, and hatreds to exploit is important to bear in mind, lest we go to the other extreme and argue that these conflicts are entirely produced from the top, an extreme toward which an overreliance on rational-

choice models may lead some analysts. Russell Hardin's otherwise excellent *One for All: The Logic of Group Conflict* (Princeton: Princeton University Press, 1995) errs, I believe, in attributing nothing but rational, self-aggrandizing motives to those leaders who stir up ethnic passions, ignoring that they, too, can be caught up in these passions. The cold rationality of leaders is itself a variable: probably Milosević fits Hardin's rational-actor model better than Tudjman, and Suharto better than Sukarno. In each case, it is an empirical question.

5. The same points could be made concerning the religious version of "ancient hatreds," such as Muslim-Hindu relations in India. However peaceful or conflictual "ancient" relations may have been (and on this issue there continues to be a great deal of controversy among historians of South Asia), the often bloody conflicts of the past ten years in India have been fueled by ambitious politicians who have seen boundless electoral opportunity in middle-class Hindu resentment toward 1) lower castes' claims that they deserve employment and education quotas, and 2) the recent prosperity of some middle-class Muslims. See the penetrating political analyses by Susanne Hoeber Rudolph and Lloyd I. Rudolph in the *New Republic*, 22 March 1993 and 14 February 1994, and a historical and ethnographic study by Peter van der Veer, *Religious Nationalism: Hindus and Muslims in India* (Berkeley: University of California Press, 1994).

6. See Ted Gurr and Barbara Harff, *Ethnic Conflict in World Politics* (Boulder, Colo.: Westview, 1994).

7. See Donald L. Horowitz, *Ethnic Groups in Conflict* (Berkeley: University of California Press, 1985), 291–364.

8. I would propose "dispersed dominance" (a situation in which each of several groups considers itself to dominate on some social or political dimension) as a second important mechanism for reducing intergroup conflict alongside the well-known "cross-cutting cleavages" (a situation in which one or more important dimensions of diversity cut across others, as religion cuts across ethnicity in many countries). "Dispersed dominance" takes into account social and cultural dimensions, such as literary preeminence or a sense of social worth stemming from putative indigenous status. It is thus broader than, but similar to, political mechanisms such as federalism, when these mechanisms are aimed at (in Donald Horowitz's phrase) "proliferating the points of power." It is the empirical correlate to the normative position articulated by Michael Walzer in *Spheres of Justice* (New York: Basic Books, 1983) that dominance in one sphere (or dimension) ought not to automatically confer dominance in others.

POSTSCRIPT

Are Cultural and Ethnic Conflicts the Defining Dimensions of Twenty-First-Century War?

Examining the causes and trends in conflict is a difficult enterprise. Perhaps no area in world politics has been studied and analyzed more than conflict. Arguing that certain forms of conflict are on the upswing or downswing is difficult because we do not have the benefit of historical context. Looking back on the twentieth century, for example, we can clearly see the increasing scope and impact of war; deaths and weapon destructiveness increased at virtually every stage. Determining the likelihood that ethnic conflicts will dominate the security agenda in the coming years, however, is difficult.

The sheer ferocity of ethnic conflicts in the past decade in places like Rwanda, Bosnia, Kosovo, Chechnya, Angola, Afghanistan, and Israel and Palestine seems to support Huntington's thesis that ethnicity and culture will define the battle lines between groups. Yet is culture at work or simply political power? The passion of the combatants may mask the more Machiavellian goals of leaders and individuals.

One element is certain. As the centralized power and influence of large imperial states like the USSR and others diminish, fault lines of conflict that may exist between people in various regions are more readily exploitable, like those in Yugoslavia by former Serb president Slobodan Milosovich. This means that such conflicts are indeed more likely, as leaders take advantage of paranoia, animosity, and fear to create an "other" against which a people may rally, at least for a short time. Whether this is part of a major global trend in ethnic conflict and cultural divides or merely political opportunism is still open to question.

Literature on this topic includes Farhang Rajaee, *Globalization on Trial: The Human Condition and the Information Civilization* (Kumarian Press, 2000); Daniel Patrick Moynihan, *Pandemonium: Ethnicity in International Politics* (Oxford University Press, 1994); and Stephen M. Walt, "Building Up a New Bogeyman," *Foreign Policy* (Spring 1997).

Contributors to This Volume

EDITORS

JAMES E. HARF is a professor of government and world affairs at the University of Tampa, where he also serves as director of the Office of International Programs. He spent most of his career at Ohio State University, where he holds the title of professor emeritus of political science. He is coauthor of *World Politics and You: A Student Study Companion to International Politics on the World Stage,* 5th ed. (Brown & Benchmark, 1995) and *The Politics of Global Resources* (Duke University Press, 1986). He also coedited a four-book series on the global issues of population, food, energy, and environment, as well as three other book series on national security education, international studies, and international business. His current research interests include the world population problem and its regime. As a staff member of the Presidential Commission on Foreign Language and International Studies, he was responsible for undergraduate education. He also served for 15 years as executive director of the Consortium for International Studies Education.

MARK OWEN LOMBARDI is vice president for academic and student affairs at the College of Santa Fe. He is also a professor of political science and coeditor of *Perspectives on Third World Sovereignty: The Post-Modern Paradox* (Macmillan, 1996). Dr. Lombardi has authored numerous articles on such topics as African political economy, U.S. foreign policy, the politics of the cold war, and collegiate curriculum reform. Dr. Lombardi is a former executive director of the U.S.-Africa Education Foundation, and he is currently a member of the International Studies Association, the African Studies Association, and the National Committee of International Studies Programs Administrators. His teaching interests include U.S. foreign policy, African politics, third world development, and globalization.

STAFF

Jeffrey L. Hahn Vice President/Publisher
Theodore Knight Managing Editor
David Brackley Senior Developmental Editor
Juliana Gribbins Developmental Editor
Rose Gleich Permissions Assistant
Brenda S. Filley Director of Production/Manufacturing
Julie Marsh Project Editor
Juliana Arbo Typesetting Supervisor
Richard Tietjen Publishing Systems Manager
Charles Vitelli Designer

AUTHORS

GRAHAM ALLISON is the Douglas Dillon Professor of Government and director of the Belfer Center for Science and International Affairs at Harvard University. During the first term of the Clinton administration, he served as assistant secretary of defense for policy and plans. His publications include *Realizing Human Rights: From Inspiration to Impact* (St. Martin's Press, 2000).

JANE T. BERTRAND is a professor in and chair of the Department of International Health and Development in the Tulane School of Public Health and Tropical Medicine. She earned her Ph.D. in sociology from the University of Chicago, and she has published over 50 peer-reviewed journal articles. Her areas of expertise include international family planning, applied research, and information-education-communication.

LEON F. BOUVIER is an adjunct professor of demography at Tulane University in New Orleans, Louisiana, and a senior fellow of Negative Population Growth. He is coauthor, with Lindsey Grant, of *How Many Americans?* (Sierra Club Books, 1994).

JOHN R. BOWEN is the Dunbar Van Cleve Professor and chair of the Social Thought and Analysis Program at Washington University in St. Louis, Missouri. His research is concerned primarily with the role of cultural forms (religious practices, aesthetic genres, legal discourse) in processes of social change. He earned his Ph.D. from the University of Chicago in 1984.

LESTER R. BROWN is founder, senior researcher, and chairman of the board of directors of the Worldwatch Institute, a research institute devoted to the analysis of global environmental issues. He earned his M.S. in agricultural economics from the University of Maryland in 1959 and his M.P.A. in public administration from Harvard University in 1962. His many books include *Eco-Economy: Building a New Economy for the Environmental Age* (W. W. Norton, 2001).

SYLVIE BRUNEL, a geographer and a specialist on development issues, is a former president of Action Contre la Faim (Action Against Hunger).

XIMING CAI is a research fellow in the Environment and Production Technology Division of the International Food Policy Research Institute. He earned his M.S. in hydrology and water resources at Tsinghua University in Beijing, China, and his Ph.D. in environmental and water resources at the University of Texas at Austin.

WILLIAM J. CARRINGTON is an economist at the U.S. Department of Labor's Bureau of Labor Statistics.

WESLEY K. CLARK is a general of the U.S. Army, now retired, whose assignments have included supreme allied commander Europe and commander in chief, United States European Command. He holds a master's degree in philosophy, politics, and economics from Oxford University, where he studied as a Rhodes scholar.

SARAH A. CLINE is a senior research assistant in the Environment and Production Technology Division of the International Food Policy Research Institute. She has also served as research assistant at Resources for the Future. She earned her M.S. in agricultural and resource economics from West Virginia University.

ROBERTA COHEN is a Foreign Policy Studies guest scholar at the Brookings Institution, where she has also been codirector of the Project on Internal Displacement. A former senior adviser to the representative of the UN secretary general on internally displaced persons, she has been a consultant to the United Nations, the World Bank, the Refugee Policy Group, and the International Human Rights Law Group. She earned her M.A. from Johns Hopkins University in 1963.

FRANCIS M. DENG is a Foreign Policy Studies senior fellow of the Brookings Institution. A former Sudanese ambassador to the United States, Scandinavia, and Canada, he has also served as the UN secretary general's special representative on internally displaced persons. He earned his LL.M. and his J.S.D. from Yale Law School in 1965 and 1968, respectively.

SETH DUNN is a research associate and a climate/energy team coleader at the Worldwatch Institute. He has also been a consultant to the Natural Resources Defense Council. He is coauthor, with Christopher Flavin and Jane A. Peterson, of *Rising Sun, Gathering Winds: Policies to Stabilize the Climate and Strengthen Economies* (Worldwatch Institute, 1997).

THOMAS L. FRIEDMAN is one of America's leading commentators on world affairs. A two-time Pulitzer Prize–winning journalist for the *New York Times,* he is the author of the international best-seller *The Lexus and the Olive Tree: Understanding Globalization* (Farrar, Straus & Giroux, 1999).

GARY GARDNER is director of research at the Worldwatch Institute. His research interests include agricultural resource degradation, materials use, global malnutrition, and the dynamics of social change. He earned an M.A. in politics from Brandeis University and an M.A. in public administration from the Monterey Institute of International Studies.

JEFFREY E. GARTEN is the William S. Beinecke Professor in the Practice of International Trade and Finance and dean of the School of Management at Yale University. An expert in finance and international trade, he was undersecretary of commerce for international trade from 1993 to 1995, and he currently writes a monthly column for *Business Week* on major challenges facing global business leaders. He is the author of *The Mind of the CEO* (Perseus Press, 2001).

MORTON H. HALPERIN, former director of policy planning at the U.S. Department of State, is a senior fellow at the Council on Foreign Relations. He worked for many years for the American Civil Liberties Union, where he directed the Center for National Security Studies.

BRIAN HALWEIL is a staff researcher at the Worldwatch Institute, where he researches issues related to food and agriculture, including organic farming, biotechnology, hunger, and water scarcity. His work has been featured in the *L.A. Times, Christian Science Monitor,* and the *New York Times.* He earned B.S. degrees in Earth systems and in biology from Stanford University.

ADAM S. HERSH is an economist at the Economic Policy Institute in Washington, D.C.

DON HINRICHSEN is an environmental reporter who specializes in covering the developing world. He is also a consultant on population for the United Nations, principally the UN Population Fund. He is the author of *Coastal Waters of the World: Trends, Threats, and Strategies* (Island Press, 1999).

KIM R. HOLMES is vice president and director of the Kathryn and Shelby Cullom Davis Institute for International Studies at the Heritage Foundation, where he is the principal spokesman on foreign and defense policy issues. He earned his M.A. and Ph.D. in history from Georgetown University in 1977 and 1982, respectively. His articles have appeared in such journals as *International Security* and *Journal of International Affairs.*

PETER HUBER is a senior fellow at the Manhattan Institute's Center for Legal Policy. His articles on science, technology, environment, and the law have appeared in such journals as the *Harvard Law Review* and the *Yale Law Journal,* and he is a regular columnist for *Forbes* magazine. He is the author of *Hard Green: Saving the Environment From the Environmentalists* (Basic Books, 2000).

SAMUEL P. HUNTINGTON is the Albert J. Weatherhead III University Professor at Harvard University, where he is also director of the John M. Olin Institute for Strategic Studies. He is the author of *The Clash of Civilizations and the Remaking of World Order* (Simon & Schuster, 1996).

INTERGOVERNMENTAL PANEL ON CLIMATE CHANGE (IPCC) was established in 1988 by the World Meteorological Organization (WMO) and the United Nations Environment Programme (UNEP). The role of the IPCC is to assess the scientific, technical, and socioeconomic information relevant to understanding the scientific basis of risk of human-induced climate change, its potential impacts, and options for adaptation and mitigation.

JOHN ISBISTER is a professor of economics at the University of California, Santa Cruz, where he has been working since 1968. His past research has been in economic development, demography, and immigration, and his current research focuses on ethical problems in economics. He is the author of *Promises Not Kept: The Betrayal of Social Change in the Third World,* 4th ed. (Kumarian Press, 1998).

JEAN M. JOHNSON is a senior analyst in the Division of Science Resources Studies at the National Science Foundation in Arlington, Virginia, an independent U.S. government agency responsible for promoting science and engineering. She is the author of *The Science and Technology Resources of Japan: A Comparison With the United States* (DIANE, 2001).

JERRY KANG is an acting professor of law at the University of California, Los Angeles. His teaching and research interests include civil procedure, Asian American jurisprudence, and cyberspace. He earned his J.D. from Harvard

Law School in 1993, and he has also worked for the National Telecommunications and Information Administration.

HISHAM KHATIB is honorary vice chairman of the World Energy Council and a former Jordanian minister of energy and minister of planning.

BJØRN LOMBORG is an associate professor of statistics in the Department of Political Science at the University of Aarhus in Denmark and a frequent participant in topical coverage in the European media. His areas of interest include the use of surveys in public administration and the use of statistics in the environmental arena. In February 2002 Lomborg was named director of Denmark's national Environmental Assessment Institute. He earned his Ph.D. from the University of Copenhagen in 1994.

N. GREGORY MANKIW is a professor of economics at Harvard University, where he has been teaching since 1985. He is also a research associate and director of the Monetary Economics Program for the National Bureau of Economic Research, associate editor of *Review of Economics and Statistics,* and a columnist for *Fortune* magazine.

DAVID MASCI is a staff writer for *CQ Researcher.*

ROBERT S. McNAMARA was president of the World Bank Group of Institutions until his retirement in 1981. A former lieutenant colonel in the U.S. Air Force, he has also taught business administration at Harvard University, where he earned his M.B.A., and he served as secretary of defense from 1961 until 1968. He has received many awards, including the Albert Einstein Peace Prize, and he is the author of *In Retrospect: The Tragedy and Lessons of Vietnam* (Random House, 1995).

EDWIN MEESE III is the Ronald Reagan Distinguished Fellow in Public Policy at the Heritage Foundation. He also served as U.S. attorney general in the Reagan administration, chairman of the Domestic Policy Council and the National Drug Policy Board, and as a member of the National Security Council. He has also been a professor of law at the University of San Diego. He earned his J.D. from the University of California, Berkeley.

ADIL NAJAM is a professor in the Department of International Relations at Boston University, where he has been teaching since 1997. He specializes in negotiation analysis to study global cooperation for sustainable development, with a particular focus on developing countries and nonstate institutions. He is the author of *Getting Beyond the Lowest Common Denominator: Developing Countries in International Environmental Negotiations* (Rowman & Littlefield, 2003).

MARK W. NOWAK is an environmental writer and a resident fellow of Negative Population Growth. His writing has appeared in national newspapers, magazines, and environmental journals, and he has contributed to several books on population, immigration, and the environment. He is a former executive director of Population-Environment Balance.

PETER G. PETERSON is chairman of the Blackstone Group, a private investment bank; chairman of the Institute for International Economics; deputy chairman of the Federal Reserve Bank of New York; and chairman of the Council on Foreign Relations. He is the author of *Gray Dawn: How the Coming Age Wave Will Transform America—And the World* (Crown, 2000).

MARK C. REGETS is a senior analyst in the Division of Science Resources Studies at the National Science Foundation in Arlington, Virginia, an independent U.S. government agency responsible for promoting science and engineering.

MARK W. ROSEGRANT is a senior fellow at the International Food Policy Research Institute. He has over 24 years of experience in research and policy analysis in agriculture and economic development, with an emphasis on critical water issues as they impact world food security and environmental sustainability. He earned his Ph.D. in public policy from the University of Michigan.

DAVID ROTHKOPF is an adjunct professor of international and public affairs at Columbia University, where he earned his B.A. He has also been managing director and a member of the board of directors of Kissinger Associates, chairman and CEO of International Media Partners, and deputy undersecretary of commerce for international trade policy development for the U.S. Department of Commerce.

ROBERT E. SCOTT is an international economist and codirector of the research department at the Economic Policy Institute in Washington, D.C. He has also been an assistant professor with the College of Business and Management of the University of Maryland at College Park. His research has been published in *The Journal of Policy Analysis and Management, The International Review of Applied Economics,* and *The Stanford Law and Policy Review.* He earned his Ph.D. from the University of California, Berkeley.

JULIAN L. SIMON (1932–1998) was a professor of economics and business administration in the College of Business and Management at the University of Maryland at College Park. His research interests focused on population economics, and his publications include *The Economics of Population: Key Modern Writings* (Edward Elgar, 1997).

MAX SINGER is president of The Potomac Organization, Inc., a public policy consulting and research firm in Chevy Chase, Maryland. He has also served as president of the World Institute in Jerusalem and of the Hudson Institute, which he cofounded. He is coauthor, with Aaron Wildavsky, of *The Real World Order: Zones of Peace/Zones of Turmoil,* rev. ed. (Seven Bridges Press, 1996).

EHUD SPRINZAK (d. 2002) was an Israeli counterterrorism specialist and expert in far-right Jewish groups and a professor of political science at Hebrew University in Jerusalem. He also served as senior scholar at the United States Institute of Peace.

PAUL STARR is a professor of sociology at Princeton University and coeditor of *The American Prospect.* He is also director of the Century Institute, founder of the Electronic Policy Network, and president of the Sandra Starr Foundation.

ANDREW STEPHEN is the U.S. editor for *New Statesman.*

JESSICA STERN is a lecturer in public policy in the John F. Kennedy School of Government at Harvard University and a faculty affiliate at the Belfer Center for Science and International Affairs. She has also served as director for Russian, Ukrainian, and Eurasian Affairs at the National Security Council, where she was responsible for national security policy toward Russia and the former Soviet states and for policies to reduce the threat of nuclear smuggling and terrorism. She is the author of *The Ultimate Terrorists* (Harvard University Press, 1999).

MICHAEL TOBIAS is the author of more than 25 books and the writer-director-producer of over 100 films, including the acclaimed 28-part series *A Parliament of Souls* and the 10-hour miniseries *Voice of the Planet.* He has also published over 200 articles and essays in such journals as *Mother Earth News, The New Scientist,* and *Discovery.*

BRIAN TUCKER a former chief of the CSIRO Division of Atmospherics Research in Melbourne, Australia, is now a senior fellow of the Institute for Public Affairs.

UNITED NATIONS ENVIRONMENT PROGRAMME (UNEP) was organized to provide leadership and to encourage partnership in caring for the environment by inspiring, informing, and enabling nations and peoples to improve their quality of life without compromising that of future generations. The executive director of UNEP is Klaus Töpfer, undersecretary general of the United Nations.

AL J. VENTER, an author and a journalist, is a Middle East correspondent for *Jane's International Defense Review.* With over 30 years' experience in that region, he has done a series of studies on nuclear and biological warfare pertaining to Iraq, Iran, Syria, Algeria, Egypt, Sudan, and other countries. He also writes on issues pertaining to water in that part of the globe.

CHRISTIAN E. WELLER is an international macroeconomist at the Economic Policy Institute in Washington, D.C. He has also worked at the Center for European Integration Studies at the University of Bonn in Germany and the Public Policy Department of the AFL-CIO in Washington, D.C. He earned his M.A. and his Ph.D. in economics from the University of Massachusetts in 1993 and 1998, respectively.

Index